Student Study Guide/ Solutions Manual

to accompany

Organic Chemistry

Fifth Edition

Janice Gorzynski Smith
University of Hawai'i at Mānoa

and

Erin R. Smith

Mc
Graw
Hill
Education

STUDENT STUDY GUIDE/SOLUTIONS MANUAL TO ACCOMPANY
ORGANIC CHEMISTRY, FIFTH EDITION

Published by McGraw-Hill Education, 2 Penn Plaza, New York, NY 10121. Copyright © 2017 by McGraw-Hill Education. All rights reserved. Printed in the United States of America. Previous editions © 2014 and 2011. No part of this publication may be reproduced or distributed in any form or by any means, or stored in a database or retrieval system, without the prior written consent of McGraw-Hill Education, including, but not limited to, in any network or other electronic storage or transmission, or broadcast for distance learning.

Some ancillaries, including electronic and print components, may not be available to customers outside the United States.

This book is printed on acid-free paper.

1 2 3 4 5 6 7 8 9 0 RHR/RHR 1 0 9 8 7 6

ISBN 978-1-259-63706-3
MHID 1-259-63706-9

mheducation.com/highered

Contents

Chapter 1: Structure and Bonding

Chapter Review

Important facts

- **The general rule of bonding:** Atoms strive to attain a complete outer shell of valence electrons (Section 1.2). H "wants" 2 electrons. Second-row elements "want" 8 electrons.

- **Formal charge** (FC) is the difference between the number of valence electrons on an atom and the number of electrons it "owns" (Section 1.3C). See Sample Problem 1.3 for a stepwise example.

- **Curved arrow notation** shows the movement of an electron pair. The tail of the arrow always begins at an electron pair, either in a bond or a lone pair. The head points to where the electron pair "moves" (Section 1.6).

Move an electron pair to O.

Use an electron pair to form a double bond.

- **Electrostatic potential plots** are color-coded maps of electron density, indicating electron rich (red) and electron deficient (blue) regions (Section 1.12).

The importance of Lewis structures (Sections 1.3–1.5)

A properly drawn Lewis structure shows the number of bonds and lone pairs present around each atom in a molecule. In a valid Lewis structure, each H has two electrons, and each second-row element has no more than eight. This is the first step needed to determine many properties of a molecule.

Resonance (Section 1.6)

The basic principles:
- Resonance occurs when a compound cannot be represented by a single Lewis structure.
- Two resonance structures differ *only* in the position of nonbonded electrons and π bonds.
- The resonance hybrid is the only accurate representation for a resonance-stabilized compound. A hybrid is more stable than any single resonance structure because electron density is delocalized.

The difference between resonance structures and isomers:
- Two **isomers** differ in the arrangement of *both* atoms and electrons.
- **Resonance structures** differ *only* in the *arrangement of electrons*.

Geometry and hybridization

The number of groups around an atom determines both its geometry (Section 1.7) and hybridization (Section 1.9).

Number of groups	Geometry	Bond angle (°)	Hybridization	Examples
2	linear	180	sp	BeH_2, $HC{\equiv}CH$
3	trigonal planar	120	sp^2	BF_3, $CH_2{=}CH_2$
4	tetrahedral	109.5	sp^3	CH_4, NH_3, H_2O

Drawing organic molecules (Section 1.8)

- Shorthand methods are used to abbreviate the structure of organic molecules.

skeletal structure　　　　　　isooctane　　　　　　condensed structure

- A carbon bonded to four atoms is tetrahedral in shape. The best way to represent a tetrahedron is to draw two bonds in the plane, one in front, and one behind.

Four equivalent drawings for CH₄

Each drawing has two solid lines, one wedge, and one dashed wedge.

Bond length

- Bond length decreases as you go from left to right across a row and increases down a column of the periodic table (Section 1.7A).

Increasing bond length　　　　　　**Increasing bond length**

- Bond length decreases as the number of electrons between two nuclei increases (Section 1.11A).

CH_3-CH_3 < $CH_2=CH_2$ < $H-C\equiv C-H$

Increasing bond length

- Bond length increases as the percent *s*-character decreases (Section 1.11B).

Increasing bond length

- Bond length and bond strength are inversely related. Shorter bonds are stronger bonds (Section 1.11).

| longest C–C bond weakest bond | —C–C— | C=C | —C≡C— | shortest C–C bond strongest bond |

Increasing bond strength

- Sigma (σ) bonds are generally stronger than π bonds (Section 1.10).

—C–C—   C=C   —C≡C—

1 strong σ bond   1 stronger σ bond   1 stronger σ bond
                         1 weaker π bond   2 weaker π bonds

Electronegativity and polarity (Sections 1.12, 1.13)

- Electronegativity increases from left to right across a row and decreases down a column of the periodic table.
- A polar bond results when two atoms with different electronegativities are bonded together. Whenever C or H is bonded to N, O, or any halogen, the bond is polar.
- A polar molecule has either one polar bond, or two or more bond dipoles that reinforce.

Drawing Lewis structures: A shortcut

Chapter 1 devotes a great deal of time to drawing valid Lewis structures. For molecules with many bonds, it may take quite awhile to find acceptable Lewis structures by using trial-and-error to place electrons. Fortunately, a shortcut can be used to figure out how many bonds are present in a molecule.

Shortcut on drawing Lewis structures—Determining the number of bonds:
  [1] Count up the number of valence electrons.
  [2] Calculate how many electrons are needed if there are no bonds between atoms and every atom has a filled shell of valence electrons; that is, hydrogen gets two electrons, and second-row elements get eight.
  [3] Subtract the number obtained in Step [1] from the sum obtained in Step [2]. **This difference tells how many electrons must be shared** to give every H two electrons and every second-row element eight. Because there are two electrons per bond, dividing this difference by two tells how many bonds are needed.

To draw the Lewis structure:
  [1] Arrange the atoms as usual.
  [2] Count up the number of valence electrons.
  [3] Use the shortcut to determine how many bonds are present.
  [4] Draw in the two-electron bonds to all the H's first. Then, draw the remaining bonds between other atoms, making sure that no second-row element gets more than eight electrons and that you use the total number of bonds determined previously.

[5] Finally, place unshared electron pairs on all atoms that do not have an octet of electrons, and calculate formal charge. You should have now used all the valence electrons determined in the first step.

Example: Draw all valid Lewis structures for CH₃NCO using the shortcut procedure.

[1] Arrange the atoms.

```
    H
H   C   N   C   O
    H
```

- In this case the arrangement of atoms is implied by the way the structure is drawn.

[2] Count up the number of valence electrons.

3 H's	x	1 electron per H	=	3 electrons	
2 C's	x	4 electrons per C	=	8 electrons	
1 N	x	5 electrons per N	=	5 electrons	
1 O	x	6 electrons per O	=	+ 6 electrons	

22 electrons total

[3] Use the shortcut to figure out how many bonds are needed.

- Number of electrons needed if there were no bonds:

3 H's	x	2 electrons per H	=	6 electrons	
4 second-row elements	x	8 electrons per element	=	+ 32 electrons	

38 electrons needed if there were no bonds

- Number of electrons that must be shared:

38 electrons
− 22 electrons

16 electrons must be shared

- Every bond requires two electrons, so 16/2 = **8 bonds are needed.**

[4] Draw all possible Lewis structures.

- Draw the bonds to the H's first (three bonds). Then add five more bonds. Arrange them between the C's, N, and O, making sure that no atom gets more than eight electrons. There are three possible arrangements of bonds; that is, there are three resonance structures.
- Add additional electron pairs to give each atom an octet and check that all 22 electrons are used.

Bonds to H's added.

All bonds drawn in.
Three arrangements possible.

Electron pairs drawn in.
Every atom has an octet.

- Calculate the formal charge on each atom.

$$\text{H-C-N}\equiv\text{C-\ddot{O}:} \longleftrightarrow \text{H-C-\ddot{N}=C=\ddot{O}} \longleftrightarrow \text{H-C-\ddot{N}-C}\equiv\text{O:}$$

        +1   −1   −1   +1

- You can evaluate the Lewis structures you have drawn. The middle structure is the best resonance structure, because it has no charged atoms.

Note: This method works for compounds that contain second-row elements in which every element gets an octet of electrons. It does NOT necessarily work for compounds with an atom that does not have an octet (such as BF_3), or compounds that have elements located in the third row and later in the periodic table.

Practice Test on Chapter Review

1.a. Which compound(s) contain a labeled atom with a +1 formal charge? All lone pairs of electrons have been drawn in.

1. (ring)$=\text{N(CH}_3)_2$ 2. (ring with $\overset{..}{O}$–CH$_3$)   3. $\text{CH}_3-\overset{..}{\text{C}}-\text{CH}_3$

4. Both (1) and (2) have labeled atoms with a +1 charge.
5. Compounds (1), (2), and (3) all contain labeled atoms with a +1 formal charge.

b. Which of the following compounds is a valid resonance structure for **A?**

(structures with $\overset{..}{O}$CH$_3$ groups)

A

1.   2.   3.

4. Both (1) and (2) are valid resonance structures for **A.**
5. Cations (1), (2), and (3) are all valid resonance structures for **A.**

c. Which species contains a labeled carbon atom that is sp^2 hybridized?

1. (ring) +   2. (ring)$=\text{C}=\text{CH}_2$ 3. (ring) −

4. Both (1) and (2) contain labeled sp^2 hybridized atoms.
5. Species (1), (2), and (3) all contain labeled sp^2 hybridized carbon atoms.

d. Which of the following compounds has a net dipole?

1. CH₃CH₂NHCH₂CH₃   4. Compounds (1) and (2) both have net dipoles.
2. CH₃CH₂CH₂OH   5. Compounds (1), (2), and (3) all have net dipoles.
3. FCH₂CH₂CH₂F

2. Rank the labeled bonds in order of increasing bond length. Label the shortest bond as **1**, the longest bond as **4**, and the bonds of intermediate length as **2** and **3**.

3. Answer the following questions about compounds **A–D**.

a. What is the hybridization of the labeled atom in **A**?
b. What is the molecular shape around the labeled atom in **B**?
c. In what type of orbital does the lone pair in **C** reside?
d. What orbitals are used to form bond [1] in **D**?
e. Which orbitals are used to form the carbon–oxygen double bond [2] in **D**?

4. Draw an acceptable Lewis structure for CH₃NO₃. Assume that the atoms are arranged as drawn.

```
      H    O
   H  C  O  N  O
      H
```

5. Follow the curved arrows and draw the product with all the needed charges and lone pairs.

Answers to Practice Test

1. a. 4   2. **A** – 1   3. a. sp^3 4. One possibility: 5.
   b. 1   **B** – 4   b. trigonal planar
   c. 1   **C** – 2   c. sp^2
   d. 5   **D** – 3   d. Csp^3–Csp^2
                     e. Csp–Osp^2,
                         Cp–Op

Answers to Problems

1.1 Two isotopes differ in the number of neutrons. Two isotopes have the same number of protons and electrons, group number, and number of valence electrons. The **mass number** is the number of protons and neutrons. The **atomic number** is the number of protons and is the same for all isotopes.

	Nitrogen-14	Nitrogen-13
a. number of protons = atomic number for N = 7	7	7
b. number of neutrons = mass number – atomic number	7	6
c. number of electrons = number of protons	7	7
d. group number	5A	5A
e. number of valence electrons	5	5

1.2 **Ionic bonds** form when an element on the far left side of the periodic table transfers an electron to an element on the far right side of the periodic table. **Covalent bonds** result when two atoms *share* electrons.

a. F–F   b. Li⁺ Br⁻   c. H–C–C–H   d. Na⁺ :N–H   e. Na⁺ :O–C–H

covalent   ionic

All C–H and C–C bonds are covalent.

ionic

Both N–H bonds are covalent.

ionic   All other bonds are covalent.

1.3 Atoms with one, two, three, or four valence electrons form one, two, three, or four bonds, respectively. Atoms with five or more valence electrons form [8 – (number of valence electrons)] bonds.

a. O $8 - 6$ valence $e^- = 2$ bonds   c. Br $8 - 7$ valence $e^- = 1$ bond

b. Al   3 valence $e^- = 3$ bonds   d. Si   4 valence $e^- = 4$ bonds

1.4 [1] Arrange the atoms with the H's on the periphery.
[2] Count the valence electrons.
[3] Arrange the electrons around the atoms. Give the H's 2 electrons first, and then fill the octets of the other atoms.
[4] Assign formal charges (Section 1.3C).

a.
[1] H H
  H C C H
     H H

[2] Count valence e⁻.
  2 C x 4 e⁻ = 8
  6 H x 1 e⁻ = 6
  total e⁻   = 14

[3] H H
  H–C–C–H
     H H

All 14 e⁻ used.
All second-row elements have an octet.

b.
[1] H
  H C N H
     H H

[2] Count valence e⁻.
  1 C x 4 e⁻ = 4
  5 H x 1 e⁻ = 5
  1 N x 5 e⁻ = 5
  total e⁻   = 14

[3] H
  H–C–N–H
     H H

12 e⁻ used.
N needs 2 more electrons for an octet.

→   H
  H–C–N–H
     H H

c.

[1] H
    H C Cl
       H

[2] Count valence e⁻.

1 C x 4 e⁻	= 4
3 H x 1 e⁻	= 3
1 Cl x 7 e⁻	= 7
total e⁻	= 14

[3]

$$H-\underset{\underset{H}{|}}{\overset{\overset{H}{|}}{C}}-Cl \longrightarrow H-\underset{\underset{H}{|}}{\overset{\overset{H}{|}}{C}}-\ddot{\underset{\cdot\cdot}{Cl}}\colon$$

8 e⁻ used.
Cl needs 6 more
electrons for an octet.

Complete octet.

1.5 Follow the directions from Answer 1.4.

a. HCN   H C N

Count valence e⁻.
1 C x 4 e⁻ =	4
1 H x 1 e⁻ =	1
1 N x 5 e⁻ =	5
total e⁻	= 10

$$H-C-N \longrightarrow H-C\equiv N\colon$$

4 e⁻ used.   Complete N and C octets.

b. H₂CO   H C O
            H

Count valence e⁻.
1 C x 4 e⁻ =	4
2 H x 1 e⁻ =	2
1 O x 6 e⁻ =	6
total e⁻	= 12

$$H-\underset{\underset{H}{|}}{C}-O \longrightarrow H-\underset{\underset{H}{|}}{C}=\ddot{O}\colon$$

6 e⁻ used.   Complete O and C octets.

c. HOCH₂CO₂H   H O C C O H
                    H

Count valence e⁻.
2 C x 4 e⁻ =	8
4 H x 1 e⁻ =	4
3 O x 6 e⁻ =	18
total e⁻	= 30

16 e⁻ used.   Complete octets.

1.6 **Formal charge** (FC) = number of valence electrons – [number of unshared electrons + (½)(number of shared electrons)]

a.

$$\left[H-\overset{\overset{H}{|}}{\underset{\underset{H}{|}}{N}}-H \right]^{+}$$

5 – [0 + (1/2)(8)] = +1

b.
5 – [0 + 1/2(8)] = +1

CH₃—N≡C:

4 – [0 + (1/2)(8)] = 0   4 – [2 + (1/2)(6)] = –1

c.
6 – [2 + 1/2(6)] = +1

:Ö=Ö–Ö:

6 – [4 + (1/2)(4)] = 0   6 – [6 + (1/2)(2)] = –1

1.7

a. CH₃O⁻   [1] H C O
                  H

[2] Count valence e⁻.
1 C x 4 e⁻ =	4
3 H x 1 e⁻ =	3
1 O x 6 e⁻ =	6
total e⁻	= 13
Add 1 for (–) charge = 14	

[3]

$$H-\underset{\underset{H}{|}}{\overset{\overset{H}{|}}{C}}-O \longrightarrow H-\underset{\underset{H}{|}}{\overset{\overset{H}{|}}{C}}-\ddot{O}\colon$$

8 e⁻ used.

[4]

$$H-\underset{\underset{H}{|}}{\overset{\overset{H}{|}}{C}}-\ddot{O}\colon^{-}$$

Assign charge.

b. HC_2^- [1] H C C   [2] Count valence e⁻.
2 C x 4 e⁻ = 8
1 H x 1 e⁻ = 1
total e⁻ = 9
Add 1 for (–) charge = 10

[3] H–C–C ⟶ H–C≡C: 4 e⁻ used.

[4] H–C≡C:⁻   Assign charge.

c. $(CH_3NH_3)^+$ [1] H H
H C N H
H H

[2] Count valence e⁻.
1 C x 4 e⁻ = 4
6 H x 1 e⁻ = 6
1 N x 5 e⁻ = 5
total e⁻ = 15
Subtract 1 for (+) charge = 14

[3]
```
    H   H
    |   |
H — C — N — H
    |   |
    H   H
```
14 e⁻ used.

[4]
```
    H   H
    |   |+
H — C — N — H
    |   |
    H   H
```
Assign charge.

d. $(CH_3NH)^-$ [1] H
H C N H
H

[2] Count valence e⁻.
1 C x 4 e⁻ = 4
4 H x 1 e⁻ = 4
1 N x 5 e⁻ = 5
total e⁻ = 13
Add 1 for (–) charge = 14

[3]
```
    H
    |
H — C — N — H
    |
    H
```
10 e⁻ used.

[4]
```
    H
    |   ..  _
H — C — N — H
    |   ..
    H
```
Complete octet and assign charge.

1.8

a. ≡O:
6 – [2 + (1/2)6] = +1

b. =O̤ –
6 – [2 + (1/2)6] = +1

c. =O̤:
6 – [4 + (1/2)4] = 0

1.9

a. $C_2H_4Cl_2$ (two isomers)

Count valence e⁻.
2 C x 4 e⁻ = 8
4 H x 1 e⁻ = 4
2 Cl x 7 e⁻ = 14
total e⁻ = 26

b. C_3H_8O (three isomers)

Count valence e⁻.
3 C x 4 e⁻ = 12
8 H x 1 e⁻ = 8
1 O x 6 e⁻ = 6
total e⁻ = 26

c. C_3H_6 (two isomers)

Count valence e⁻.
3 C x 4 e⁻ = 12
6 H x 1 e⁻ = 6
total e⁻ = 18

1.10 Two different definitions:
- **Isomers** have the same molecular formula and a *different* arrangement of atoms.
- **Resonance structures** have the same molecular formula and the *same* arrangement of atoms.

a. :N̈=C=Ö: and   :C≡N⁺–Ö:⁻

N at the end   N in the middle

different arrangement of atoms = **isomers**

b. (2 lone pairs) and (3 lone pairs)

same arrangement of atoms =
resonance structures

1.11 **Isomers** have the same molecular formula and a *different* arrangement of atoms.
Resonance structures have the same molecular formula and the *same* arrangement of atoms.

a. 2 lone pairs / 3 lone pairs

A   **B**

same arrangement of atoms =
resonance structures

c. CH₃ bonded to C=O / H bonded to C=O

A   **D**

different arrangement of atoms = **isomers**

b. **A**   **C**

$C_2H_4O_2$ $(C_2H_5O_2)^-$

different molecular formulas = **neither**

d. **B**   **D**

different arrangement of atoms = **isomers**

1.12 Curved arrow notation shows the movement of an electron pair. The tail begins at an electron pair (a bond or a lone pair) and the head points to where the electron pair moves.

a.

The net charge is the same
in both resonance structures.

b.

The net charge is the same
in both resonance structures.

1.13 Compare the resonance structures to see what electrons have "moved." **Use one curved arrow to show the movement of each electron pair.**

a.

One electron pair moves:
one curved arrow.

b.

Two electron pairs move:
two curved arrows.

1.14 To draw another resonance structure, **move electrons only in multiple bonds and lone pairs** and keep the number of unpaired electrons constant.

a.

b.

c.

or

1.15 **A "better" resonance structure is one that has more bonds and fewer charges.** The better structure is the major contributor and all others are minor contributors. To draw the resonance hybrid, use dashed lines for bonds that are in only one resonance structure, and use partial charges when the charge is on different atoms in the resonance structures.

a.

All atoms have octets.
one more bond
major contributor

hybrid:

b.

These two resonance structures are equivalent. They both have one charge and the same number of bonds. They are **equal contributors** to the hybrid.

hybrid:

1.16

a.

A

b. The N atom in **B** has four atoms and no lone pairs, so there is no way to move the electrons to give the N another bond.

1.17 To predict the geometry around an atom, **count the number of groups (atoms + lone pairs),** making sure to draw in any needed lone pairs or hydrogens: 2 groups = linear, 3 groups = trigonal planar, 4 groups = tetrahedral.

1.18 To predict the bond angle around an atom, **count the number of groups (atoms + lone pairs),** making sure to draw in any needed lone pairs or hydrogens: 2 groups = 180°, 3 groups = 120°, 4 groups = 109.5°.

a. 2 groups = 180°   CH₃—C≡C—C̈l: 2 groups = 180°

b. This C has 3 groups, so both angles are 120°.   CH₂=C—C̈l:

c. CH₃—C—C̈l: This C has 4 groups, so both angles are 109.5°.

1.19 To predict the geometry around an atom, use the rules in Answer 1.17.

enanthotoxin

1.20 Reading from left-to-right, draw the molecule as a Lewis structure. Always check that carbon has four bonds and all heteroatoms have an octet by adding any needed lone pairs.

$(CH_3)_3CCH(OH)CH_2CH_3$

$(CH_3)_2CHCHO$

$(HOCH_2)_2CH(CH_2)_3C(CH_3)_2CH_2CH_3$

a.

b.

c.

double bond
needed to give
C an octet

d.

$CH_3(CH_2)_4CH(CH_3)_2$

1.21 Draw the Lewis structure of lactic acid.

$CH_3CH(OH)CO_2H \longrightarrow$

1.22 In shorthand or skeletal drawings, **all line junctions or ends of lines represent carbon atoms.**
The carbons are all tetravalent.

1 H

3 H's

1 H

a.

0 H's

octinoxate
(2-ethylhexyl 4-methoxycinnamate)
$C_{18}H_{26}O_3$

1 H

b.

0 H's   0 H's

avobenzone
$C_{20}H_{22}O_3$

3 H's

1.23 In shorthand or skeletal drawings, **all line junctions or ends of lines represent carbon atoms**.
Convert by writing in all carbons, and then adding hydrogen atoms to make the carbons
tetravalent.

a.

b.

c.

d.

1.24

HO

CH_3O

quinine

$C_{20}H_{24}N_2O_2$
molecular formula

1.25 A charge on a carbon atom takes the place of one hydrogen atom. **A negatively charged C has one lone pair, and a positively charged C has none.**

a. positive charge
no lone pairs
no H's needed

b. negative charge
one lone pair
one H needed

c. positive charge
no lone pairs
one H needed

d. negative charge
one lone pair
one H needed

1.26 Draw each indicated structure. Recall that in the skeletal drawings, a carbon atom is located at the intersection of any two lines and at the end of any line.

a. $(CH_3)_2C=CH(CH_2)_4CH_3$ =

c. = $(CH_3)_2CH(CH_2)_2CONHCH_3$

b. $CH_3-\overset{H}{\underset{H}{C}}-\overset{H}{\underset{|}{C}}-CH_2CH_2Cl$, $H_2N-\overset{|}{\underset{H}{C}}-\overset{|}{\underset{H}{C}}-H$ =

d. HO ... = $HO(CH_2)_2CH=CHCO_2CH(CH_3)_2$

1.27 To determine the orbitals used in bonding, **count the number of groups** (atoms + lone pairs):
4 groups = sp^3, 3 groups = sp^2, 2 groups = sp, H atom = $1s$ (no hybridization).
All covalent single bonds are σ, and all double bonds contain one σ and one π bond.

Each H uses a **1s orbital**.

All single bonds are σ bonds.

Each C has 4 groups and is **sp^3 hybridized.**

Each C–C bond is C_{sp^3}–C_{sp^3}.
Each C–H bond is C_{sp^3}–H_{1s}.

Total of 10 σ bonds

1.28 [1] Draw a valid Lewis structure for each molecule.
[2] **Count the number of groups** around each atom: 4 groups = sp^3, 3 groups = sp^2, 2 groups = sp,
H atom = $1s$ (no hybridization).

Note: **Be and B** (Groups 2A and 3A) do not have enough valence e⁻ to form an octet, **and do not form an octet in neutral molecules.**

a. [1] $H-\overset{H}{\underset{H}{C}}-Be-H$

Be has 2 bonds.

[2] Count groups around each atom:

$H-\overset{H}{\underset{H}{C}}-Be-H$

4 groups
sp^3

2 groups
sp

[3] All C–H bonds: C_{sp^3}–H_{1s}
C–Be bond: C_{sp^3}–Be_{sp}
Be–H bond: Be_{sp}–H_{1s}

b. [1] CH₃–B–CH₃ (with CH₃ above B)
B forms 3 bonds.

[2] Count groups around each atom:
CH₃–B–CH₃ (with CH₃ above B)
4 groups sp^3 3 groups sp^2

[3] All C–H bonds: C_{sp^3}–H_{1s}
C–B bonds: C_{sp^3}–B_{sp^2}

c. [1] H–C–O–C–H (with H above and below each C, lone pairs on O)

[2] Count groups around each atom:
H–C–O–C–H
4 groups sp^3
4 groups sp^3

[3] All C–H bonds: C_{sp^3}–H_{1s}
C–O bonds: C_{sp^3}–O_{sp^3}

1.29 To determine the hybridization, **count the number of groups** around each atom: 4 groups = sp^3, 3 groups = sp^2, 2 groups = sp, H atom = $1s$ (no hybridization).

a. CH₃–C≡CH
4 groups sp^3 2 groups sp

b. (cyclohexane ring)=N–CH₃
3 groups sp^2 3 groups sp^2

c. CH₂=C=CH₂
3 groups sp^2 2 groups sp

1.30

a. (structure with :O: and H)

b. O has three groups (one atom + two lone pairs), so it is sp^2 hybridized.

c. C_{sp^2}–O_{sp^2} C_p–O_p

Five sp^2 hybridized C's are labeled.

d, e. Draw in all H atoms to count σ bonds. Each C–H and C–C single bond is a σ bond. Each double bond has one σ bond and one π bond.

H–C–C–C=C–C–C–C=C–C–H (with H atoms and O)
23 σ bonds
3 π bonds

1.31 Single bonds are weaker and longer than double bonds, which are weaker and longer than triple bonds. Increasing percent *s*-character increases bond strength and decreases bond length.

a.   triple bond
      The triple bond is shorter
      than the double bond.

b.   single bond
      double bond
      The C=N bond is shorter
      than the C–N bond.

c. C_{sp^2}–H_{1s}
      33% s-character
      shorter bond
      or
      C_{sp^3}–H_{1s}
      25% s-character

d. N_{sp^2}–H_{1s}
      33% s-character
      shorter bond
      or
      N_{sp^3}–H_{1s}
      25% s-character

1.32 Electronegativity increases from left-to-right across a row of the periodic table and decreases down a column. Look at the relative position of the atoms to determine their relative electronegativity.

a.   most electropositive
      most electronegative
      $Se < S < O$
      increasing electronegativity

b.   most electropositive
      most electronegative
      $Na < P < Cl$
      increasing electronegativity

c.   most electropositive
      most electronegative
      $S < Cl < F$
      increasing electronegativity

d.   most electropositive
      most electronegative
      $P < N < O$
      increasing electronegativity

1.33 Dipoles result from unequal sharing of electrons in covalent bonds. More electronegative atoms "pull" electron density towards them, making a dipole. **Dipole arrows point towards the atom of higher electron density.**

a. $\delta+$ $\delta-$
      H–F

b. $\delta+$ B–C $\delta-$

c. $\delta-$ C–Li $\delta+$

d. $\delta+$ C–Cl $\delta-$

1.34 Polar molecules result from a net dipole. To determine polarity, draw the molecule in three dimensions around any polar bonds, draw in the dipoles, and look to see whether the dipoles cancel or reinforce.

Electronegative atom pulls e⁻ density.

a. All C–H bonds have no dipole. one polar bond net dipole = **polar molecule**

b. re-draw. **Note:** You must draw the molecule in three dimensions to observe the net dipole. In the Lewis structure, it appears the dipoles would cancel out, when in fact they add to make a polar molecule.

c. no resulting dipole = **nonpolar molecule**. Four polar bonds cancel.

d. resulting dipole = **polar molecule**

e. no resulting dipole = **nonpolar molecule**. Two polar bonds are **equal and opposite** and cancel.

1.35

a. The two circled C's are sp^3 hybridized.

b. All the C–H bonds are nonpolar. All H's bonded to O and N bear a partial positive charge (δ+).

c.

one possibility

1.36

a.

skeletal structure

b. Circled carbons are sp^3 hybridized. All others are sp^2 hybridized.

c. Each N is surrounded by three atoms and a lone pair, making it sp^3 hybridized and trigonal pyramidal in molecular shape.

d.

11 polar bonds shown in bold

1.37

a, b, c.

$C_6H_8O_7$
14 lone pairs, 2 lone pairs on each O

d. Each C that is part of a C=O is sp^2 hybridized, so there are three sp^2 C's.
e. Orbitals:

[1] C=O, Csp^2–Osp^2 and Cp–Op
[2] C–C, Csp^3–Csp^2
[3] O–H, Osp^3–$H1s$
[4] C–O, Csp^3–Osp^3

1.38

a, b, c.

$C_{11}H_{14}O_3$
6 lone pairs, 2 lone pairs on each O

d. The sp^2 hybridized C's (seven) are labeled with circles.
e. Orbitals:

[1] C–C, Csp^3–Csp^2
[2] C–C, Csp^3–Csp^3
[3] C–H, Csp^3–$H1s$
[4] C–H, Csp^2–$H1s$

1.39 Formal charge (FC) = number of valence electrons – [number of unshared electrons + ($\frac{1}{2}$)(number of shared electrons)]. C is in group 4A.

a. CH_2=$\ddot{C}H$
4 – [0 + (1/2)(8)] = 0   4 – [2 + (1/2)(6)] = –1

b.   H–$\ddot{C}$–H

c.   H–$\dot{C}$–H
4 – [2 + (1/2)(4)] = 0   4 – [1 + (1/2)(6)] = 0

d.   H–C–C
4 – [0 + (1/2)(8)] = 0   4 – [0 + (1/2)(6)] = +1

1.40 Formal charge (FC) = number of valence electrons – [number of unshared electrons + ($\frac{1}{2}$)(number of shared electrons)]. N is in group 5A and O is in group 6A.

5 – [0 + (1/2)(8)] = +1

a.

b.   :$\ddot{N}$=N=$\ddot{N}$:

c.

d.

5 – [2 + (1/2)(6)] = 0

5 – [4 + (1/2)(4)] = –1   5 – [4 + (1/2)(4)] = –1

6 – [2 + (1/2)(6)] = +1   6 – [4 + (1/2)(4)] = 0

5 – [4 + (1/2)(4)] = –1

1.41 Follow the steps in Answer 1.4 to draw Lewis structures.

a. CH$_2$N$_2$

H–C=N=N:
│
H
or

H–C–N≡N:
│
H

valence e⁻

1 C x 4 e⁻ = 4
2 H x 1 e⁻ = 2
2 N x 5 e⁻ = 10
total e⁻ = 16

c. CH$_3$CNO

valence e⁻

2 C x 4 e⁻ = 8
3 H x 1 e⁻ = 3
1 N x 5 e⁻ = 5
1 O x 6 e⁻ = 6
total e⁻ = 22

b. CH$_3$NO$_2$

valence e⁻

1 C x 4 e⁻ = 4
3 H x 1 e⁻ = 3
1 N x 5 e⁻ = 5
2 O x 6 e⁻ = 12
total e⁻ = 24

d. ⁻CH$_2$CN

valence e⁻

2 C x 4 e⁻ = 8
2 H x 1 e⁻ = 2
1 N x 5 e⁻ = 5
1 for (–) charge = 1
total e⁻ = 16

1.42 Follow the steps in Answer 1.4 to draw Lewis structures.

a. (CH$_3$CH$_2$)$_2$O

[2] Count valence e⁻.
1 O x 6 e⁻ = 6
10 H x 1 e⁻ = 10
4 C x 4 e⁻ = 16
total e⁻ = 32

[3] 28 e⁻ used.

[4] Add lone pairs.

b. CH$_2$CHCN

[2] Count valence e⁻.
1 N x 5 e⁻ = 5
3 H x 1 e⁻ = 3
3 C x 4 e⁻ = 12
total e⁻ = 20

[3] 12 e⁻ used.

[4] Add lone pairs and π bonds.

c. (HOCH$_2$)$_2$CO

[2] Count valence e⁻.
3 O x 6 e⁻ = 18
6 H x 1 e⁻ = 6
3 C x 4 e⁻ = 12
total e⁻ = 36

[3] 22 e⁻ used.

[4] Add lone pairs and π bonds.

d. (CH$_3$CO)$_2$O

[2] Count valence e⁻.
3 O x 6 e⁻ = 18
6 H x 1 e⁻ = 6
4 C x 4 e⁻ = 16
total e⁻ = 40

[3] 24 e⁻ used.

[4] Add lone pairs and π bonds.

1.43

a.

b. Two of the possible resonance structures:

1.44 Isomers must have a different arrangement of atoms. Compounds are drawn as Lewis structures with no implied geometry.

a. Two isomers of molecular formula C_3H_7Cl

$$
\begin{array}{ccc}
& :\!\ddot{Cl}\!: & H & H \\
& | & | & | \\
H- & C- & C- & C-H \\
& | & | & | \\
& H & H & H
\end{array}
\qquad
\begin{array}{ccc}
& H & :\!\ddot{Cl}\!: & H \\
& | & | & | \\
H- & C- & C- & C-H \\
& | & | & | \\
& H & H & H
\end{array}
$$

c. Four isomers of molecular formula C_3H_9N

b. Three isomers of molecular formula C_2H_4O

1.45

Nine isomers of C_3H_6O:

1.46 Use the definition of isomers and resonance structures in Answer 1.10.

:O:

C_6H_8O

A

a. C_6H_8O **isomers**

b. $C_6H_{10}O$ different molecular formula **neither** isomers nor resonance structures

c. C_6H_8O same arrangement of atoms **resonance structures**

d. 5 membered ring C_6H_8O different arrangement of atoms **isomers**

1.47 Use the definitions of isomers and resonance structures in Answer 1.9.

$(C_8H_{11})^+$ **B**

a. $(C_8H_{11})^+$ different arrangement of atoms **isomers**

b. $(C_8H_{11})^+$ same arrangement of atoms **resonance structures**

c. $(C_8H_{11})^+$ different arrangement of atoms **isomers**

d. $(C_8H_9)^+$ different molecular formula **neither**

1.48 Use the definitions of isomers and resonance structures in Answer 1.9.

a.

two C–O bonds   one O–H bond
different arrangement of atoms
Both have molecular formula C_3H_8O = **isomers**

b.

ring   double bond
different arrangement of atoms
Both have molecular formula C_4H_8 = **isomers**

c.

same arrangement of atoms
Both have molecular formula $(C_4H_7)^-$.
Different arrangement of electrons = **resonance structures**

d. $CH_3CH_2CH_3$ and $CH_3CH_2\ddot{C}H_2$

molecular formula C_3H_8 molecular formula $(C_3H_7)^-$

different molecular formulas = **neither**

1.49 Compare the resonance structures to see what electrons have "moved." **Use one curved arrow to show the movement of each electron pair.**

a.

One electron pair moves = one arrow

b.

Four electron pairs move = four arrows

1.50 Curved arrow notation shows the movement of an electron pair. The tail begins at an electron pair (a bond or a lone pair) and the head points to where the electron pair moves.

a.

b.

c.

1.51 Use the rules in Answer 1.14.

a.

Two electron pairs move = two arrows

Charge is on both O's.

hybrid

Double bond can be in two locations.

b.

Two electron pairs move = two arrows

Charge is on both atoms.

Double bond can be in two locations.

hybrid

c.

One electron pair moves = one arrow

C–O bond has
partial double bond character.

1.52 For the compounds where the arrangement of atoms is not given, first draw a Lewis structure. Then use the rules in Answer 1.14.

a. O_3

Count valence e⁻.
3 O x 6 e⁻ = 18
total e⁻   = 18

b. NO_3^- (a central N atom)

Count valence e⁻.
1 N x 5 e⁻   = 5
3 O x 6 e⁻   = 18
(–) charge   = 1
total e⁻   = 24

c. N_3^-

Count valence e⁻.
3 N x 5 e⁻   = 15
(–) charge   = 1
total e⁻   = 16

d.

e.

1.53 a. No additional Lewis structures can be drawn for **B.**
  b. Additional Lewis structures can be drawn for **A, C,** and **D,** which are all examples of X=Y–Z* resonance.

A

C

D

1.54 To draw the **resonance hybrid,** use the rules in Answer 1.15.

resonance hybrid

1.55 A "better" resonance structure is one that has more bonds and fewer charges. The better structure is the major contributor and all others are minor contributors.

3 C–O bonds	2 C–O bonds	3 C–O bonds
no charges	2 charges	2 charges
contributes the most	contributes the least	

1.56

This C would have 5 bonds.

invalid

invalid

[Note: The pentavalent C's in (a) and (d) bear a (–1) formal charge.]

This C would have 5 bonds.

1.57 Use the rules in Answer 1.18.

a. CH₃Cl

4 groups = 109.5°

c. 3 groups = 120°

120°

4 groups = 109.5°

3 groups = 120°

e.

All C atoms have
3 groups = 120°.

120°

b. 4 groups = ~109.5°

H–N–O–H

4 groups = ~109.5°

d. 4 groups = 109.5°

HC≡C–C–OH

both C's surrounded by
2 groups = 180°

4 groups = ~109.5°

1.58 To predict the geometry around an atom, use the rules in Answer 1.17.

a. 4 groups
(4 atoms)
tetrahedral

b. (CH₃)₂N:

4 groups
(2 atoms, 2 lone pairs)
tetrahedral
(bent molecular shape)

c. 3 groups
(3 atoms)
trigonal planar

d. 3 groups
(3 atoms)
trigonal planar

e. (CH₃)₃N:

4 groups
(3 atoms, 1 lone pair)
tetrahedral
(trigonal pyramidal molecular shape)

1.59 In shorthand or skeletal drawings, **all line junctions or ends of lines represent carbon atoms.**
The C's are all tetravalent. All H's bonded to C's are drawn in the following structures. C's
labeled with (*) have no H's bonded to them.

a.

b.

1.60 In shorthand or skeletal drawings, **all line junctions or ends of lines represent carbon atoms.**
Convert by writing in all C's, and then adding H's to make the C's tetravalent.

a.

menthol
(isolated from peppermint oil)

c.

ethambutol
(drug used to treat tuberculosis)

d.

estradiol
(a female sex hormone)

b.

myrcene
(isolated from bayberry)

1.61 In skeletal formulas, leave out all C's and H's, except H's bonded to heteroatoms.

a. $(CH_3)_2CHCH_2CH_2CH(CH_3)_2$

c. $CH_3(CH_2)_2C(CH_3)_2CH(CH_3)CH(CH_3)CH(Br)CH_3$

b. $CH_3CH(Cl)CH(OH)CH_3$

d.

limonene
(oil of lemon)

1.62 In skeletal formulas, leave out all C's and H's, except H's bonded to heteroatoms.

a. $CH_3CONHCH_3$

c. $(CH_3)_3COH$

e. $CH_3COCH_2CO_2H$

b. CH_3COCH_2Br

d. CH_3COCl

f. $HO_2CCH(OH)CO_2H$

1.63 A charge on a C atom takes the place of one H atom. A negatively charged C has one lone pair, and a positively charged C has none.

a.

b.

c.

d. $CH_3-C≡\overset{+}{N}H$

e.

1.64 To determine the hybridization around the labeled atoms, use the procedure in Answer 1.29.

a.
4 groups
(3 atoms, 1 lone pair)
sp³, tetrahedral
(trigonal pyramidal
molecular shape)

b.
4 groups
(3 atoms, 1 lone pair)
sp³, tetrahedral
(trigonal pyramidal
molecular shape)

c.
4 groups
(4 atoms)
sp³, tetrahedral

d.
2 groups
(2 atoms)
sp, linear

e.
3 groups
(2 atoms, 1 lone pair)
sp², trigonal planar

1.65 To determine what orbitals are involved in bonding, use the procedure in Answer 1.27.

a.
C_{sp^2}–H_{1s}
σ: C_{sp^2}–C_{sp^2}
π: C_p–C_p
C_{sp^2}–C_{sp^3}

b.
C_{sp^2}–C_{sp^3}
:O:
σ: C_{sp^2}–O_{sp^2}
π: C_p–O_p

c. H–C≡C–C=N–CH₃
C_{sp}–C_{sp^2}
σ: C_{sp^3}–N_{sp^2}
C_{sp}–H_{1s}
σ: C_{sp}–C_{sp}
π: C_p–C_p
π: C_p–C_p

1.66

ketene
σ: C_{sp}–C_{sp^2}
π: C_p–C_p
sp   sp²
CH_2=C=O
sp²
σ: C_{sp}–O_{sp^2}
π: C_p–O_p

σ bonds
1s
sp
sp²   sp²
sp²

π bonds
π bond
π bond
H
H

[For clarity, only the large bonding lobes of the hybrid orbitals are drawn.]

1.67 To determine relative bond length, use the rules in Answer 1.31. All C's and H's are drawn in for emphasis.

H–C≡C–CH₂–C=C–C–H

C_{sp}–H_{1s}
highest
% s-character
shortest

C_{sp^2}–H_{1s}
middle
% s-character
middle

C_{sp^3}–H_{1s}
lowest
% s-character
longest

1.68

b. and d.   bond (1)
longest,
weakest C–C
single bond

c. shortest, **strongest** C–C bond

bond (2)

a. **shortest** C–C
single bond

e. strongest C–H bond
f. Bond (1) is a C_{sp^3}–C_{sp^3} bond, and bond (2) is a C_{sp^3}–C_{sp^2} bond. Bond (2) is shorter due to the increased percent s-character in the sp² hybridized carbon.

1.69 Percent *s*-character determines the strength of a bond. **The higher percent *s*-character of an orbital used to form a bond, the stronger the bond.**

vinyl chloride

$CH_2=CH-Cl$

33% *s*-character
higher percent *s*-character
stronger bond

C_{sp^2}

chloroethane (ethyl chloride)

CH_3-CH_2-Cl

25% *s*-character

C_{sp^3}

1.70 Dipoles result from unequal sharing of electrons in covalent bonds. More electronegative atoms "pull" electron density towards them, making a dipole.

a. $\overset{\delta+}{NH_2}-\overset{\delta-}{OH}$

b. $\delta+$ —$\overset{\delta-}{NH_2}$

c. $\delta-$ —Li $\delta+$

1.71 Use the directions from Answer 1.34.

a. CHBr₃

H
C
Br Br
Br **net dipole**

b. CH₃CH₂OCH₂CH₃

net dipole

c.
Br
Br
net dipole

d.
Cl
Cl
no net dipole

1.72

aspirin

a. molecular formula C₉H₈O₄
b. eight lone pairs total
c. C labeled with a circle is *sp³* hybridized. All other C's are *sp²* hybridized.
d. three possible resonance structures:

caffeine

a. molecular formula C₈H₁₀N₄O₂
b. eight lone pairs total
c. C's labeled with a circle are *sp³* hybridized. All other C's are *sp²* hybridized.
d. three possible resonance structures:

1.73

1 σ and 2 π bonds
polar bond

σ
(essentially) nonpolar

tetrahedral ——→ CH₃—C≡N:
sp^3 hybridized

linear
sp hybridized

linear
sp hybridized
The lone pair is in an
sp hybrid orbital.

All C–H bonds are nonpolar σ bonds.
All H's use a 1s orbital in bonding.

1.74 a. sp^2

b. Each C is trigonal planar; the ring is flat, drawn as a hexagon.

c.

d. Benzene is stable because of its two resonance structures that contribute equally to the hybrid. (This is only part of the story. We'll learn more about benzene's unusual stability in Chapter 17.)

1.75

4 groups
sp^3
tetrahedral
(trigonal pyramidal
molecular shape)

3 groups
sp^2
trigonal planar

3 groups
sp^2
trigonal planar

a.

4 groups
sp^3
tetrahedral

3 groups
sp^2
trigonal planar

b.

All C–O, C–N, C–S, N–H, and O–H bonds are polar
and labeled with arrows.
All partial positive charges lie on the C.
All partial negative charges lie on the O, N, or S.
In OH and NH bonds, H bears a δ+.

c.

6 π bonds

d. **33% s-character = sp^2 hybridized**

These C–H bonds have a C atom
with 33% s-character.

1.76

a, b, c.

4 groups
*sp*³
tetrahedral
(trigonal pyramidal molecular shape)
lone pair in *sp*³ orbital

3 groups
*sp*²
trigonal planar
lone pair in *sp*² orbital

d.

constitutional isomer

e.

resonance structure

1.77 a.

longest C–N bond

shortest C–N bond

longest C–C bond

b. The C–C bonds in the CH₂CH₃ groups are the longest because they are formed from *sp*³ hybridized C's.

c. The shortest C–C bond is labeled with an asterisk (*) because it is formed from orbitals with the highest percent *s*-character (C*sp*–C*sp*²).

d. The longest C–N bond is formed from the *sp*³ hybridized C atom bonded to a N atom [labeled in part (a)].

e. The shortest C–N bond is the triple bond (C≡N); increasing the number of electrons between atoms decreases bond length.

f.

g.

1.78

⁺CH₃

3 groups
*sp*² trigonal planar
plot **A**

The blue region is evidence of the electron-poor cation.

⁻:CH₃

4 groups
*sp*³ tetrahedral
(The molecular shape is trigonal pyramidal.)
plot **B**

The red region is evidence of the electron-rich anion.

1.79 If the N atom is sp^2 hybridized, the lone pair occupies a p orbital, which can overlap with the π bond of the adjacent C=O. This allows electron density to delocalize, which is a stabilizing feature.

σ bonds in –CONH₂

C, O, and N use sp^2 hybrid orbitals to form σ bonds.

π bond formed by overlap of C$_p$–O$_p$

π bonds in –CONH₂

The lone pair occupies the unhybridized p orbital.

1.80

a.

[1]

CH₃—OH

sp^3

25% s-character
The lower percent s-character
makes this bond longer.

[2]

CH₃—C(=O)—OH

sp^2

33% s-character

b.

[3]

CH₃—C(=O)—O⁻   ←→   CH₃—C(O⁻)=O

[4]

two resonance structures

CH₃—C(:O: δ−)=Ö: δ−

hybrid

Bonds [3] and [4] are both equivalent in length, because the anion is resonance stabilized, and the C–O bond of the hybrid is a composite of one single bond and one double bond. Both resonance structures contribute equally to the hybrid. Since each C–O bond in the hybrid has partial double bond character, it is shorter than the C–O bond labeled [2].

1.81 Ten additional resonance structures are drawn. (There are more possibilities.)

1.82 Polar bonds result from unequal sharing of electrons in covalent bonds. Normally we think of more electronegative atoms "pulling" more of the electron density towards them, making a dipole. In looking at a C_{sp^2}–C_{sp^3} bond, the atom with a higher percent s-character will "pull" more of the electron density towards it, creating a small dipole.

33% s-character
higher percent s-character
pulls more electron density
more electronegative

$\delta-$ $\delta+$
C_{sp^2} — C_{sp^3} —

25% s-character

1.83

Isomers of C_4H_8:

These two compounds are different because of restricted rotation around the C=C (Section 8.2B).

1.84 Carbocation **A** is more stable than carbocation **B** because resonance distributes the positive charge over two carbons. Delocalizing electron density is stabilizing. **B** has no possibility of resonance delocalization.

1.85

Chapter 2: Acids and Bases

Chapter Review

A comparison of Brønsted–Lowry and Lewis acids and bases

Type	Definition	Structural feature	Examples
Brønsted–Lowry acid (2.1)	proton donor	a proton	HCl, H_2SO_4, H_2O, CH_3CO_2H, TsOH
Brønsted–Lowry base (2.1)	proton acceptor	a lone pair or a π bond	^-OH, $^-OCH_3$, H^-, $^-NH_2$, $CH_2=CH_2$
Lewis acid (2.8)	electron pair acceptor	a proton, or an unfilled valence shell, or a partial (+) charge	BF_3, $AlCl_3$, HCl, CH_3CO_2H, H_2O
Lewis base (2.8)	electron pair donor	a lone pair or a π bond	^-OH, $^-OCH_3$, H^-, $^-NH_2$, $CH_2=CH_2$

Acid–base reactions

[1] A Brønsted–Lowry acid donates a proton to a Brønsted–Lowry base (2.2).

             acid   **base**   **conjugate base**   **conjugate acid**

    proton donor   proton acceptor

[2] A Lewis base donates an electron pair to a Lewis acid (2.8).

    Lewis acid   Lewis base

    electrophile   nucleophile

- Electron-rich species react with electron-poor ones.
- Nucleophiles react with electrophiles.

Important facts

- Definition: **$pK_a = -\log K_a$**. The **lower the pK_a**, the **stronger** the acid (2.3).

             NH_3 versus H_2O

           $pK_a = 38$ $pK_a = 15.7$

                                               lower pK_a = stronger acid

- The stronger the acid, the weaker the conjugate base (2.3).

- In proton transfer reactions, equilibrium favors the weaker acid and the weaker base (2.4).

- An acid can be deprotonated by the conjugate base of any acid having a **higher pK$_a$** (2.4).

Acid	pK$_a$	Conjugate base	
CH_3CO_2–H	4.8	$CH_3CO_2^-$	
CH_3CH_2O–H	16	$CH_3CH_2O^-$	These bases
HC≡CH	25	HC≡C$^-$	can deprotonate
H–H	35	H$^-$	CH_3CO_2–H.
	higher pK$_a$ than CH_3CO_2–H		

Factors that determine acidity (2.5)

[1] **Element effects** (2.5A)   The acidity of H–A increases both left-to-right across a row and down a column of the periodic table.

[2] Inductive effects (2.5B)

The acidity of H–A increases with the presence of electron-withdrawing groups in A.

CH₃CH₂OH

weaker acid

⟶ CH₃CH₂O⁻

No additional electronegative atoms stabilize the conjugate base.

CF₃CH₂OH

stronger acid

⟶

CF₃ withdraws electron density, stabilizing the conjugate base.

[3] Resonance effects (2.5C)

The acidity of H–A increases when the conjugate base A:⁻ is resonance stabilized.

CH₃CH₂Ö–H ⟶ CH₃CH₂Ö:⁻

ethanol

ethoxide

conjugate base

only **one** Lewis structure

acetic acid

more acidic

acetate

conjugate base

two resonance structures

[4] Hybridization effects (2.5D)

The acidity of H–A increases as the percent *s*-character of the A:⁻ increases.

CH₃CH₃

ethane

pK_a = 50

CH₂=CH₂

ethylene

pK_a = 44

H–C≡C–H

acetylene

pK_a = 25

Increasing acidity

Practice Test on Chapter Review

1.a. Given the pK_a data, which of the following bases is strong enough to deprotonate C_6H_5OH
   ($pK_a = 10$) so that the equilibrium lies to the right?

Compound	pK_a
H_3O^+	−1.7
NH_4^+	9.4
H_2O	15.7
NH_3	38

1. NaOH
2. $NaNH_2$
3. NH_3
4. Compounds (1) and (2) are strong enough to deprotonate C_6H_5OH.
5. Compounds (1), (2), and (3) are all strong enough to deprotonate C_6H_5OH.

   b. Which of the following statements is true about pK_a, acidity, and basicity?
      1. A higher pK_a means the acid is less acidic.
      2. In an acid–base reaction, the equilibrium lies on the side of the acid with the higher pK_a.
      3. A lower pK_a value for the acid means the conjugate base is more basic.
      4. Statements (1) and (2) are both true.
      5. Statements (1), (2), and (3) are all true.

   c. Which of the following species can be Lewis acids?
      1. BCl_3
      2. CH_3OH
      3. $(CH_3)_3C^+$
      4. Both (1) and (2) can be Lewis acids.
      5. Species (1), (2), and (3) can all be Lewis acids.

2. Answer the following questions about compounds **A–D.**

      A   B   C   D

   a. Which compound is the strongest acid?
   b. Which compound forms the strongest conjugate base?
   c. The conjugate base of **C** is strong enough to remove a proton on which compound(s), so that the equilibrium favors the products?

3. (a) Which compound is the strongest Brønsted–Lowry acid? (b) Which compound is the weakest Brønsted–Lowry acid?

      A   B   C   D

4. Draw all the products formed in the following reactions.

a. [structure with NHCH₃] + HBr ⟶   b. [structure with Cl, HO, NO₂] + Na⁺ ⁻NH₂ ⟶

5. Draw the product(s) formed in the following Lewis acid–base reaction.

(CH₃)₂ĊH + CH₃OH ⟶

Answers to Practice Test

1. a. 4   2. a. **D**   3. a. **C**   4. a.   5.
   b. 4   b. **A**   b. **B**
   c. 5   c. **B, D**

Answers to Problems

2.1 Brønsted–Lowry acids are **proton donors** and must contain a hydrogen atom.
Brønsted–Lowry bases are **proton acceptors** and must have an available electron pair (either a lone pair or a π bond).

a. HB̈r: N̈H₃   CCl₄
    acid   **acid**   not an acid—no H

b. CH₃CH₃   (CH₃)₃CÖ:⁻   H–C≡C–H
    no lone pairs lone pairs
    or π bonds   on O
    not a base   **base**   base— π bonds

c. CH₃CH₂ÖH   CH₃CH₂CH₂CH₃   CH₃–C(=Ö:)–:ÖCH₃
    base—lone pairs on O   not a base—no lone pairs   **base**—lone pairs on O's, π bond
    acid—contains H atoms   or π bonds   **acid**—contains H atoms
                              acid—contains H atoms

2.2 A Brønsted–Lowry base accepts a proton to form the conjugate acid. A Brønsted–Lowry acid loses a proton to form the conjugate base.

a. NH₃ ⟶ NH₄⁺   b. HBr ⟶ Br⁻
   Cl⁻ ⟶ HCl   HSO₄⁻ ⟶ SO₄²⁻
   (CH₃)₂C=O ⟶ (CH₃)₂C=ÖH⁺   CH₃OH ⟶ CH₃O⁻

2.3 a. True.

$$CH_2{=}CH_2 \;+\; H^+ \longrightarrow CH_3CH_2^+$$
base   conjugate acid

b. False. $CH_3CH_2^-$ cannot be the conjugate base of $CH_3CH_2^+$ because they both have the same number of H's and a conjugate base must have one fewer H.

c. False. $CH_2{=}CH_2$ and $CH_3CH_2^-$ differ by the presence of H^-, not H^+.

d. True.

$$CH_2{=}CH_2 \;\xrightarrow{\text{Remove } H^+}\; CH_2{=}\overset{-}{C}H$$
acid   conjugate base

e. True.

$$CH_3CH_2^- \;+\; H^+ \longrightarrow CH_3CH_3$$
base   conjugate acid

2.4 The Brønsted–Lowry base accepts a proton to form the conjugate acid. The Brønsted–Lowry acid loses a proton to form the conjugate base. Use curved arrows to show the movement of electrons (**NOT protons**). Re-draw the starting materials if necessary to clarify the electron movement.

a.

H–Cl: +   (ether)   ⇌   :Cl:⁻   +   (protonated ether)

acid   base   conjugate base   conjugate acid

b.

(ketone)–H   +   :O⁻ (alkoxide)   ⇌   (enolate)   +   H–O (alcohol)

acid   base   conjugate base   conjugate acid

2.5 To draw the products:

[1] Find the acid and base.

[2] Transfer a proton from the acid to the base.

[3] Check that the charges on each side of the arrows are balanced.

a.

(ester with Cl and O–H) + :OCH₃⁻ ⇌ (carboxylate) + HO–CH₃   (–)1 charge on each side

acid   base

b.

(cyclohexyl–C≡C–H) + H:⁻ ⇌ (cyclohexyl–C≡C:⁻) + H₂   (–)1 charge on each side

acid   base

c.

(cyclopentyl)–NH₂ + H–Cl: ⇌ (cyclopentyl)–NH₃⁺ + :Cl:⁻   net neutral on each side

base   acid

d.

(propyl)–OH + H–OSO₃H ⇌ (propyl)–OH₂⁺ + ⁻OSO₃H   net neutral on each side

base   acid

2.6 Draw the products in each reaction as in Answer 2.4.

2.7 The smaller the pK_a, the stronger the acid. The larger the K_a, the stronger the acid.

a. CH₃CH₂CH₃   or   CH₃CH₂OH

   pK_a = 50   pK_a = 16
                          ↑
                    smaller pK_a
                    stronger acid

b. [phenol, OH] or [toluene, CH₃]

   K_a = 10⁻¹⁰ K_a = 10⁻⁴¹
   ↑
   larger K_a
   stronger acid

2.8 To convert from K_a to pK_a, take (–) the log of the K_a; **pK_a = –log K_a.**
To convert pK_a to K_a, take the antilog of (–) the pK_a.

a. K_a = 10⁻¹⁰ K_a = 10⁻²¹ K_a = 5.2 x 10⁻⁵   b. pK_a = 7   pK_a = 11   pK_a = 3.2
    ↓   ↓   ↓   ↓   ↓   ↓
    pK_a = 10   pK_a = 21   pK_a = 4.3 K_a = 10⁻⁷ K_a = 10⁻¹¹ K_a = 6.3 x 10⁻⁴

2.9 Since **strong acids form weak conjugate bases,** the basicity of conjugate bases increases with increasing pK_a of their acids. Find the pK_a of each acid from Table 2.1 and then rank the acids in order of increasing pK_a. This will also be the order of increasing basicity of their conjugate bases.

2.10 Use the definitions in Answer 2.9 to compare the acids. The smaller the pK_a, the larger the K_a and the stronger the acid. When a stronger acid dissolves in water, the equilibrium lies further to the right.

HCO₂H   (CH₃)₃CCO₂H
formic acid   **pivalic acid**
pK_a = 3.8   pK_a = 5.0

a. smaller pK_a = larger K_a c. weaker acid = stronger conjugate base
b. smaller pK_a = stronger acid
d. stronger acid = equilibrium further to the right

2.11 To estimate the pK_a of the indicated bond, find a similar bond in the pK_a table (H bonded to the same atom with the same hybridization).

a.

For NH$_3$, pK_a is 38.
estimated pK_a = 38

b.

For CH$_3$CH$_2$OH,
pK_a is 16.
estimated pK_a = 16

c.

For CH$_3$COOH, pK_a is 4.8.
estimated pK_a = 5

2.12 Label the acid and the base and then transfer a proton from the acid to the base. To determine if the reaction will proceed as written, compare the pK_a of the acid on the left with the conjugate acid on the right. **The equilibrium always favors the formation of the weaker acid and the weaker base.**

a. CH$_2$=CH$_2$ + H:⁻ ⇌ CH$_2$=ĊH + H$_2$

   acid   base   conjugate base conjugate acid

   pK_a = 44   pK_a = 35

   weaker acid

   Equilibrium favors the **starting materials.**

b. CH$_4$ + ⁻:ÖH ⇌ :CH$_3$ + H$_2$Ö:

   acid base   conjugate base conjugate acid

   pK_a = 50   pK_a = 15.7

   weaker acid

   Equilibrium favors the **starting materials.**

c. CH$_3$CO$_2$H + CH$_3$CH$_2$Ö:⁻ ⇌ CH$_3$CO$_2$⁻ + CH$_3$CH$_2$ÖH

   acid   base   conjugate base conjugate acid

   pK_a = 4.8   pK_a = 16

                    weaker acid

   Equilibrium favors the **products.**

d. :Ċl:⁻ + CH$_3$CH$_2$ÖH ⇌ HĊl: + CH$_3$CH$_2$Ö:⁻

   base   acid   conjugate acid conjugate base

       pK_a = 16   pK_a = –7

      weaker acid

   Equilibrium favors the **starting materials.**

2.13 An acid can be deprotonated by the conjugate base of any acid with a higher pK_a.

CH$_3$CN
pK_a = 25
Any base having a conjugate acid with a pK_a higher than 25 can deprotonate this acid.

Base	Conjugate acid	pK_a
NaH	H$_2$	35
Na$_2$CO$_3$	HCO$_3$⁻	10.2
NaOH	H$_2$O	15.7
NaNH$_2$	NH$_3$	38
NaHCO$_3$	H$_2$CO$_3$	6.4

Only NaH and NaNH$_2$ are strong enough to deprotonate acetonitrile.

2.14 The acidity of H–Z **increases left-to-right across a row and down a column** of the periodic table.

a.   or H$_2$O

  only C–H bonds   O–H bond

    O is farther to the right in the periodic table, so H$_2$O is the stronger acid.

b.   or H$_2$S

         S–H bond

  only C–H bonds

  Because acidity increases across a row and down a column, H$_2$S is the stronger acid.

2.15

Because acidity increases across a row and down a column, the order of acidity is N–H < O–H < S–H.

2.16 Look at the element bonded to the acidic H and decide its acidity based on the periodic trends. **Farther to the right and down the periodic table is more acidic.**

a.
most acidic

Molecule contains C–H and O–H bonds.
O is farther right; therefore, O–H hydrogen is the most acidic.

b.
most acidic

Molecule contains C–H, N–H, and O–H bonds.
O is farthest right; therefore, O–H hydrogen is the most acidic.

c.
most acidic

Molecule contains C–H and N–H bonds.
N is farther right; therefore, N–H hydrogen is the most acidic.

2.17 The acidity of HA increases left-to-right across the periodic table. Pseudoephedrine contains C–H, N–H, and O–H bonds. The O–H bond is most acidic.

pseudoephedrine
skeletal structure

2.18 Compare the most acidic protons in each compound to determine the stronger acid.

a.   or

N–H bond

N is farther to the right in the periodic table.
stronger acid

C–H bond

b.   or

C–H bond

O–H bond

O is farther to the right in the periodic table.
stronger acid

2.19 **More electronegative atoms stabilize the conjugate base, making the acid stronger.**
Compare the electron-withdrawing groups on the acids below to decide which is a stronger acid **(more electronegative groups = more acidic).**

a.   or

more acidic

F is more electronegative than Cl, making the O–H bond in the acid on the right **more acidic.**

b.   or

Cl is closer to the acidic O–H bond.
more acidic

Cl is farther from the O–H bond.

c.

more acidic
NO$_2$ is electron withdrawing, making the
O–H bond in the acid on the
right **more acidic.**

2.20   More electronegative groups stabilize the conjugate base, making the acid stronger.

HOCH$_2$CO$_2$H   CH$_3$CO$_2$H

an α-hydroxy acid   acetic acid
The extra OH group contains an electronegative O, which
stabilizes the conjugate base.
stronger acid

2.21   HBr is a stronger acid than HCl because Br is farther down a column of the periodic table, and the larger Br⁻ anion is more stable than the smaller Cl⁻ anion. In these acids the H is bonded directly to the halogen. In HOCl and HOBr, the H is bonded to O, and the halogens Cl and Br exert an inductive effect. In this case, the more electronegative Cl stabilizes ⁻OCl more than the less electronegative Br stabilizes ⁻OBr. Thus, HOCl forms the more stable conjugate base, making it the stronger acid.

2.22   The acidity of an acid increases when the conjugate base is resonance stabilized. Compare the conjugate bases of acetone and propane to explain why acetone is more acidic.

acetone
pK$_a$ = 19.2

One resonance structure places the (–) charge on the more
electronegative O atom. This is especially good.

CH$_3$CH$_2$CH$_3$ →   CH$_3$CH$_2$CH$_2$ only one Lewis structure
less stable conjugate base

propane
pK$_a$ = 50   (Any C–H bond in the starting
material can be removed.)

2.23   The acidity of an acid increases when the conjugate base is resonance stabilized. Acetonitrile has a resonance-stabilized conjugate base, which accounts for its acidity.

The negative charge is stabilized by
delocalization on the C and N atoms.

acetonitrile
(one Lewis structure)   Having the (–) charge on the electronegative N atom adds stability.

2.24   Increasing percent s-character makes an acid more acidic. Compare the percent s-character of the carbon atoms in each of the C–H bonds in question. A stronger acid has a weaker conjugate base.

2.25 To compare the acids, first **look for element effects.** Then identify electron-withdrawing groups, resonance, or hybridization differences.

a.

C is farthest left in the periodic table. **CH bond is least acidic.**

intermediate acidity

O is farthest right in the periodic table. **OH bond is most acidic.**

c.

C is farthest left in the periodic table. **CH bond is least acidic.**

intermediate acidity

O is farthest right in the periodic table. **OH bond is most acidic.**

b.

OH group **least acidic**

intermediate acidity

Br is electron withdrawing and the conjugate base is resonance stabilized. **most acidic**

2.26 Look at the element bonded to the acidic H and decide its acidity based on the periodic trends. **Farther to the right and down the periodic table is more acidic.**

a.

most acidic

THC
tetrahydrocannabinol
The molecule contains C–H and O–H bonds.
O is farther right; therefore,
O–H hydrogen is the most acidic.

b.

most acidic

ketoprofen
The molecule contains C–H and O–H bonds.
O is farther right; therefore,
O–H hydrogen is the most acidic.

2.27

A is resonance stabilized so A is the weaker base.

The negative charge is localized on O in **B**, so **B** is the stronger base.

A   **B**

2.28 Draw the products of proton transfer from the acid to the base.

a. $(CH_3)_2CHO{-}H$ + $Na^+ H:^-$ ⇌ $(CH_3)_2CHO:^- Na^+$ + H_2
     acid   base   conjugate base   conjugate acid

b. $(CH_3)_2CHO{-}H$ + $H{-}OSO_3H$ ⇌ $(CH_3)_2CHOH_2^+$ + HSO_4^-
     base   acid   conjugate acid   conjugate base

c. $(CH_3)_2CHO{-}H$ + $Li^+ \ {}^-N[CH(CH_3)_2]_2$ ⇌ $(CH_3)_2CHO:^- Li^+$ + $HN[CH(CH_3)_2]_2$
     acid   base   conjugate base   conjugate acid

d. $(CH_3)_2CHO{-}H$ + $H{-}OCOCH_3$ ⇌ $(CH_3)_2CHOH_2^+$ + $^-OCOCH_3$
     base   acid   conjugate acid   conjugate base

2.29 To cross a cell membrane, amphetamine must be in its neutral (not ionic) form.

absorption here
in the neutral form

$:NH_2$ protonation → $^+NH_3$ deprotonation → $:NH_2$
         by HCl in   in the intestines
         the stomach

amphetamine

2.30 **Lewis bases are electron pair donors:** they contain a lone pair or a π bond.

a. $\ddot{N}H_3$ b. $CH_3CH_2CH_3$ c. $H:^-$ d. $H-C{\equiv}C-H$

yes—has   **no**—no lone pair   **yes**—has   **yes**—has
lone pair   or π bond   lone pair   2 π bonds

2.31 **Lewis acids are electron pair acceptors.** Most Lewis acids contain a proton or an unfilled valence shell of electrons.

a. BBr_3 b. CH_3CH_2OH c. $(CH_3)_3C^+$ d. Br^-

yes   **yes**   **yes**   **no**
unfilled valence shell   contains a proton   unfilled valence shell   no proton
on B   on C   no unfilled valence shell

2.32 Label the Lewis acid and Lewis base and then draw the curved arrows.

a.

Lewis acid
unfilled valence shell
on B

Lewis base
lone pairs
on O

new bond

b.

Lewis acid
unfilled valence
shell on C

Lewis base
lone pairs
on O

2.33 A Lewis acid is also called an **electrophile**. When a Lewis base reacts with an electrophile other than a proton, it is called a **nucleophile**. Label the electrophile and nucleophile in the starting materials and then draw the products.

a.

Lewis base
nucleophile
lone pairs
on O

BBr₃

Lewis acid
electrophile
unfilled valence shell
on B

b.

Lewis base
nucleophile
lone pairs
on O

AlCl₃

Lewis acid
electrophile
unfilled valence shell
on Al

2.34 Draw the product of each reaction by using an electron pair of the Lewis base to form a new bond to the Lewis acid.

a. CH₃CH₂—N̈—CH₂CH₃ + B(CH₃)₃ ⟶ CH₃CH₂—N⁺—CH₂CH₃
   | |
   CH₂CH₃   CH₂CH₃
                                               with CH₃—B⁻—CH₃ / CH₃ group

Lewis base
nucleophile
lone pair
on N

Lewis acid
electrophile
unfilled valence shell
on B

b. CH₃CH₂—N̈—CH₂CH₃ + ⁺C(CH₃)₃ ⟶ CH₃CH₂—N⁺—CH₂CH₃
   | |
   CH₂CH₃   CH₂CH₃
                                               with CH₃—C—CH₃ / CH₃ group

Lewis base
nucleophile
lone pair
on N

Lewis acid
electrophile
unfilled valence shell
on C

c. CH₃CH₂—N̈—CH₂CH₃ + AlCl₃ ⟶ CH₃CH₂—N⁺—CH₂CH₃
   | |
   CH₂CH₃   CH₂CH₃
                                               with Cl—Al⁻—Cl / Cl group

Lewis base
nucleophile
lone pair
on N

Lewis acid
electrophile
unfilled valence shell
on Al

2.35 Curved arrows begin at the Lewis base and point towards the Lewis acid.

Lewis base
contains a lone pair

Lewis acid
contains a proton

2.36 a, b. Since acidity increases from left-to-right across a row of the periodic table and propranolol has C–H, N–H, and O–H bonds, the O–H bond is most acidic. NaH is a base that removes the most acidic OH proton.

most acidic

propranolol
skeletal structure

c, d. Of the atoms with lone pairs (N and O), N is to the left in the periodic table, making it the most basic site. HCl is an acid, which protonates the most basic site.

2.37 a, b. Using periodic trends, the N–H bond of amphetamine is most acidic. NaH is a base that removes an acidic proton on N, forming H_2 in the process.

amphetamine
skeletal structure   **most acidic**

c. HCl protonates the lone pair on N (the most basic site).

2.38 To draw the conjugate acid of a Brønsted–Lowry base, **add a proton to the base.**

a. $HCO_3^- \xrightarrow{H^+} H_2CO_3$

2.39 To draw the conjugate base of a Brønsted–Lowry acid, **remove a proton from the acid.**

a. $HCO_3^- \xrightarrow{-H^+} CO_3^{2-}$

c.

b.

d.

2.40 To draw the products of an acid–base reaction, transfer a proton from the acid (H_2SO_4 in this case) to the base.

a.

b.

c.

d.

2.41 To draw the products of an acid–base reaction, transfer a proton from the acid to the base (^-OH in this case).

a.

b.

c.

d.

2.42 Label the Brønsted–Lowry acid and Brønsted–Lowry base in the starting materials and **transfer a proton from the acid to the base** for the products.

a.

b.

base   acid   conjugate acid   conjugate base

c.

acid   base   conjugate base   conjugate acid

d.

base   acid   conjugate acid   conjugate base

2.43 Draw the products of proton transfer from acid to base.

a.

acid   base

b.

base   acid

2.44 Draw the products of proton transfer from acid to base.

base   acid

base   acid

2.45 To convert pK_a to K_a, take the antilog of (−) the pK_a.

a. H_2S
  $pK_a = 7.0$
  $K_a = 10^{-7}$

b. $ClCH_2COOH$
  $pK_a = 2.8$
  $K_a = 1.6 \times 10^{-3}$

c. HCN
  $pK_a = 9.1$
  $K_a = 7.9 \times 10^{-10}$

2.46 To convert from K_a to pK_a, take (–) the log of the K_a; $\mathbf{pK_a = -\log K_a.}$

a. [structure: benzene ring]–$CH_2\overset{+}{N}H_3$

$K_a = 4.7 \times 10^{-10}$
$pK_a = 9.3$

b. [structure: benzene ring]–$\overset{+}{N}H_3$

$K_a = 2.3 \times 10^{-5}$
$pK_a = 4.6$

c. CF_3COOH

$K_a = 5.9 \times 10^{-1}$
$pK_a = 0.23$

2.47 An acid can be deprotonated by the conjugate base of any acid with a higher pK_a.

$CH_3CH_2CH_2C\equiv CH$
$\mathbf{pK_a = 25}$
Any base having a conjugate
acid with a pK_a higher than
25 can deprotonate this acid.

Base	Conjugate acid	pK_a
H_2O	H_3O^+	–1.7
NaOH	H_2O	15.7
$NaNH_2$	NH_3	38
NH_3	NH_4^+	9.4
NaH	H_2	35
CH_3Li	CH_4	50

Only $NaNH_2$, NaH, and CH_3Li
are strong enough to
deprotonate the acid.

2.48 ⁻OH can deprotonate any acid with a $pK_a < 15.7$.

a. HCOOH

$pK_a = 3.8$
stronger acid
deprotonated

b. H_2S

$pK_a = 7.0$
stronger acid
deprotonated

c. [benzene ring]–CH_3

$pK_a = 41$
weaker acid

d. CH_3NH_2

$pK_a = 40$
weaker acid

These acids are too weak to be
deprotonated by ⁻OH.

2.49 Draw the products and then compare the pK_a of the acid on the left and the conjugate acid on the right. **The equilibrium lies towards the side having the acid with a higher pK_a (weaker acid).**

a. $CH_3\ddot{N}H_2$ + H–OSO_3H ⇌ $CH_3\overset{+}{N}H_3$ + HSO_4^- **products favored**

$pK_a = -9$ $pK_a = 10.7$

b. [propanoic acid structure] $pK_a = \sim5$ + Na^+ :$\ddot{Cl}$:⁻ ⇌ [propanoate structure] $\ddot{O}$:⁻ Na^+ + $H\ddot{Cl}$: $pK_a = -7$ **starting material favored**

c. [phenol structure with Ö–H] $pK_a = 10$ + Na^+ HCO_3^- ⇌ [phenoxide structure] $\ddot{O}$:⁻ Na^+ + H_2CO_3 $pK_a = 6.4$ **starting material favored**

d. $H-C\equiv C-H$ + $CH_3CH_2^-Li^+$ ⇌ $H-C\equiv C$:⁻ Li^+ + CH_3CH_3 **products favored**

$pK_a = 25$ $pK_a = 50$

2.50 Compare element effects first and then resonance, hybridization, and electron-withdrawing groups to determine the relative strengths of the acids.

a.
acidity increases across a row
weakest acid ...... **strongest acid**

b. (CH–Cl, OH, SH structures)
only C–H bonds / **weakest acid** ...... O–H bond ...... S–H bond / **strongest acid**

c.
only C–H bonds / **weakest acid** ...... O–H bond ...... O–H bond and electron-withdrawing Cl / **strongest acid**

d.
all sp^3 C–H / **weakest acid** ...... sp^2 C–H ...... sp C–H / **strongest acid**

2.51 The strongest acid has the weakest conjugate base.

a. Draw the conjugate acid.
   Increasing acidity of conjugate acids:
   $CH_3CH_3 < CH_3NH_2 < CH_3OH$

   increasing basicity: $CH_3O^- < CH_3\overline{N}H < CH_3\overline{C}H_2$

b. Draw the conjugate acid.
   Increasing acidity of conjugate acids:
   $CH_4 < H_2O < HBr$

   increasing basicity: $Br^- < HO^- < CH_3^-$

c. Draw the conjugate acid.
   Increasing acidity of conjugate acids:

   (CH₃CH₂OH < CH₃CO–OH < ClCH₂CO–OH)

   increasing basicity:

   (ClCH₂CO–O⁻ < CH₃CO–O⁻ < CH₃CH₂O⁻)

d. Draw the conjugate acid.
   Increasing acidity of conjugate acids:

   (cyclohexyl-ethyl < cyclohexyl-vinyl < cyclohexyl-ethynyl)

   increasing basicity:

   (cyclohexyl–C≡C:⁻ < cyclohexyl-vinyl anion < cyclohexyl-ethyl anion)

2.52 **More electronegative atoms stabilize the conjugate base by an electron-withdrawing inductive effect, making the acid stronger.** Thus, an O atom increases the acidity of an acid.

(piperidinium $^+NH_2$)
$pK_a = 11.1$

(morpholinium O $^+NH_2$)
The O atom makes this cation the stronger acid.
$pK_a = 8.33$

2.53

(pentan-2-one structure with H H_b, H H_a labels)
pentan-2-one

H_a is more acidic than H_b because loss of H_a forms a resonance-stabilized conjugate base.

(mechanism: B: abstracts H_a → resonance-stabilized conjugate base)
resonance-stabilized conjugate base

(B: abstracts H_b → only one Lewis structure)
only one Lewis structure

2.54

The negative charge on O is good. This makes this resonance structure especially good.

pK$_a$ = 50

pK$_a$ = 43

pK$_a$ = 19.2

conjugate base:

one Lewis structure
weakest acid

two resonance structures
negative charge delocalized
on two carbons

two resonance structures
negative charge delocalized
on one O and one C
strongest acid

2.55 To draw the conjugate acid, look for the most basic site and protonate it. To draw the conjugate base, look for the most acidic site and remove a proton.

conjugate acid   most basic site   **A**   most acidic proton   conjugate base

2.56 Estimate the pK$_a$ of **B** as 16. A difference of 10^5 in acidity is a difference of 5 pK$_a$ units.

a.

most acidic H
pK$_a$ ~ 5

b.

most acidic
pK$_a$ ~ 25

c.

most acidic
pK$_a$ ~ 16

2.57 Remove the most acidic proton to form the conjugate base. Protonate the most basic electron pair to form the conjugate acid.

a.

only O–H bond
most acidic proton

ibuprofen

conjugate base:

b.

most basic electron pair

Increasing basicity:

—O—   —N—

conjugate acid:

cocaine

2.58   Compare the isomers.

dimethyl ether   CH$_3$—O—CH$_3$ CH$_3$CH$_2$OH   ethanol

All H's are on C.

One O–H bond
O–H bonds are more acidic
than C–H bonds.
more acidic

2.59   Compare the Lewis structures of the conjugate bases when each H is removed. The more stable base makes the proton more acidic.

The negative charge on N is stabilized by resonance.   This conjugate base is more stable, so it is formed by removal of the more acidic H.

more acidic proton

less acidic proton

no resonance stabilization
formed from the weaker acid

2.60   Look at the element bonded to the acidic H and decide its acidity based on the periodic trends. **Farther to the right across a row and down a column of the periodic table is more acidic.**

a.

CO$_2$H

most acidic

The molecule contains C–H and O–H bonds. O is farther right in the periodic table; therefore, the O–H hydrogen is the most acidic.

b.

F

most acidic

The molecule contains C–H and N–H bonds. N is farther right in the periodic table; therefore, the N–H hydrogen is the most acidic.

c.

OH ◄— **most acidic**

CH$_3$O

The molecule contains C–H, N–H, and O–H bonds. O is farthest right in the periodic table; therefore, the O–H hydrogen is the most acidic.

2.61   Use element effects, inductive effects, and resonance to determine which protons are the most acidic. The H's of the CH$_3$ group are least acidic, because they are bonded to an sp^3 hybridized C and the conjugate base formed by their removal is not resonance stabilized.

lactic acid

c > b > a
Both O–H protons [(b) and (c)] are more acidic than the C–H proton (a) by the element effect. The most acidic proton has added resonance stabilization when it is removed, making its conjugate base the most stable.

conjugate base by loss of (c):

resonance stabilization
negative charge on O in both resonance structures
This makes (c) most acidic.

conjugate base by loss of (b):

no resonance stabilization, but negative charge on O, an electronegative atom

conjugate base by loss of (a):

This conjugate base has two resonance structures, but one places a negative charge on C.

2.62

This lone pair is localized on N, so it is more basic.

bupivacaine

This lone pair is delocalized by resonance, so it is less available to donate to an acid, making it less basic.

2.63 ***Lewis bases* are electron pair donors:** they contain a lone pair or a π bond. ***Brønsted–Lowry bases* are proton acceptors:** to accept a proton they need a lone pair or a π bond. This means Lewis bases are also Brønsted–Lowry bases.

a.   lone pairs on O
both

b. $CH_3CH_2-\ddot{C}l:$ ← lone pairs on Cl
both

c.   **neither** = no lone pairs or π bonds

d.   π bonds
both

2.64 **A *Lewis acid* is an electron pair acceptor** and usually contains a proton or an unfilled valence shell of electrons. A ***Brønsted–Lowry acid* is a proton donor** and must contain a hydrogen atom. All Brønsted–Lowry acids are Lewis acids, though the reverse may not be true.

a. H_3O^+	b. Cl_3C^+	c. BCl_3	d. BF_4^-
both contains a H	**Lewis acid** unfilled valence shell on C	**Lewis acid** unfilled valence shell on B	**neither** no H or unfilled valence shell

2.65 Label the Lewis acid and Lewis base and then draw the products.

a. :Cl:⁻ + BCl₃ ⟶ Cl–B–Cl (new bond)

Lewis base   **Lewis acid**

b. (reaction with :OH⁻ giving product with new bond) **Lewis base** ... **Lewis acid** ... **new bond**

2.66 A Lewis acid is also called an **electrophile.** When a Lewis base reacts with an electrophile other than a proton, it is called a **nucleophile.** Label the electrophile and nucleophile in the starting materials and then draw the products.

a.   (compound) + AlCl₃ ⟶ (product with ⁻AlCl₃)

nucleophile   **electrophile**

b.   (compound) + BF₃ ⟶ (product O–BF₃)

nucleophile   **electrophile**

c.   (cyclohexyl cation) + H₂O ⟶ (product OH₂⁺)

electrophile   **nucleophile**

2.67 Draw the product of each reaction.

a.   (cation) + CH₃ÖH ⟶ (product O with CH₃ and H)

b.   (cation) + (CH₃)₂Ö ⟶ (product O with CH₃ and CH₃)

c.   (cation) + (CH₃)₂N̈H ⟶ (product N⁺ with CH₃, H, CH₃)

2.68

A   B   C   D

a. A conjugate acid–base pair differ in the presence of a proton. **A** and **B** (or **B** and **D**) represent a conjugate acid–base pair.

b. **A** and **D** are resonance structures because the atom placement is the same, but the electron pairs are placed differently.

c. **A** and **C** (or **C** and **D**) are constitutional isomers because they have the same molecular formula (C_5H_9NO), but the bonds are different. **A** and **D** have an N–H bond, and **C** has an O–H bond.

2.69

2.70 Draw the products of each reaction. In part (a), ⁻OH pulls off a proton and thus acts as a Brønsted–Lowry base. In part (b), ⁻OH attacks a carbon and thus acts as a Lewis base.

2.71 Answer each question about esmolol.

a.

most acidic

Esmolol contains C–H, N–H, and O–H bonds. Since acidity increases across a row of the periodic table, the OH bond is most acidic.

b.

c.

d, e, f.

All sp^2 C's are indicated with an arrow.
The N is the only trigonal pyramidal atom.
The δ+ C's are indicated with a (*).

2.72

[1] → no additional stabilization

or

[2]

more basic N   resonance-stabilized cation

Path [2] is favored because a resonance-stabilized conjugate acid is formed. The N that is part of the C=N is therefore more basic.

2.73 Draw the product of protonation of either O or N and compare the conjugate acids. When acetamide reacts with an acid, the O atom is protonated because it results in a resonance-stabilized conjugate acid.

protonate O → resonance stabilization of the + charge O is more readily protonated because the product is resonance stabilized.

acetamide

protonate N → no other resonance structure

2.74

$pK_a = 2.86$ $pK_a = 5.70$

This group destabilizes the second negative charge.

δ+ stabilizes the (–) charge of the conjugate base.

The nearby COOH group serves as an electron-**withdrawing** group to stabilize the negative charge. This makes the first proton **more** acidic than CH_3COOH.

COO⁻ now acts as an electron-**donor** group which destabilizes the conjugate base, making removal of the second proton more difficult and thus it is **less** acidic than CH_3COOH.

2.75 The COOH group of glycine gives up a proton to the basic NH₂ group to form the zwitterion.

a.   acts as a base →

glycine   acts as an acid

proton transfer →

zwitterion form

b. H₃N⁺–CH₂–C(=O)–Ö:⁻ + H–Cl → H₃N⁺–CH₂–C(=O)–OH + Cl⁻

most basic site

c. most acidic site

H₃N⁺–CH₂–C(=O)–O⁻ + Na⁺ :ÖH⁻ → H₂N–CH₂–C(=O)–O⁻ + Na⁺ + H₂Ö:

2.76 Use curved arrows to show how the reaction occurs.

[1]

[2]

Protonate the negative charge on this carbon to form the product.

2.77 Compare the OH bonds in vitamin C and decide which one is the most acidic.

vitamin C
ascorbic acid

This is the most acidic proton, because the conjugate base is most resonance stabilized.

loss of H⁺

Removal of either of these H's does not give a resonance-stabilized anion.

The most delocalized anion with 3 resonance structures.

loss of H⁺

only 2 resonance structures

This proton is less acidic, because its conjugate base is less resonance stabilized.

2.78

N has two resonance structures with the same number of bonds and charges, so both contribute approximately equally to the hybrid. This makes **N** more resonance stabilized than its conjugate base, and less willing to give up a proton than **M,** which has no similar resonance stabilization. Thus **M** is a stronger acid than **N.** (Resonance structures that break the C=O bond are not drawn in this solution, because they are possible for both compounds.)

Chapter 3: Introduction to Organic Molecules and Functional Groups

Chapter Review

Classifying carbon atoms, hydrogen atoms, alcohols, alkyl halides, amines, and amides (3.2)

- Carbon atoms are classified by the number of carbons bonded to them; a 1° carbon is bonded to one other carbon, and so forth.
- Hydrogen atoms are classified by the type of carbon atom to which they are bonded; a 1° hydrogen is bonded to a 1° carbon, and so forth.
- Alkyl halides and alcohols are classified by the type of carbon to which the OH or X group is bonded; a 1° alcohol has an OH group bonded to a 1° carbon, and so forth.
- Amines and amides are classified by the number of carbons bonded to the nitrogen atom; a 1° amine has one carbon–nitrogen bond, and so forth.

Types of intermolecular forces (3.3)

Type of force	Cause	Examples
van der Waals (VDW)	Due to the interaction of temporary dipoles • Larger surface area, stronger forces • Larger, more polarizable atoms, stronger forces	All organic compounds
dipole–dipole (DD)	Due to the interaction of permanent dipoles	$(CH_3)_2C=O$, H_2O
hydrogen bonding (HB or H-bonding)	Due to the electrostatic interaction of a H atom in an O–H, N–H, or H–F bond with another N, O, or F atom.	H_2O
ion–ion	Due to the interaction of two ions	NaCl, LiF

Increasing strength (↓)

Physical properties

Property	Observation
Boiling point (3.4A)	• For compounds of comparable molecular weight, the stronger the forces the higher the bp.

VDW
MW = 72
bp = 36 °C

VDW, DD
MW = 72
bp = 76 °C

VDW, DD, HB
MW = 74
bp = 118 °C

→ **Increasing strength of intermolecular forces
Increasing boiling point**

- For compounds with similar functional groups, the larger the surface area, the higher the bp.

bp = 0 °C   bp = 36 °C

Increasing surface area
Increasing boiling point

- For compounds with similar functional groups, the more polarizable the atoms, the higher the bp.

CH$_3$F   CH$_3$I
bp = –78 °C   bp = 42 °C

Increasing polarizability
Increasing boiling point

| **Melting point (3.4B)** | • For compounds of comparable molecular weight, the stronger the forces the higher the mp. |

VDW   VDW, DD   VDW, DD, HB
MW = 72   MW = 72   MW = 74
mp = –130 °C   mp = –96 °C   mp = –90 °C

Increasing strength of intermolecular forces
Increasing melting point

- For compounds with similar functional groups, the more symmetrical the compound, the higher the mp.

mp = –160 °C   mp = –17 °C

Increasing symmetry
Increasing melting point

Solubility (3.4C)	Types of water-soluble compounds:
	• Ionic compounds
	• Organic compounds having ≤ 5 C's, and an O or N atom for hydrogen bonding (for a compound with one functional group)
	Types of compounds soluble in organic solvents:
	• Organic compounds regardless of size or functional group
	• Examples:
	Key: VDW = van der Waals, DD = dipole–dipole, HB = hydrogen bonding MW = molecular weight

Reactivity (3.8)

- **Nucleophiles react with electrophiles.**
- Electronegative heteroatoms create electrophilic carbon atoms that react with nucleophiles.
- Lone pairs and π bonds are nucleophilic sites that react with electrophiles.

Practice Test on Chapter Review

1.a. Which of the following compounds exhibits dipole–dipole interactions?

1. CH_2Cl_2
2. $(CH_3)_2O$
3. $CH_3CH_2CH_2—OH$

4. Compounds (1) and (2) both exhibit dipole–dipole interactions.

5. Compounds (1), (2), and (3) all exhibit dipole–dipole interactions.

b. Which of the following compounds can hydrogen bond both to another molecule of itself and to water?

1.

2.

3. $CH_3C≡N$

4. Compounds (1) and (2) can each hydrogen bond to another molecule like itself and water.

5. Each of the three compounds, (1), (2), and (3), can hydrogen bond to another molecule like itself and to water.

c. Which of the following compounds has the highest boiling point?

1. $CH_3CH_2CH_2OCH_2CH_3$
2. $CH_3(CH_2)_4CH_3$
3. $CH_3(CH_2)_4NH_2$
4. $(CH_3)_2CHCH(CH_3)NH_2$
5. $(CH_3)_2CHCH(CH_3)_2$

d. Which statement(s) is (are) true about the following compounds?

A

C

B

D

1. The boiling point of **A** is higher than the boiling point of **B**.
2. The melting point of **B** is higher than the melting point of **C**.
3. The boiling point of **A** is higher than the boiling point of **D**.
4. Statements (1) and (2) are both true.
5. Statements (1), (2), and (3) are all true.

2. Consider the anti-hypertensive agent atenolol drawn below.

a. What is the hybridization of the N atom labeled with **A**?
b. What is the shape around the C atom labeled with **F**?
c. What orbitals are used to form the bond labeled with **E**?
d. Which bond (**B, C,** or **D**) is the longest bond?
e. Would you predict atenolol to be soluble in water?
f. Which of the labeled H atoms (H_a, H_b, or H_c) is most acidic?

Answers to Practice Test

1. a. 5
   b. 2
   c. 3
   d. 5

2. a. sp^3
   b. trigonal planar
   c. C_{sp2}–C_{sp3}
   d. **D**
   e. yes
   f. H_b

Answers to Problems

3.1

3.2 To classify a carbon atom as 1°, 2°, 3°, or 4°, **determine how many carbon atoms it is bonded to** (**1° C** = bonded to **one** other C, **2° C** = bonded to **two** other C's, **3° C** = bonded to **three** other C's, **4° C** = bonded to **four** other C's). Re-draw if necessary to see each carbon clearly.

To classify a hydrogen atom as 1°, 2°, or 3°, **determine if it is bonded to a 1°, 2°, or 3° C (a 1° H is bonded to a 1° C; a 2° H is bonded to a 2° C; a 3° H is bonded to a 3° C).** Re-draw if necessary.

3.3 Use the definition of 1°, 2°, 3°, or 4° carbon atoms from Answer 3.2.

3.4 A 1° ROH or RX has the functional group bonded to a 1° C; a 2° ROH or RX has the functional group bonded to a 2° C; a 3° ROH or RX has the functional group bonded to a 3° C.

3.5 Use the definitions from Answer 3.4.

dexamethasone

3.6 Amines are classified as 1°, 2°, or 3° by the number of alkyl groups bonded to the *nitrogen* atom.

3.7

3.8 A 1° amide has one carbon–nitrogen bond; a 2° amide has two carbon–nitrogen bonds; a 3° amide has three carbon–nitrogen bonds.

ction to Organic Molecules and Functional Groups 3–7

3.9 Identify the functional groups based on Tables 3.1, 3.2, and 3.3.

3.10 One possible structure for each functional group:

a. aldehyde =

b. ketone =

c. carboxylic acid =

d. ester =

3.11 One possible structure for each description:

a. $C_5H_{10}O$

aldehyde   ketone

b. $C_6H_{10}O$

ketone   alkene

ketone   alkene

3.12 Summary of forces:
- **All compounds exhibit van der Waals forces (VDW).**
- **Polar molecules have dipole–dipole forces (DD).**
- **Hydrogen bonding (H-bonding)** can occur only when a **H is bonded to an O, N, or F.**

a.
- only nonpolar C–C and C–H bonds
- **VDW only**

b.
- **VDW forces**
- 2 polar C–O bonds and a net dipole – **DD**
- no H on O, so no H-bonding

c.
- **VDW forces**
- polar C–N bonds – **DD**
- no H on N, so no H-bonding

d.
- **VDW forces**
- polar C–Cl bond – **DD**

e.
- **VDW forces**
- polar C–O bonds and a net dipole – **DD**
- H bonded to O – **H-bonding**

f.
- only nonpolar C–H and C–C bonds
- **VDW only**

3.13 One principle governs boiling point:

- **Stronger intermolecular forces = higher bp.**
  Increasing intermolecular forces: van der Waals < dipole–dipole < hydrogen bonding

Two factors affect the strength of van der Waals forces, and thus affect bp:

- **Increasing surface area = increasing bp.**
  Longer molecules have a larger surface area. Any branching decreases the surface area of a molecule.
- **Increasing polarizability = increasing bp.**

a.

only VDW
VDW and DD
polar, stronger intermolecular forces
higher boiling point

c.

longer molecule, more surface area
higher boiling point

b.

VDW, DD, and H-bonding
stronger intermolecular forces
higher boiling point

no H-bonding

d.

I is more polarizable.
higher boiling point

3.14 Increasing intermolecular forces: van der Waals < dipole–dipole < hydrogen bonding

CH_3CH_2—C(=O)—NH_2

N–H bonds allow for hydrogen bonding.
stronger intermolecular forces
higher boiling point

H—C(=O)—N(CH_3)(CH_3)

no hydrogen bonding
weaker intermolecular forces

3.15

a.

or

NH_2

more polar
stronger intermolecular forces
(H-bonding)
higher mp

b.

or

more spherical
packs better
higher mp

3.16 Compare the intermolecular forces to explain why sodium acetate has a higher melting point than acetic acid.

CH_3—C(=O)—OH
acetic acid

CH_3—C(=O)—O^- Na^+
sodium acetate

a. VDW, DD, and H-bonding
b. not ionic, lower melting point

a. VDW, DD, ionic bonds
b. Ionic bonds are the strongest: **higher melting point.**

3.17 A compound is water soluble if it is ionic or if it has an O or N atom and ≤ 5 C's.

a.

an O atom that
can H-bond with water
≤ 5 C's
water soluble

b.

nonpolar
not water soluble

c.

an N atom that can
H-bond to H_2O, but
> 5 C's
not water soluble

3.18 Hydrophobic portions will primarily be hydrocarbon chains. **Hydrophilic** portions will be polar. Circled regions are **hydrophilic** because they are polar.

All other regions are **hydrophobic** because they have only C and H.

can H-bond to itself

can H-bond to H_2O

can H-bond to H_2O

C
norethindrone

D
arachidonic acid

3.19 Like dissolves like.

- To be **soluble in water**, a molecule must be ionic, or have a polar functional group capable of H-bonding for every 5 C's.

- Organic compounds are generally **soluble in organic solvents** regardless of size or functional group.

a.

polar
polar
polar

vitamin B₃
(niacin)

soluble in water due to
two polar functional groups
and only 6 C's in the molecule

b.

nonpolar
long hydrocarbon chain

nonpolar

vitamin K₁
(phylloquinone)

soluble in organic solvents
two polar C–O bonds but the
compound has > 10 C's
water insoluble

3.20 a.

alcohol

OH

HO

amide

carboxylic acid

b. The amide, carboxylic acid, and both alcohols can all hydrogen bond with water.

c. Since pantothenic acid has only nine carbons with four functional groups that can hydrogen bond, pantothenic acid is a water-soluble vitamin.

3.21 A soap contains both a long hydrocarbon chain and a carboxylic acid salt.

a.

short chain
carboxylic acid salt

c.

no salt

b.

ionic salt

long chain
This is a soap because it contains both a
long chain and a carboxylic acid salt.

3.22 Detergents have a polar head consisting of oppositely charged ions, and a nonpolar tail consisting of C–C and C–H bonds, just like soaps do. Detergents clean by having the **hydrophobic ends of molecules surround grease**, while the **hydrophilic portion of the molecule interacts with the polar solvent** (usually water).

a detergent

SO_3^- Na$^+$

nonpolar tail
hydrophobic
This end interacts with the
grease to dissolve it.

polar head
ionic – hydrophilic
This end interacts with the water solvent
to maintain the micelle's solubility in water.

3.23

a.

HO

HO

morphine
VDW forces
dipole–dipole interactions
hydrogen bonding due to the OH groups

heroin
VDW forces
dipole–dipole interactions
NO hydrogen bonding

b. Heroin can cross the blood–brain barrier more readily because it is less polar than morphine, and therefore more soluble in the nonpolar interior of the cell membrane.

3.24 The noble gas xenon consists of uncharged atoms that exhibit only van der Waals interactions, so it is very soluble in the nonpolar interior of the cell membrane. It can thus cross the blood–brain barrier and act as an anesthetic.

3.25 Because the interior of a cell membrane is nonpolar, aspirin crosses a cell membrane as a neutral carboxylic acid, by the general rule that "like dissolves like."

3.26 Electronegative heteroatoms like N, O, or X make a carbon atom an *electrophile*.
A lone pair on a heteroatom makes it basic and nucleophilic.
Pi (π) bonds create *nucleophilic* sites and are more easily broken than σ bonds.

3.27 Electrophiles and nucleophiles react with each other.

3.28

H's that are boxed in can hydrogen bond to O of H_2O.
Atoms labeled with (*) can hydrogen bond to H of H_2O.

3.29

three ethers

Electrophilic carbons are labeled with (*).

b. $HOCH_2$

$HOCH_2$

CH_2OH

Three OH's allow for H-bonding.
Stronger intermolecular forces mean a
higher bp and mp.

3.30

a, c. — alkene, aldehyde
The most electrophilic C is labeled with *.

b. — OH
This isomer has an OH, giving more opportunities for H-bonding (through both O and H atoms), and probably making it more H_2O soluble.

3.31 Use the rules from Answer 3.2.

a. [1] 1°, 3°, 2°, 4°, 1°   [2] 2°, 3°, 2°, 1°

b. [1] CH_3 CH_3 H H, CH_3—C—C—CH_3, H CH_3
All CH_3's have 1° H's.
All CH_2's have 2° H's.
All CH's have 3° H's.

[2] H H, H—C—C—CH_3, H—C—C—H, CH_3—C—CH_2, CH_3 H

3.32 Identify the functional groups based on Tables 3.1, 3.2, and 3.3.

a. 3° amine, aromatic rings, ester
Darvon

b. 1° amine, carboxylic acid
pregabalin

c. carboxylic acid, aromatic ring
ibuprofen

d. alkyne, alkene, 2° amine, 2° alcohol
histrionicotoxin

e. aromatic ring, 2° amide, sulfide, 3° amide, HO, carboxylic acid
penicillin G

f. alkene, alkene, ester, ketone, alkene, alkenes
pyrethrin I

3.33 A cyclic ester is called a lactone. A cyclic amide is called a lactam.

a. N–CH_3
amine

b. O
ether

c. ester
lactone

d. NH
amide
lactam

3.34

salinosporamide A

3.35 Draw the constitutional isomers and identify the functional groups.

3.36 Use the rules from Answer 3.12.

a.

VDW
dipole–dipole
H-bonding (O–H bond)

b.

VDW
dipole–dipole
no H-bonding (no O–H bond)

c.

VDW
dipole–dipole
no H-bonding (no N–H bond)

d.

VDW
no dipole–dipole
(nonpolar C–C, C–H bonds)
no H-bonding (no O, N, F)

3.37 Increasing intermolecular forces: van der Waals < dipole–dipole < H-bonding

a. **increasing intermolecular forces:**

VDW

VDW
dipole–dipole

VDW
dipole–dipole
H-bonding

b. **increasing intermolecular forces:**

VDW

VDW
dipole–dipole

VDW
dipole–dipole
H-bonding

3.38

indinavir

Indinavir can hydrogen bond to another molecule of itself at boxed-in sites. Indinavir can hydrogen bond to water at both the boxed-in sites and the sites labeled with an arrow.

3.39 **A** = VDW forces; **B** = H-bonding; **C** = ion–ion interactions; **D** = H-bonding; **E** = H-bonding; **F** = VDW forces.

3.40

a.

aldehyde   ketone   ether   alcohol

b. The alcohol is the highest boiling, because it is the only isomer that can hydrogen bond with another molecule of itself.

3.41 Use the principles from Answer 3.13.

a.

smallest surface area
most branching   largest surface area

Increasing boiling point

b.

VDW   VDW   VDW
          dipole–dipole   dipole–dipole
                    H-bonding

Increasing boiling point

3.42 In $CH_3CH_2NHCH_3$, there is a N–H bond, so the molecules exhibit intermolecular hydrogen bonding, whereas in $(CH_3)_3N$ the N is bonded only to C, so there is no hydrogen bonding. The hydrogen bonding in $CH_3CH_2NHCH_3$ makes it have much **stronger intermolecular forces** than $(CH_3)_3N$. As intermolecular forces increase, the boiling point of a molecule of the same molecular weight increases.

3.43   Stronger forces, higher mp.

menthone
VDW
dipole–dipole
lower melting point

menthol
VDW
dipole–dipole
H-bonding
stronger forces
higher melting point

3.44   Stronger forces, higher mp.

VDW

VDW
DD

VDW
DD
H-bonding

Increasing intermolecular forces →

Increasing melting point

3.45   **Boiling point is determined solely by the strength of the intermolecular forces.** Because benzene has a smaller size, it has less surface area and weaker VDW interactions and therefore a lower boiling point than toluene. The increased melting point for benzene can be explained by symmetry: benzene is much more symmetrical than toluene. More symmetrical molecules can pack more tightly together, increasing their melting point. Symmetry has no effect on boiling point.

benzene
bp = 80 °C
mp = 5 °C

very symmetrical
closer packing in solid form
higher mp

and

toluene
bp = 111 °C
mp = −93 °C

less symmetrical
lower mp

3.46 Increasing polarity = increasing water solubility.

—Br

polar
no H-bonding

=O

polar
H-bonding to H_2O,
not itself

—OH

polar and
H-bonding
More opportunities
for H-bonding with its
O atom and its H on O.

3.47 Look for two things:
- To H-bond to another molecule of itself, the molecule must contain a **H bonded to O, N, or F.**
- To H-bond with water, a molecule needs **only to contain an O, N, or F.**

Only (c) can H-bond to another molecule like itself.
    c. $CH_3CH_2CONH_2$

These molecules can H-bond with water. All of these molecules have an O or N atom.
    b. $(CH_3CH_2)_3N$
    c. $CH_3CH_2CONH_2$
    d. $CH_3CH_2COOCH_3$

3.48 Draw the molecules in question and look at the intermolecular forces involved.

no H bonded to O

diethyl ether

VDW forces
dipole–dipole forces

H bonded to O:
hydrogen bonding

butan-1-ol

VDW forces
dipole–dipole forces
H-bonding

- Both have ≤ 5 C's and an electronegative O atom, so they can H-bond to water, making them soluble in water.
- Only butan-1-ol can H-bond to another molecule of itself, and this increases its boiling point.

3.49 Use the solubility rule from Answer 3.19.

a.

caffeine

many polar bonds with N and O atoms
many opportunities for H-bonding
water soluble

c.

sucrose
many polar bonds with O
11 O's and 12 C's
many opportunities for H-bonding with H_2O
water soluble

b.

mestranol

2 polar functional groups
but > 10 C's
not water soluble

d.

carotatoxin
1 polar functional group
but > 10 C's
not water soluble

3.50 Water solubility is determined by polarity. Polar molecules are soluble in water, while nonpolar molecules are soluble in organic solvents.

Arrows indicate polar functional groups.

a.

vitamin E
only 2 polar functional groups
many nonpolar C–C and C–H bonds (29 C's)
soluble in organic solvents
insoluble in H₂O

b.

pyridoxine
vitamin B₆

many polar bonds and few nonpolar bonds
soluble in H₂O
It is also **soluble in organic solvents** because it
is organic, but is probably more soluble in H₂O.

3.51 Compare the functional groups in the two components of sunscreen. Dioxybenzone will most likely be washed off in water because it contains two hydroxy groups and is more water soluble.

avobenzone
two ketones
one ether

dioxybenzone
two hydroxy groups
one ketone
one ether
more water soluble

3.52 Because of the O atoms, PEG is capable of hydrogen bonding with water, which makes PEG water soluble and suitable for a product like shampoo. PVC cannot hydrogen bond to water, so PVC is water insoluble, even though it has many polar bonds. Because PVC is water insoluble, it can be used to transport and hold water.

H-bond

poly(ethylene glycol)
PEG
water soluble

no H-bonding

poly(vinyl chloride)
PVC
water insoluble

3.53 Molecules that dissolve in water are readily excreted from the body in urine, whereas less polar molecules that dissolve in organic solvents are soluble in fatty tissue and are retained for longer periods. Compare the solubility properties of THC and ethanol to determine why drug screenings can detect THC and not ethanol weeks after introduction to the body.

tetrahydrocannabinol
THC

THC has relatively few polar
bonds compared to the number
of nonpolar bonds, making it
soluble in organic solvents
and therefore **soluble in fatty tissue.**

ethanol

Ethanol has 1 O atom and
only 2 C's, making it
soluble in water.

Due to their solubilities, **THC is retained much longer in the fatty tissue of the body,** being slowly excreted over many weeks, while ethanol is excreted rapidly in urine after ingestion.

3.54 Compare the intermolecular forces of crack and cocaine hydrochloride. Stronger intermolecular forces increase both the boiling point and the water solubility.

cocaine (crack)
neutral organic molecule

cocaine hydrochloride
a salt

The molecules are identical except for the ionic bond in cocaine hydrochloride. Ionic forces are extremely strong forces, and therefore the cocaine hydrochloride salt has a much **higher boiling point and is more water soluble.** Because the salt is highly water soluble, it can be injected directly into the bloodstream, where it dissolves. Crack is smoked because it can dissolve in the organic tissues of the nasal passages and lungs.

3.55

a.

b.

five functional groups that have
many opportunities for H-bonding
water soluble

ionic salt
more water soluble

c. Because the hydrochloride salt is ionic and therefore more water soluble, it is more readily transported in the bloodstream.

3.56   Use the rules from Answer 3.26.

3.57

3.58   More rigid cell membranes have phospholipids with *fewer* C=C's. Each C=C introduces a bend in the molecule, making the phospholipids pack less tightly. Phospholipids without C=C's can pack very tightly, making the membrane less fluid and more rigid.

The double bonds introduce kinks in the chain, making packing of the hydrocarbon chains less efficient. This makes the cell membrane formed from them more fluid.

3.59   **B** is ionic, so it does not "dissolve" in the nonpolar interior of a cell membrane and it cannot cross the blood–brain barrier. **A** is a neutral organic molecule with polar bonds, so it can enter the nonpolar interior of the cell membrane and cross the blood–brain barrier.

3.60

a, b.

quinapril

quinapril

c. Quinapril can hydrogen bond to water at all N and O atoms (boxed in).
d. Quinapril can hydrogen bond to acetone at its O–H and N–H bonds (shown with arrows).
e. The most acidic H is the lone H bonded to O, which is part of a carboxylic acid.
f. The most basic site is the N atom of the amine (circled).

3.61

a, b.

oxycodone

c. The lone OH proton is the most acidic site.
d. The lone N atom is the most basic site.
e. The N atom is surrounded by three atoms and a lone pair (four groups), so it is sp^3 hybridized.
f. Oxycodone contains seven sp^2 hybridized C's labeled with gray circles.

3.62 Because the O atom in tetrahydrofuran is in a ring, the C atoms bonded to it are kept away from the lone pairs on O. This allows the O atom to more readily hydrogen bond with water, thus increasing its solubility in water.

Electron pairs are more exposed. This facilitates H-bonding.

3.63

a.

amide amine

aromatic ring

aromatic ring

e. An isomer that can hydrogen bond should have a higher boiling point.

hydrogen bonding possible

b, c, f, g.

most basic atom

The N atom of the amide is less basic because the lone pair is part of resonance with the C=O.

Atoms that can hydrogen bond to H_2O are boxed in. Electrophilic carbons are labeled with (*).

most acidic H, $pK_a \sim 25$
All H's are bonded to C's. Removal of the H on the C adjacent to the C=O results in a resonance-stabilized anion.

d. Fentanyl exhibits van der Waals and dipole–dipole interactions but no hydrogen bonding, because there is no H bonded to O or N.

3.64

A

The OH and CHO groups are close enough that they can intramolecularly H-bond to each other. Because the two polar functional groups are involved in intramolecular H-bonding, they are less available for H-bonding to H_2O.
This makes **A** less H_2O soluble than **B,** whose two functional groups are both available for H-bonding to the H_2O solvent.

B

The OH and the CHO are too far apart to intramolecularly H-bond to each other, leaving more opportunity to H-bond with solvent.

3.65

a. melting point

fumaric acid

Fumaric acid has its two larger COOH groups on opposite ends of the molecule, and in this way it can pack better in a lattice than maleic acid, giving it a **higher mp.**

b. solubility

maleic acid

Maleic acid is more polar, giving it greater **H_2O solubility.** The bond dipoles in fumaric acid cancel.

c. removal of the first proton (pK_{a1})

In maleic acid, intramolecular H-bonding
stabilizes the conjugate base after one H is
removed, making maleic acid more acidic
than fumaric acid.

Intramolecular H-bonding
is not possible here.

d. removal of the second proton (pK_{a2})

Now the dianion is held in close proximity
in maleic acid, and this destabilizes the conjugate
base. Thus, removing the second H in maleic
acid is harder, making it a weaker acid than
fumaric acid for removal of the second proton.

The two negative charges are much farther
apart. This makes the dianion from fumaric
acid more stable and thus pK_{a2} is lower for
fumaric acid than maleic acid.

Chapter 4: Alkanes

Chapter Review

General facts about alkanes (4.1–4.3)

- Alkanes are composed of **tetrahedral, sp^3** hybridized C's.
- There are two types of alkanes: acyclic alkanes having molecular formula C_nH_{2n+2}, and cycloalkanes having molecular formula C_nH_{2n}.
- Alkanes have only **nonpolar C–C and C–H bonds** and no functional group, so they undergo few reactions.
- Alkanes are named with the suffix **-ane**.

Names of alkyl groups (4.4A)

Conformations in acyclic alkanes (4.9, 4.10)

- Alkane conformations can be classified as **staggered, eclipsed, anti,** or **gauche** depending on the relative orientation of the groups on adjacent carbons.

eclipsed	staggered	anti	gauche

- Dihedral angle = 0°
- Dihedral angle = 60°
- Dihedral angle of 2 CH₃'s = 180°
- Dihedral angle of 2 CH₃'s = 60°

- A staggered conformation is **lower in energy** than an eclipsed conformation.
- An anti conformation is **lower in energy** than a gauche conformation.

Types of strain

- **Torsional strain**—an increase in energy due to eclipsing interactions (4.9).
- **Steric strain**—an increase in energy when atoms are forced too close to each other (4.10).
- **Angle strain**—an increase in energy when tetrahedral bond angles deviate from $109.5°$ (4.11).

Two types of isomers

[1] **Constitutional isomers**—isomers that differ in the way the atoms are connected to each other (4.1A).

[2] **Stereoisomers**—isomers that differ only in the way atoms are oriented in space (4.13B).

Conformations in cyclohexane (4.12, 4.13)

- Cyclohexane exists as **two chair conformations** in rapid equilibrium at room temperature.
- Each carbon atom on a cyclohexane ring has **one axial** and **one equatorial hydrogen.** Ring-flipping converts axial H's to equatorial H's, and vice versa.

- In substituted cyclohexanes, groups larger than hydrogen are more stable in the **more roomy equatorial position.**

- Disubstituted cyclohexanes with substituents on different atoms exist as two possible stereoisomers.
  - The **cis** isomer has two groups on the **same side** of the ring, either both up or both down.
  - The **trans** isomer has two groups on **opposite sides** of the ring, one up and one down.

Oxidation–reduction reactions (4.14)

- **Oxidation** results in an **increase in the number of C–Z bonds** or a **decrease in the number of C–H bonds.**

$CH_3CH_2-OH \longrightarrow$ ethanol, acetic acid

| Increase in C–O bonds = **oxidation** |

- **Reduction** results in a **decrease in the number of C–Z bonds** or an **increase in the number of C–H bonds.**

ethylene → ethane

| Increase in C–H bonds = **reduction** |

Practice Test on Chapter Review

1.a. Which statement is true about compounds **A–D** below?

    A   **B**   **C**   **D**

    1. **A** and **C** are stereoisomers.   4. Statements (1) and (2) are both true.
    2. **B** and **D** are identical.   5. Statements (1), (2), and (3) are all true.
    3. **A** and **B** are stereoisomers.

b. Which of the following statements is true about *cis*-1-isopropyl-2-methylcyclohexane?
1. The more stable conformation has the isopropyl group in the equatorial position and the methyl group in the axial position.
2. The more stable conformation has both the methyl and isopropyl groups in the equatorial position.
3. *cis*-1-Isopropyl-2-methylcyclohexane is a stereoisomer of *trans*-1-isopopyl-3-methyl-cyclohexane.
4. Statements (1) and (2) are true.
5. Statements (1), (2), and (3) are all true.

c. Rank the following conformations in order of *increasing energy*.

A   B   C

1. C < B < A   3. A < C < B   5. A < B < C
2. C < A < B   4. B < A < C

2. Give the IUPAC name for each of the following compounds.

a.

b.

3. How are the molecules in each pair related? Are they constitutional isomers, stereoisomers, identical, or not isomers?

a.

and

b.

and

c.

and

4. Rank the following conformations in order of increasing energy. Label the conformation of lowest energy as **1**, the highest energy as **4**, and the conformations of intermediate energy as **2** and **3**.

A   B   C   D

5. Consider the following disubstituted cyclohexane drawn below:

   a. Draw the more stable chair conformation for the cis isomer.
   b. Draw the more stable chair conformation for the trans isomer.

Answers to Practice Test

1. a. 4	2. a. 5-isobutyl-2,6-dimethyl-6-propyldecane	3. a. identical	4. **A**–3 **B**–4 **C**–2 **D**–1	5. a.
b. 1	b. 1-*sec*-butyl-4-propylcyclooctane	b. constitutional isomers		
c. 2		c. stereoisomers		b.

Answers to Problems

4.1 The general molecular formula for an acyclic alkane is C_nH_{2n+2}.

Number of C atoms = n	$2n + 2$	Number of H atoms
23	2(23) + 2 =	48
25	2(25) + 2 =	52
27	2(27) + 2 =	56

4.2 2-Methylbutane has 4 C's in a row with a 1 C branch.

a.

2-methylbutane

b.

2-methylbutane

c.

re-draw

2-methylbutane

d.

5 C's in a row
pentane

4.3 Constitutional isomers differ in the way the atoms are connected to each other. To draw all the constitutional isomers:

[1] Draw all of the C's in a long chain.

[2] Take off one C and use it as a substituent. (Don't add it to the end carbon: this re-makes the long chain.)

[3] Take off two C's and use these as substituents, etc.

Five **constitutional isomers** of molecular formula C_6H_{14}:

[1] long chain   [2] with one C as a substituent   [3] using two C's as substituents

4.4 Draw each alkane to satisfy the requirements.

a.   4° C

b.   1° C   1° C
      All other C's are 2° C's.

c.   H
      1° H   3° H   2° H

4.5

A
6 C chain
CH_3 group on C3

B
identical

C
identical

D
isomer
CH_3 bonded to C2

E
identical

F
isomer
7 C chain

4.6 Use the steps from Answer 4.3 to draw the constitutional isomers.

Five **constitutional isomers** of molecular formula C_5H_{10} having one ring:

[1]   [2]   [3]

4.7 Follow these steps to name an alkane:

[1] **Name the parent chain** by finding the longest C chain.

[2] **Number the chain** so that the first substituent gets the lower number. Then **name and number all substituents,** giving like substituents a prefix (di, tri, etc.).

[3] **Combine all parts,** alphabetizing the substituents, ignoring all prefixes except *iso*.

[1]

a.

8 carbons = **octane**

[2] 4-methyl

4-*tert*-butyl

[3] **4-*tert*-butyl-4-methyloctane**

[1]

b.

6 carbons = **hexane**

[2] 4-methyl

2-methyl

[3] **2,4-dimethylhexane**

[1]

c.

9 carbons = **nonane**

[2] 6-isopropyl

3-methyl

[3] **6-isopropyl-3-methylnonane**

[1]

d.

7 carbons = **heptane**

[2]

2-methyl

4-methyl

[3] **2,4-dimethylheptane**

4.8 Use the steps in Answer 4.7 to name each alkane.

a. (CH₃)₃CCH₂CH(CH₂CH₃)₂

[1]   ↓ re-draw

$$CH_3-\underset{\underset{CH_3}{|}}{\overset{\overset{CH_3}{|}}{C}}-CH_2-\underset{\underset{CH_2CH_3}{|}}{CH}-CH_2CH_3$$

6 carbons = **hexane**

[2] 2   4

$$CH_3-\underset{\underset{CH_3}{|}}{\overset{\overset{CH_3}{|}}{C}}-CH_2-\underset{\underset{CH_2CH_3}{|}}{CH}-CH_2CH_3$$

2,2-dimethyl

4-ethyl

[3] **4-ethyl-2,2-dimethylhexane**

[1]

b.

or

[2] 2-methyl

3-ethyl   5-methyl

[3] **3-ethyl-2,5-dimethylheptane**

longest chain = 7 carbons = **heptane**
Number so there are more substituents.
Pick the upper option.

c. $CH_3(CH_2)_3CH(CH_2CH_2CH_3)CH(CH_3)_2$

[1]

↓ re-draw

$CH_3CH_2CH_2CH_2-CH-CH-CH_3$ with CH₃ branch and CH₂CH₂CH₃ branch

8 carbons = **octane**

[2]

$\overset{8}{C}H_3\overset{7}{C}H_2CH_2CH_2-\overset{4}{C}H-CH-CH_3$ with CH₃ branch and CH₂CH₂CH₃ branch (labeled 1)

4-isopropyl

[3] **4-isopropyloctane**

d. [1]

10 carbons = **decane**

[2]

2-methyl **5-sec-butyl** **3-ethyl** **7-methyl**

[3] **5-sec-butyl-3-ethyl-2,7-dimethyldecane**

4.9 To work backwards from a name to a structure:

[1] Find the parent name and draw that number of C's. Use the suffix to identify the functional group **(-ane = alkane).**

[2] Arbitrarily number the C's in the chain or ring. Add the substituents to the appropriate C's.

a. 3-methyl**hexane**

[1] 6 carbon alkane

[2] methyl on C3

b. 3,3-dimethyl**pentane**

[1] 5 carbon alkane

[2] methyl groups on C3

c. 3,5,5-trimethyl**octane**

[1] 8 carbon alkane

[2] methyl groups on C3 and C5

d. 3-ethyl-4-methyl**hexane**

[1] 6 carbon alkane

[2] methyl group on C4 — ethyl group on C3

e. 3-ethyl-5-isobutyl**nonane**

[1] 9 carbon alkane

[2] ethyl group on C3   isobutyl group on C5

4.10 Use the steps in Answer 4.7 to name each alkane.

[1] 6 carbons = **hexane**   [2] no substituents   [3] **hexane**

[1] 5 carbons = **pentane**   [2] 2-methyl   [3] **2-methylpentane**

[1] 5 carbons = **pentane**   [2] 3-methyl   [3] **3-methylpentane**

[1] 4 carbons = **butane**   [2] 2,2-dimethyl   [3] **2,2-dimethylbutane**

[1] 4 carbons = **butane**   [2] 2,3-dimethyl   [3] **2,3-dimethylbutane**

4.11 Follow these steps to name a cycloalkane:

[1] **Name the parent cycloalkane** by counting the C's in the ring and adding cyclo-.

[2] **Numbering:**

     a. **Number around the ring** beginning at a substituent and giving the second substituent the lower number.

     b. **Number to assign the lower number to the substituents alphabetically.**

     c. **Name and number all substituents,** giving like substituents a prefix (di, tri, etc.).

[3] **Combine all parts,** alphabetizing the substituents, ignoring all prefixes except *iso.*

     (Remember: If a carbon chain has more C's than the ring, the chain is the parent, and the ring is a substituent.)

a. [1]   6 carbons in ring = **cyclohexane**

[2] →1,1-dimethyl Number so the substituents are at C1.

[3] **1,1-dimethylcyclohexane**

b. [1]   5 carbons in ring = **cyclopentane**

[2] 1,2,3-trimethyl Number so the first substituent is at C1, second at C2.

[3] **1,2,3-trimethylcyclopentane**

c. [1]   6 carbons in ring = **cyclohexane**

[2] **1-butyl** **4-methyl** Number so the earlier alphabetical substituent is at C1, **b**utyl before **m**ethyl.

[3] **1-butyl-4-methylcyclohexane**

d. [1]   6 carbons in ring = **cyclohexane**

[2] **1-*sec*-butyl** **2-isopropyl** Number so the earlier alphabetical substituent is at C1, **b**utyl before **i**sopropyl.

[3] **1-*sec*-butyl-2-isopropylcyclohexane**

e. [1]   longest chain = 5 carbons = **pentane**

[2] **1-cyclopropyl** Number so the cyclopropyl is at C1.

[3] **1-cyclopropylpentane**

f. [1]   6 carbons in ring = **cyclohexane**

[2] **1,1-dimethyl** **3-butyl** Number so the two methyls are at C1.

[3] **3-butyl-1,1-dimethylcyclohexane**

4.12 To draw the structures, use the steps in Answer 4.9.

a. 1,2-dimethyl**cyclobutane**

[1]   4 carbon cycloalkane

[2]   methyl groups on C1 and C2

[3]

b. 1,1,2-trimethyl**cyclopropane**

[1]   3 carbon cycloalkane

[2]   3 CH₃'s

[3]

c. 4-ethyl-1,2-dimethyl**cyclohexane**

[1]   6 carbon cycloalkane

[2]   ethyl on C4   2 CH₃'s

[3]

d. 1-*sec*-butyl-3-isopropyl**cyclopentane**

[1]   5 carbon cycloalkane

[2]   isopropyl   *sec*-butyl

[3]

e. 1,1,2,3,4-pentamethyl**cycloheptane**

[1]   7 carbon cycloalkane

[2]   5 CH₃'s

[3]

4.13 Compare the number of C's and surface area to determine relative boiling points. Rules:
[1] Increasing number of C's = increasing boiling point.
[2] Increasing surface area = increasing boiling point (branching decreases surface area).

8 C's linear largest number of C's no branching **highest bp**

7 C's linear

7 C's one branch

7 C's three branches

increasing branching decreasing surface area **decreasing bp**

Increasing boiling point: $(CH_3)_3CCH(CH_3)_2$ < $CH_3CH_2CH_2CH_2CH(CH_3)_2$ < $CH_3(CH_2)_5CH_3$ < $CH_3(CH_2)_6CH_3$

4.14   To draw a Newman projection, visualize the carbons as one in front and one in back of each other. The C–C bond is not drawn. There is only one staggered and one eclipsed conformation.

4.15   Re-draw each Newman projection as a skeletal structure, remembering that a Newman projection shows the substituents bonded to two C's but not the C's themselves.

2-methylpentane

A

CH₃—C(CH₃)(H)—CH₂CH₂CH₃

C in back

2-methylpentane

B

CH₃—C(CH₃)(H)—C(H)(H)—CH₂CH₃

C in back

2-methylpentane

C

CH₃CH₂—C(CH₃)(H)—C(H)(H)—CH₃

C in back

3-methylpentane

4.16

H,H eclipsing
4.0 kJ/mol of destabilization

4.0 kJ/mol

4.0 kJ/mol

To calculate H,CH₃ destabilization:

14 kJ/mol (total) –
8.0 kJ/mol for 2 H,H eclipsing interactions
= **6 kJ/mol** for one H,CH₃ eclipsing interaction

4.17   To determine the energy of conformations, keep two things in mind:
[1] Staggered conformations are more stable than eclipsed conformations.
[2] Minimize steric interactions: keep large groups away from each other.
The highest energy conformation is the eclipsed conformation in which the two largest groups are eclipsed. The lowest energy conformation is the staggered conformation in which the two largest groups are anti.

rotation here

1
staggered
most stable

60°

2
eclipsed

60°

3
staggered
most stable

60°

6
eclipsed
least stable

60°

5
staggered

60°

4
eclipsed
least stable

4.18 Use the criteria in problem 4.17 to determine the relative energy.

A

B

C

D

Staggered conformations **A** and **D** are lower in energy than eclipsed conformations **B** and **C**. **D** is lower in energy than **A** because it has two gauche interactions and **A** has three. **C** is higher in energy than **B** because all the larger groups (CH_3 and CH_2CH_3) are eclipsed.
Ranking: **D < A < B < C**

4.19 To determine the most and least stable conformations, use the rules from Answer 4.17.

a. **1,2-dichloroethane**

$ClCH_2$—CH_2Cl

rotation here

1
staggered, anti

60°

2
eclipsed

60°

3
staggered, gauche

60°

6
eclipsed

60°

5
staggered, gauche

60°

4
eclipsed

4.20 Add the energy increase for each eclipsing interaction to determine the destabilization.

a.

b.

1 H,H interaction =   4.0 kJ/mol
2 H,CH₃ interactions
  (2 x 6.0 kJ/mol) = 12.0 kJ/mol
_____

Total destabilization = 16 kJ/mol

3 H,CH₃ interactions
  (3 x 6.0 kJ/mol) = **18 kJ/mol**

Total destabilization

4.21 Two points:
  • Axial bonds point up or down, while equatorial bonds point out.
  • An *up* carbon has an axial *up* bond, and a *down* carbon has an axial *down* bond.

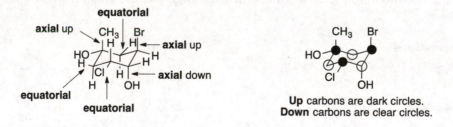

4.22

a.

b.

c.

4.23 Draw the second chair conformation by flipping the ring.

- **The *up* carbons become *down* carbons, and the axial bonds become equatorial bonds.**
- **Axial bonds become equatorial, but *up* bonds stay *up*;** that is, an axial *up* bond becomes an equatorial *up* bond.
- The conformation with **larger groups equatorial is the more stable** conformation and is present in higher concentration at equilibrium.

a.

more stable
Br is equatorial.

Draw in the H
and label the C
as up or down.

axial
H eq
Br

Axial bond is up =
up carbon

Draw second conformation.
Up carbons switch
to down carbons.

H ← eq
Br
↑
axial

Axial bond is down =
down carbon

b.

Cl

Draw in the H
and label the C
as up or down.

axial
Cl
H ↑
eq

Axial bond is up =
up carbon

Draw second conformation.
Up carbons switch
to down carbons.

eq
Cl
axial → H

Axial bond is down =
down carbon

more stable
Cl is equatorial.

c.

CH₂CH₃

more stable
CH₂CH₃ is equatorial.

Draw in the H
and label the C
as up or down.

Axial bond is up =
up carbon
H
CH₂CH₃
eq

Draw second conformation.
Up carbons switch
to down carbons.

eq
H
CH₂CH₃

Axial bond is down =
down carbon

4.24

CH₃ ax
CH₂CH₃
eq

more stable
The larger ethyl group is in the
roomier equatorial position.

CH₃ eq
CH₂CH₃
ax

4.25 Wedges represent "up" groups in front of the page, whereas dashed wedges are "down" groups in back of the page. Cis groups are on the same side of the ring, whereas trans groups are on opposite sides of the ring.

a.   ***cis*-1,2-dimethylcyclopropane**

or

cis = same side of the ring
both groups on wedges or
both on dashed wedges

b.   ***trans*-1-ethyl-2-methylcyclopentane**

or

trans = opposite sides of the ring
one group on a wedge,
one group on a dashed wedge

4.26 Cis and trans isomers are stereoisomers.

cis-1,3-diethylcyclobutane

cis = same side of the ring
both groups on wedges or
both on dashed wedges

a. *trans*-1,3-diethylcyclobutane

trans = opposite sides of the ring
one group on a wedge,
one group on a dashed wedge

b. *cis*-1,2-diethylcyclobutane

constitutional isomer
different arrangement of atoms

4.27 To classify a compound as a cis or trans isomer, **classify each non-hydrogen group as up or down. Groups on the same side = cis isomer; groups on opposite sides = trans isomer.**

a.
down bond
(equatorial)
(up)
H (up)
HO
HO
down bond
(equatorial)
both groups down =
cis isomer

H
H
OH
OH

b.
down bond
(equatorial)
(up)
H
Cl
Cl
H
(down)
up bond
(equatorial)
one group up, one down =
trans isomer

up bond
(axial)
Cl
H
H
Cl

c.
H
(down)
up bond
(axial)
(up)
Br
H
Br
down bond
(equatorial)
one group up, one down =
trans isomer

Br
H
H
Br

4.28

a.
groups on same side
cis isomer

groups on opposite sides
trans isomer
(one possibility)

c.
trans:
CH₃
H
H
CH₃
H
CH₃

both groups equatorial
more stable
two chair conformations for the **trans isomer**

b.
cis:
CH₃
H
CH₃
H

CH₃
CH₃
H
H

two chair conformations for the **cis isomer**

Same stability because they both have
one equatorial, one axial CH₃ group.

d. The **trans isomer is more stable** because
it can have both methyl groups in the more roomy
equatorial position.

4.29

a.
CH₂CH₃ ← axial CH₂CH₃
CH₃
1,1-disubstituted

c.
up → CH₃CH₂   H
H   CH₃ ← down
(equatorial)
trans-1,3-disubstituted

b.
CH₃ ← up (axial)
H
CH₂CH₃ ← up
H
cis-1,2-disubstituted

d.
H
up → CH₃CH₂   CH₃ ← down
(equatorial)   H
trans-1,4-disubstituted

4.30 *Oxidation* results in an *increase* in the number of C–Z bonds, or a *decrease* in the number of C–H bonds.

Reduction results in a *decrease* in the number of C–Z bonds, or an *increase* in the number of C–H bonds.

a.

Decrease in the number of C–H bonds.
Increase in the number of C–O bonds.
Oxidation

c.

No change in the number of C–O
or C–H bonds. **Neither**

b.

Decrease in the number of C–O bonds.
Increase in the number of C–H bonds.
Reduction

d.

Decrease in the number of C–O bonds.
Increase in the number of C–H bonds.
Reduction

4.31 The products of a combustion reaction of a hydrocarbon are always the same: **CO_2 and H_2O.**

a. $CH_3CH_2CH_3$ + $5 O_2$ $\xrightarrow{\text{flame}}$ $3 CO_2$ + $4 H_2O$ +   heat

b.   (cyclohexane)   + $9 O_2$ $\xrightarrow{\text{flame}}$ $6 CO_2$ + $6 H_2O$ +   heat

4.32 "Like dissolves like." Beeswax is a lipid, so it will be more soluble in nonpolar solvents. H_2O is very polar, ethanol is slightly less polar, and chloroform is least polar. Beeswax is most soluble in the least polar solvent.

Increasing polarity

H_2O CH_3CH_2OH $CHCl_3$

Increasing solubility of beeswax

4.33 Re-draw the model as a skeletal structure. The longest chain has 15 C's, making it a derivative of pentadecane. Numbering from either direction gives the same numbers.

15 C's - - - - ▸ pentadecane
4 methyl groups at C2, C6, C10, and C14
2,6,10,14-tetramethylpentadecane

4.34 Re-draw each alkane as a skeletal structure, and use the steps in Answer 4.7 to name each compound. Classify C's as in Section 3.2.

a.

7 C's in longest chain = heptane
Number from left to right to put two
substituents on C3.

● = 2° C
□ = 3° C   1° C's unlabeled
■ = 4° C

Four CH₃ groups ----→ tetramethyl

Answer: **3,3,4,5-tetramethylheptane**

b.

8 C's in longest chain = octane
Number from right to left to put first
substituent at C3.

● = 2° C
□ = 3° C   1° C's unlabeled
■ = 4° C

ethyl at C3; two methyls (C5) ----→ dimethyl

Answer: **3-ethyl-5,5-dimethyloctane**

4.35 Re-draw the ball-and-stick model as a chair form.

a.

equatorial
$CH(CH_3)_2$
equatorial
CH_3
axial

b. CH_3 on C1 and Br on C2 are both down, making them cis.

c. Br on C2 and $CH(CH_3)_2$ on C4 are both down, making them cis.

d. Second chair form:

CH_3
Br $CH(CH_3)_2$

This chair form is less stable because two groups
[Br and $CH(CH_3)_2$] are in the more crowded axial
position.

4.36

a.

Br
Cl   H
H   H
H

b.

Cl
H   H
H   H
CH_3

c.

Cl
H   Cl
H   H
CH_3

4.37

a. Five constitutional isomers of molecular formula C_4H_8:

b. Nine constitutional isomers of molecular formula C$_7$H$_{16}$:

c. Twelve constitutional isomers of molecular formula C$_6$H$_{12}$ containing one ring:

4.38 Use the steps in Answers 4.7 and 4.11 to name the alkanes.

a.
[1] 10 carbons = **decane**
[2] **3-methyl** · **6-isopropyl**
[3] **6-isopropyl-3-methyldecane**

b.
[1] 10 carbons = **decane**
[2] **4-isopropyl** · **8-ethyl** · **2,6-dimethyl**
[3] **8-ethyl-4-isopropyl-2,6-dimethyldecane**

c.
[1] 8 carbons = **octane**
[2] **5-propyl** · **4-methyl** · **3-methyl** · **3-ethyl**
[3] **3-ethyl-3,4-dimethyl-5-propyloctane**

d.
[1] 9 carbons = **nonane**
[2] **7-methyl** · **6-ethyl** · **3,3-diethyl**
[3] **3,3,6-triethyl-7-methylnonane**

e.

3-cyclobutyl

3-cyclobutylpentane

f.

1-*sec*-butyl

2-isopropyl

1-*sec*-butyl-2-isopropylcyclopentane

g.

1-isobutyl

3-isopropyl

1-isobutyl-3-isopropylcyclohexane

h.

4-isopropyl

2-methyl

6-propyl

5-isobutyl

5-isobutyl-4-isopropyl-2-methyl-6-propyldecane

i.

2-*sec*-butyl

1,1-dimethyl

5-ethyl

2-*sec*-butyl-5-ethyl-1,1-dimethylcyclohexane

j.

9-ethyl

8-ethyl

7-isopropyl

8,9-diethyl-7-isopropyl-4-methyltridecane

4-methyl

4.39

← 2,2-dimethyl

2,2-dimethylheptane

3,3-dimethyl →

3,3-dimethylheptane

← 4,4-dimethyl

4,4-dimethylheptane

3,4-dimethylheptane

2,4-dimethylheptane   **2,5-dimethylheptane**   **2,3-dimethylheptane**   **3,5-dimethylheptane**   **2,6-dimethylheptane**

4.40 Use the steps in Answer 4.9 to draw the structures.

a. 3-ethyl-2-methyl**hexane**

[1]   6 C chain

[2]

methyl on C2   ← ethyl on C3

b. *sec*-butyl**cyclopentane**

[1]   5 C ring

[2]

c. 4-isopropyl-2,4,5-trimethyl**undecane**

[1]   11 C chain

[2]   isopropyl on C4

methyls on C2, C4, and C5

d. cyclobutylcycloheptane

[1]   7 C cycloalkane

[2]

e. 3-ethyl-1,1-dimethyl**cyclohexane**

[1]   6 C cycloalkane

[2]

ethyl on C3   ← 2 methyl groups on C1

f. 4-butyl-1,1-diethyl**cyclooctane**

[1]   8 C cycloalkane

[2]   2 ethyl groups

g. 6-isopropyl-2,3-dimethyl**dodecane**

[1]   12 C alkane

[2]

methyl on C2   isopropyl on C6

← methyl on C3

h. 2,2,6,6,7-pentamethyl**octane**

[1]   8 C alkane

[2]   5 methyl groups

i. *cis*-1-ethyl-3-methyl**cyclopentane**

[1]   5 C ring

[2]   ← ethyl on C1

← methyl on C3   or

j. *trans*-1-*tert*-butyl-4-ethyl**cyclohexane**

[1]   6 C ring

[2]

4.41 Draw the compounds.

a. 2,2-dimethyl-4-ethylheptane

alphabetized incorrectly
ethyl before **m**ethyl   →

4-ethyl-2,2-dimethylheptane

b. 5-ethyl-2-methylhexane

Longest chain was not
chosen = **heptane**   →

2,5-dimethylheptane

c. 2-methyl-2-isopropylheptane

Longest chain was not
chosen = **octane**   →

2,3,3-trimethyloctane

d. 1,5-dimethylcyclohexane

Numbered incorrectly.   →
Re-number so methyls
are at C1 and C3.

1,3-dimethylcyclohexane

e. 1-ethyl-2,6-dimethylcycloheptane

Numbered incorrectly.
Re-number so methyls   →
are at C1 and C4.

2-ethyl-1,4-dimethylcycloheptane

f. 5,5,6-trimethyloctane

Numbered incorrectly.   →
Re-number so methyls
are at C3 and C4.

3,4,4-trimethyloctane

g. 3-butyl-2,2-dimethylhexane

Longest chain not   →
chosen = **octane**

4-*tert*-butyloctane

h. 1,3-dimethylbutane

Longest chain not   →
chosen = **pentane**

2-methylpentane

4.42

a.

CH₃

H   CH₂CH₂CH₃

CH₃   H

CH₂CH₂CH₃

↓ re-draw

4

4-isopropylheptane

b.

CH₃

CH₃   CH₂CH₃

H   H

CH₂CH₃

↓ re-draw

3

3-ethyl-3-methylpentane

c.

CH₃

CH₃CH₂   CH₂CH₂CH₃

CH₃CH₂CH₂   H

CH₂CH₃

↓ re-draw

4   5

4,4-diethyl-5-methyloctane

4.43 Use the rules from Answer 4.13.

most branching
lowest boiling point

least branching
highest boiling point

4.44 a.

$CH_3(CH_2)_6CH_3$
no branching = higher surface area
higher boiling point

$(CH_3)_3CC(CH_3)_3$
branching = lower surface area
lower boiling point
more spherical, better packing =
higher melting point

b. There is a 159 °C difference in the melting points, but only a 20 °C difference in the boiling points because the symmetry in $(CH_3)_3CC(CH_3)_3$ allows it to pack more tightly in the solid, thus requiring more energy to melt. In contrast, once the compounds are in the liquid state, symmetry is no longer a factor, the compounds are isomeric alkanes, and the boiling points are closer together.

4.45

a.

CH₃
H   CH₃
H   H
CH₃

1 gauche CH_3,CH_3
= 3.8 kJ/mol
of destabilization

or

H
H   CH₃
H   CH₃
CH₃

higher energy
2 gauche CH_3,CH_3
3.8 kJ/mol x 2 = 7.6 kJ/mol
of destabilization

Energy difference =
7.6 kJ/mol − 3.8 kJ/mol =   **3.8 kJ/mol**

b.

H
H   H
CH₃   CH₃
CH₃

2 gauche CH_3,CH_3
3.8 kJ/mol x 2 =
7.6 kJ/mol
of destabilization

or

CH₃
H   H CH₃
H
CH₃

higher energy
3 eclipsed H,CH_3
6 kJ/mol x 3 = 18 kJ/mol
of destabilization

Energy difference =
18 kJ/mol − 7.6 kJ/mol =   **10.4 kJ/mol**

4.46 Use the rules from Answer 4.17 to determine the most and least stable conformations.

a.

staggered
most stable

eclipsed
least stable

All staggered conformations are equal in energy.
All eclipsed conformations are equal in energy.

b.

staggered
ethyl groups anti
most stable

eclipsed
ethyl groups eclipsed
least stable

4.47

a.

b.

c.

4.48

[1]

1 →60°→ 2 →60°→ 3

6 ←60°← 5 ←60°← 4

1 ↑60° ... 60° ... 60°

Energy vs Dihedral angle between two alkyl groups

least stable
4

Eclipsed forms are
higher in energy.

2   6

3   5

Staggered forms
are lower in energy.

1
most stable

1
most stable

180° 120° 60° 0° 60° 120° 180°

Dihedral angle between two alkyl groups

[2]

most stable
CH₃CH₂

1 → 60° → 2 → 60° → 3 → 60° → 4 (**least stable**)

4 → 60° → 5 → 60° → 6 → 60° → 1

Energy vs. Dihedral angle graph:

Eclipsed forms are higher in energy.

Staggered forms are lower in energy.

180° 120° 60° 0° 60° 120° 180°

Dihedral angle
(between CH₃CH₂ in back and CH₃ in front)

most stable — 1 (at 180°)
2 — eclipsed
3 — staggered
least stable 4
5 — staggered
6 — eclipsed
most stable — 1 (at 180°)

4.49 Two types of strain:
- *Torsional strain* is due to eclipsed groups on adjacent carbon atoms.
- *Steric strain* is due to overlapping electron clouds of large groups (e.g. gauche interactions).

a.
two sites
three bulky methyl groups close =
steric strain

b.
eclipsed conformation =
torsional strain

c.
two bulky ethyl groups close =
steric strain
eclipsed conformation =
torsional strain

4.50 The barrier to rotation is equal to the difference in energy between the highest energy eclipsed and lowest energy staggered conformations of the molecule.

a.

most stable

least stable

Destabilization energy =

2 H,CH₃ eclipsing interactions
$$2(6.0 \text{ kJ/mol}) = 12.0 \text{ kJ/mol}$$
$$1 \text{ H,H eclipsing interaction} = 4.0 \text{ kJ/mol}$$

Total destabilization = **16.0 kJ/mol**

16.0 kJ/mol = rotation barrier

b.

most stable

least stable

Destabilization energy =

3 H,CH₃ eclipsing interactions
$$3(6.0 \text{ kJ/mol}) = 18.0 \text{ kJ/mol}$$

Total destabilization = **18.0 kJ/mol**

18.0 kJ/mol = rotation barrier

4.51

most stable

least stable

2 H,H eclipsing interactions = 2(4.0 kJ/mol) = 8.0 kJ/mol

Since the barrier to rotation is 15 kJ/mol, the difference between this value and the destabilization due to H,H eclipsing is the destabilization due to H,Cl eclipsing.

15.0 kJ/mol – 8.0 kJ/mol = 7.0 kJ/mol

destabilization due to H,Cl eclipsing

4.52 The gauche conformation can intramolecularly hydrogen bond, making it the more stable conformation.

$HOCH_2{-}CH_2OH$

rotation here

anti

hydrogen bonding

gauche

Hydrogen bonding can occur only in the gauche conformation, making it **more stable.**

4.53

a.
[1]

axial
axial
HO
eq

eq

b.

down HO

H **up**

one up, one down =
trans

c.

HO,,,..

OH

d.

ax
H ax
OH
eq HO

H
eq

⇌

H eq
eq HO

OH ax
H
ax

[2] a. **axial** Br, H **eq**, CH₃ **eq**, H **axial**
b. **up** Br, H, CH₃ **up**, H — both up = **cis**
c. Br, CH₃
d. ax Br, H eq, CH₃ eq, H ax ⇌ ax CH₃, eq Br, H eq, H ax

[3] a. **axial** H, HO **eq**, OH **eq**, H **axial**
b. H, HO **up**, OH **down**, H — one up, one down = **trans**
c. OH, HO
d. ax H, eq HO, OH eq, H ax ⇌ eq H, ax OH, H eq, OH ax

4.54

ax H, ax H, CH₃ eq, CH₃ eq ⇌ eq H, H eq, CH₃ ax, CH₃ ax

both groups equatorial
more stable

4.55 A **cis isomer** has two groups on the **same side** of the ring. The two groups can be drawn both up or both down. Only one possibility is drawn. A **trans isomer** has one group on one side of the ring and one group on the other side. Either group can be drawn on either side. Only one possibility is drawn.

[1]

a. cis   trans

[2]

a. cis   trans

[3]

a. cis   trans

b. cis isomer

both groups equatorial
more stable

b. cis isomer

larger group equatorial
more stable

b. cis isomer

larger group equatorial
more stable

c. trans isomer

larger group equatorial
more stable

c. trans isomer

both groups equatorial
more stable

c. trans isomer

both groups equatorial
more stable

d.

The cis isomer is more
stable than the trans
because one conformation has
both groups equatorial.

d.

The trans isomer is more
stable than the cis
because one conformation has
both groups equatorial.

d.

The trans isomer is more
stable than the cis
because one conformation has
both groups equatorial.

4.56

Re-draw to see
axial and equatorial.

all equatorial
menthol

4.57 a.

CH$_3$ is axial and up. OH is equatorial
and down, so the two groups are trans.

b. A substituent (**A**) on C$_a$ that is cis to the existing CH$_3$ must be up, so it is equatorial.

c. An equatorial Br on C$_b$ is down, so it is cis to the OH group.

d. A H on C$_c$ must be axial and down, making it cis to the OH group.

e. A substituent (**D**) on C$_d$ that is trans to the OH must be up and axial.

4.58

a. **most stable**
All groups are equatorial.

c. **constitutional isomer**

b.

d.

4.59

a.

1 down, 1 up =
trans

1 down, 1 up =
trans

same arrangement in three dimensions
identical

c.

1 down, 1 up =
trans

both down =
cis

different arrangement in three dimensions
stereoisomers

b.

same molecular formula C$_{10}$H$_{20}$
different connectivity
constitutional isomers

d.

both up = **cis**

both up = **cis**

same arrangement in three dimensions
identical

4.60

a.

3-ethyl-2-methylpentane   3-ethyl-2-methylpentane

same molecular formula
same name
identical molecules

b.

and

same molecular formula
different arrangement of atoms
constitutional isomers

re-draw

4.61

a. **A** and **B**: constitutional isomers
   A and **C**: identical
   B and **D**: stereoisomers

b.

A	B	C	D
cis	trans	cis	cis

c.

B

d.

stereoisomer of **A**

4.62

Three constitutional isomers of C$_7$H$_{14}$:

1,1-dimethylcyclopentane   1,2-dimethylcyclopentane   1,3-dimethylcyclopentane

or   trans   cis   or   trans   cis

4.63 Use the definitions from Answer 4.30 to classify the reactions.

a.

Increase in the number of C–Z
bonds. **Oxidation**

b.

Loss of one C–O
bond *and* one C–H
bond. **Neither**

4.64 Use the rule from Answer 4.31.

a.

$$\xrightarrow[\text{11 O}_2]{\text{flame}} 7\,CO_2 + 8\,H_2O + \text{heat}$$

b.

$$\xrightarrow[\text{(13/2) O}_2]{\text{flame}} 4\,CO_2 + 5\,H_2O + \text{heat}$$

4.65

a.

benzene   [1] →   an arene oxide
2 C–O bonds

[2] →   phenol

1 C–O bond
1 C–H bond

[1]   increase in C–O bonds
oxidation reaction

[2]   loss of 1 C–O bond,
loss of 1 C–H bond
neither

b. Phenol is more water soluble than benzene because it is **polar (contains an O–H group) and
can hydrogen bond with water,** whereas benzene is nonpolar and cannot hydrogen bond.

4.66

cholic acid
a bile acid

a bile salt

This polar part of the molecule
interacts with water.

This nonpolar part of the molecule
can **interact with lipids** to create
micelles that allow for transport
of lipids through aqueous environments.

4.67 The mineral oil can prevent the body's absorption of important fat-soluble vitamins. The vitamins
dissolve in the mineral oil, and are thus not absorbed. Instead, they are expelled with the mineral
oil.

4.68 Cyclopropane has larger angle strain than cyclobutane because the internal angles in the three-
membered ring (60°) are smaller than they are in cyclobutane. Although cyclobutane is not flat, as
shown in Figure 4.11, there are more C–H bonds than there are in cyclopropane, so there are more
sites of torsional strain. Thus cyclopropane has more angle strain but less torsional strain. The
result is that both cyclopropane and cyclobutane have roughly similar strain energies.

4.69 The amide in the four-membered ring has 90° bond angles giving it angle strain, which makes it more reactive.

amide

penicillin G

strained amide
more reactive

4.70

Example:

Although I is a much bigger atom than Cl, the C–I bond is also much longer than the C–Cl bond. As a result the eclipsing interaction of the H and I atoms is not very much different in magnitude from the H,Cl eclipsing interaction.

longer bond

4.71

decalin

trans-decalin

cis-decalin

trans

The trans isomer is more stable because the carbon groups at the ring junction are both in the favorable equatorial position.

1,3-diaxial interaction

cis

This bond is axial, creating unfavorable 1,3-diaxial interactions.

4.72 Re-draw the ball-and-stick model using chair forms.

a.

axial
(above)

CH₃ H

HO

equatorial
(above)

H   H   H
OH ← axial
(below)

A

b. All bonds above the ring are on wedges and all bonds below the ring are on dashed wedges.

HO

H

HO   H   OH

4.73

a, b.

B

c. The circled H's at one ring fusion are cis. The boxed in CH₃ and H at the second ring fusion are trans.

4.74

pentylcyclopentane

(1,1-dimethylpropyl)cyclopentane

(2-methylbutyl)cyclopentane

(2,2-dimethylpropyl)cyclopentane

(1-methylbutyl)cyclopentane

(1-ethylpropyl)cyclopentane

(1,2-dimethylpropyl)cyclopentane

(3-methylbutyl)cyclopentane

4.75

a.

2,3-dimethylbicyclo[3.1.1]heptane

b.

2-ethyl-7,7-dimethylbicyclo[2.2.1]heptane

c.

1-methyl-7-propylbicyclo[3.2.1]octane

d.

6-ethyl-3,3-dimethylbicyclo[3.2.0]heptane

Chapter 5: Stereochemistry

Chapter Review

Isomers are different compounds with the same molecular formula (5.2, 5.11).

[1] **Constitutional isomers**—isomers that differ in the way the atoms are connected to each other. They have:
- different IUPAC names
- the same or different functional groups
- different physical and chemical properties.

[2] **Stereoisomers**—isomers that differ only in the way atoms are oriented in space. They have the same functional group and the same IUPAC name except for prefixes such as cis, trans, *R*, and *S*.
- **Enantiomers**—stereoisomers that are nonsuperimposable mirror images of each other (5.4).
- **Diastereomers**—stereoisomers that are not mirror images of each other (5.7).

A and B are diastereomers of C and D.

Assigning priority (5.6)

- Assign priorities (1, 2, 3, or 4) to the atoms bonded directly to the stereogenic center in order of decreasing atomic number. The atom of *highest* atomic number gets the *highest* priority (1).
- If two atoms on a stereogenic center are the *same*, assign priority based on the atomic number of the atoms bonded to these atoms. *One* atom of higher atomic number determines a higher priority.
- If two isotopes are bonded to the stereogenic center, assign priorities in order of decreasing *mass* number.
- To assign a priority to an atom that is part of a multiple bond, treat a multiply bonded atom as an equivalent number of singly bonded atoms.

- The stereogenic center is bonded to Br, Cl, C, and H.
- The stereogenic center is *not* bonded directly to I.

- $CH(CH_3)_2$ gets the highest priority because the C is bonded to two other C's.

- OH gets the highest priority because O has the highest atomic number.
- CO_2H (three bonds to O) gets higher priority than CH_2OH (one bond to O).

Some basic principles

- When a compound and its mirror image are **superimposable,** they are **identical achiral compounds.** A plane of symmetry in one conformation makes a compound achiral (5.3).
- When a compound and its mirror image are **not superimposable,** they are **different chiral compounds** called **enantiomers.** A chiral compound has no plane of symmetry in any conformation (5.3).
- A **tetrahedral stereogenic center** is a carbon atom bonded to four different groups (5.4, 5.5).
- For *n* **stereogenic centers,** the maximum number of stereoisomers is 2^n (5.7).

no stereogenic centers	one stereogenic center	two stereogenic centers	two stereogenic centers
achiral	chiral	chiral	achiral

Optical activity is the ability of a compound to rotate plane-polarized light (5.12).

- An optically active solution contains a chiral compound.
- An optically inactive solution contains one of the following:
  - an achiral compound with no stereogenic centers.
  - a meso compound—an achiral compound with two or more stereogenic centers.
  - a racemic mixture—an equal amount of two enantiomers.

The prefixes *R* and *S* compared with *d* and *l*

The prefixes *R* and *S* are labels used in nomenclature. Rules on assigning *R,S* are found in Section 5.6.
- An enantiomer has every stereogenic center opposite in configuration.
- A diastereomer of this same compound has one stereogenic center with the same configuration and one that is opposite.

The prefixes *d* (or +) and *l* (or –) tell the direction a compound rotates plane-polarized light (5.12).
- *d* (or +) stands for dextrorotatory, rotating polarized light clockwise.
- *l* (or –) stands for levorotatory, rotating polarized light counterclockwise.

The physical properties of isomers compared (5.12)

Type of isomer	Physical properties
Constitutional isomers	Different
Enantiomers	Identical except the direction of rotation of polarized light
Diastereomers	Different
Racemic mixture	Possibly different from either enantiomer

Equations

- Specific rotation (5.12C):

$$\text{specific rotation} = [\alpha] = \frac{\alpha}{l \times c}$$

α = observed rotation (°)
l = length of sample tube (dm)
c = concentration (g/mL)

[dm = decimeter
1 dm = 10 cm]

- Enantiomeric excess (5.12D):

$$ee = \text{% of one enantiomer} - \text{% of other enantiomer}$$

$$= \frac{[\alpha] \text{ mixture}}{[\alpha] \text{ pure enantiomer}} \times 100\%$$

Practice Test on Chapter Review

1.a. Which of the following statements is true for compounds **A–D** below?

    A   **B**   **C**   **D**

1. **A** and **B** are separable by physical methods such as distillation.
2. **A** and **C** are separable by physical methods such as distillation.
3. **A** and **D** are separable by physical methods such as distillation.
4. Statements (1) and (2) are both true.
5. Statements (1), (2), and (3) are all true.

b. Which of the following statements is true about compounds **A–C** below?

    A   **B**   **C**

1. **A** and **B** are enantiomers.
2. **A** and **C** are enantiomers.
3. An equal mixture of **B** and **C** is optically active.
4. Statements (1) and (2) are true.
5. Statements (1), (2), and (3) are all true.

c. Which compound is a diastereomer of **A?**

| A | B | C | D |

1. **B** only
2. **C** only
3. **D** only
4. Both **B** and **C**
5. Compounds **B**, **C**, and **D**

2. Rank the following four groups around a stereogenic center in order of decreasing priority. Rank the highest priority group as **1,** the lowest priority group as **4,** and the two groups of intermediate priority as **2** and **3.**

—CH₂Cl   —CH₂CH₂Br   —COOH   —CH₂OH

A   B   C   D

3. Label each stereogenic center in the following compound as *R* or *S*.

4. State how the compounds in each pair are related to each other. Choose from constitutional isomers, enantiomers, diastereomers, or identical compounds.

a.   and

b.   and

c. and

5. The enantiomeric excess of a mixture of **A** and **B** is 62% with **A** in excess. How much of **A** and **B** are present in the mixture?

Answers to Practice Test

1. a. 1	2. **A**–1	3. a. *S*	4. a. diastereomers	5. 81% **A**
b. 2	**B**–4	b. *S*	b. enantiomers	19% **B**
c. 3	**C**–2		c. identical	
	D–3			

Answers to Problems

5.1 Cellulose consists of long chains held together by intermolecular hydrogen bonds forming sheets that stack in extensive three-dimensional arrays. Most of the OH groups in cellulose are in the interior of this three-dimensional network, unavailable for hydrogen bonding to water. Thus, even though cellulose has many OH groups, its three-dimensional structure prevents many of the OH groups from hydrogen bonding with the solvent and this makes it water insoluble.

5.2 **Constitutional isomers** have atoms bonded to different atoms.
Stereoisomers differ only in the three-dimensional arrangement of atoms.

5.3 Draw the mirror image of each molecule by drawing a mirror plane and then drawing the molecule's reflection. **A chiral molecule is one that is not superimposable on its mirror image.** A molecule with one stereogenic center is always chiral. A molecule with zero stereogenic centers is not chiral (in general).

5.4 A plane of symmetry cuts the molecule into **two identical halves.**

5.5 Rotate around a C–C bond to draw an eclipsed conformation.

5.6 To locate a stereogenic center, omit all C's with two or more H's, all *sp* and *sp²* hybridized atoms, and all heteroatoms. [In Chapter 25, we will learn that the N atoms of ammonium salts ($R_4N^+ X^-$) can sometimes be stereogenic centers.] Then evaluate any remaining atoms. A tetrahedral stereogenic center has a carbon bonded to **four different groups.**

a.

b.

c.

d.

e.

f.

* = stereogenic center

5.7 Use the directions from Answer 5.6 to locate the stereogenic centers.

aliskiren

4 C's bonded to 4 different groups
4 stereogenic centers

5.8 Find the C bonded to four different groups in each molecule. At the stereogenic center, draw two bonds in the plane of the page, one in front (on a wedge), and one behind (on a dashed wedge). Then draw the mirror image (enantiomer).

a. stereogenic center
mirror images
nonsuperimposable
enantiomers

b. stereogenic center
mirror images
nonsuperimposable
enantiomers

c. stereogenic center
mirror images
nonsuperimposable
enantiomers

5.9 Use the directions from Answer 5.6 to locate the stereogenic centers.

a. C bonded to H and 3 different C's: **1 stereogenic center**

c. 4 C's bonded to 4 different groups: **4 stereogenic centers**

e. C bonded to H, 2 different O's and 1 C: **1 stereogenic center**

b. Each labeled C is bonded to: H, Cl, CH₂, CHCl: **2 stereogenic centers**

d. **no stereogenic centers**

f. 3 C's bonded to 4 different groups: **3 stereogenic centers**

5.10

a. cholesterol

b. simvastatin

All stereogenic C's are circled. Each C is *sp*³ hybridized and bonded to 4 different groups.

5.11 Assign priority based on atomic number: atoms with a higher atomic number get a higher priority. If two atoms are the same, look at what they are bonded to and assign priority based on the atomic number of these atoms.

a. –CH₃, –CH₂CH₃
higher priority

c. –H, –D
higher mass
higher priority

e. –CH₂CH₂Cl, –CH₂CH(CH₃)₂
higher priority

b. –I, –Br
higher priority

d. –CH₂Br, –CH₂CH₂Br
higher priority

f. –CH₂OH, –CHO = —C=O = —C–O
2 H's, 1 O   **2 O's, 1 H**   2 C–O bonds

C bonded to 2 O's has **higher priority.**

5.12 Rank by decreasing priority. Lower atomic number = lower priority.

Highest priority = 1, Lowest priority = 4

		priority
a. –COOH	C = second lowest atomic number	3
–H	H = lowest atomic number	4
–NH₂	N = second highest atomic number	2
–OH	O = highest atomic number	1

decreasing priority: **–OH, –NH₂, –COOH, –H**

		priority
b. –H	H = lowest atomic number	4
–CH₃	C bonded to 3 H's	3
–Cl	Cl = highest atomic number	1
–CH₂Cl	C bonded to 2 H's + **1 Cl**	2

decreasing priority: **–Cl, –CH₂Cl, –CH₃, –H**

c. –CH$_2$CH$_3$ C bonded to 2 H's + **1 C**   **priority 2**

    –CH$_3$ C bonded to 3 H's   3

    –H   H = lowest atomic number   4

    –CH(CH$_3$)$_2$ C bonded to 1 H + **2 C's**   1

decreasing priority: –CH(CH$_3$)$_2$, –CH$_2$CH$_3$, –CH$_3$, –H

d. –CH=CH$_2$ C bonded to 1 H + **2 C's**   **priority 2**

    –CH$_3$ C bonded to 3 H's   3

    –C≡CH   C bonded to **3 C's**   1

    –H   H = lowest atomic number   4

decreasing priority: –C≡CH, –CH=CH$_2$, –CH$_3$, –H

5.13 To assign *R* or *S* to the molecule, first rank the groups. The lowest priority group must be oriented behind the page. If tracing a circle from (1) → (2) → (3) proceeds in the clockwise direction, then the stereogenic center is labeled *R;* if the circle is counterclockwise, then it is labeled *S*.

a.

b. lowest priority forward
clockwise
It looks like *R*, but reverse answer
R → S

c.

d. lowest priority forward
clockwise
It looks like *R*, but reverse answer
R → S

5.14

clopidogrel

clockwise
R isomer

counterclockwise
S isomer
Plavix

5.15 a, b. Re-draw lisinopril as a skeletal structure, locate the stereogenic centers, and assign *R,S*.

three stereogenic centers
All have the *S* configuration.

5.16 The maximum number of stereoisomers = 2^n where n = the number of stereogenic centers.

a. 3 stereogenic centers
2^3 = 8 stereoisomers

b. 8 stereogenic centers
2^8 = 256 stereoisomers

5.17

5.18

5.19

5.20 **A meso compound must have at least two stereogenic centers. Usually a meso compound has a plane of symmetry.** You may have to rotate around a C–C bond to see the plane of symmetry clearly.

a.

2 stereogenic centers
plane of symmetry
meso compound

b.

2 stereogenic centers
no plane of symmetry
not a meso compound

c. rotate

2 stereogenic centers
plane of symmetry
meso compound

5.21 Use the definition in Answer 5.20 to draw the meso compounds.

a.

plane of symmetry

b.

plane of symmetry

c.

plane of symmetry

5.22 The enantiomer has the exact opposite R,S designations. Diastereomers with two stereogenic centers have one center the same and one different.

If a compound is **R,S:**

Its enantiomer is: **S,R** ◄—————— Exact opposite: R and S interchanged.

Its diastereomers are: **R,R and S,S** ◄——— One designation remains the same, the other changes.

5.23 The enantiomer has the exact opposite R,S designations. For diastereomers, at least one of the R,S designations is the same, but not all of them.

a. (2R,3S)-hexane-2,3-diol and (2R,3R)-hexane-2,3-diol
      One changes; one remains the same:
            diastereomers
b. (2R,3R)-hexane-2,3-diol and (2S,3S)-hexane-2,3-diol
      Both R's change to S's:
            enantiomers
c. (2R,3S,4R)-hexane-2,3,4-triol and (2S,3R,4R)-hexane-2,3,4-triol
      Two change; one remains the same:
            diastereomers

5.24 The enantiomer must have the exact opposite R,S designations. For diastereomers, at least one of the R,S designations is the same, but not all of them.

a.

sorbitol

b.

A
One changes; three remain the same.
diastereomer

c.

B
All stereogenic centers change.
enantiomer

5.25 Meso compounds generally have a plane of symmetry. They cannot have just one stereogenic center.

a.

no plane of symmetry
not a meso compound

b.

plane of symmetry
meso compound

c.

no plane of symmetry
not a meso compound

5.26

a.
2 stereogenic centers =
4 stereoisomers maximum

c.

Draw the cis and trans isomers:

cis

A

identical

trans

B   **C**

Pair of enantiomers: **B** and **C**.
Pairs of diastereomers: **A** and **B**, **A** and **C**.

Only 3 stereoisomers exist.

Draw the cis and trans isomers:

A

identical

B

identical

Pair of diastereomers: **A** and **B**.

Only 2 stereoisomers exist.

b.
HO
2 stereogenic centers =
4 stereoisomers maximum

Draw the cis and trans isomers:

cis

OH   HO

A   **B**

trans

OH   HO

C   **D**

Pairs of enantiomers: **A** and **B**, **C** and **D**.
Pairs of diastereomers: **A** and **C**, **A** and **D**,
B and **C**, **B** and **D**.

All 4 stereoisomers exist.

5.27 Four facts:
- **Enantiomers** are mirror image isomers.
- **Diastereomers** are stereoisomers that are not mirror images.
- **Constitutional isomers** have the same molecular formula but the atoms are bonded to different atoms.
- **Cis and trans isomers** are always diastereomers.

a.
same molecular formula
same *R,S* designation:
identical

b.
same molecular formula,
opposite configuration at one
stereogenic center
enantiomers

c.
1,4- isomer   1,3-isomer
constitutional isomers

d.
trans   **cis**
Both 1,3 isomers,
cis and trans:
diastereomers

5.28

(S)-alanine

$[\alpha] = +8.5$

mp = 297 °C

a. Mp = same as the *S* isomer.
b. The mp of a racemic mixture is often different from the melting point of the enantiomers.
c. –8.5, same as *S* but opposite sign
d. Zero. A racemic mixture is optically inactive.
e. Solution of pure (*S*)-alanine: **optically active**
   Equal mixture of (*R*)- and (*S*)-alanine: **optically inactive**
   75% (*S*)- and 25% (*R*)-alanine: **optically active**

5.29

$$[\alpha] = \frac{\alpha}{l \times c}$$

α = observed rotation
l = length of tube (dm)
c = concentration (g/mL)

$$[\alpha] = \frac{10°}{1 \text{ dm} \times (1 \text{ g/10 mL})} = +100 = \textbf{specific rotation}$$

5.30 Enantiomeric excess = *ee* = % of one enantiomer – % of other enantiomer.

a. 95% – 5% = **90% *ee***   b. 85% – 15% = **70% *ee***

5.31
a. 90% *ee* means 90% excess of **A** and 10% racemic mixture of **A** and **B** (5% each); therefore, **95% A and 5% B.**
b. 99% *ee* means 99% excess of **A** and 1% racemic mixture of **A** and **B** (0.5% each); therefore, **99.5% A and 0.5% B.**
c. 60% *ee* means 60% excess of **A** and 40% racemic mixture of **A** and **B** (20% each); therefore, **80% A and 20% B.**

5.32

$$ee = \frac{[\alpha] \text{ mixture}}{[\alpha] \text{ pure enantiomer}} \times 100\%$$

a. $\dfrac{+10}{+24} \times 100\% = 42\%\ ee$

b. $\dfrac{[\alpha] \text{ solution}}{+24} \times 100\% = 80\%\ ee$

[α] solution = +19.2

5.33

a. $\dfrac{[\alpha] \text{ mixture}}{+3.8} \times 100\% = 60\%\ ee$

[α] mixture = +2.3

b. % one enantiomer – % other enantiomer = *ee*
80% – 20% = 60% *ee*

80% dextrorotatory (+) enantiomer
20% levorotatory (–) enantiomer

5.34• Enantiomers have the same physical properties (mp, bp, solubility), and rotate the plane of polarized light to an equal extent, but in opposite directions.
 • **Diastereomers have different physical properties.**
 • **A racemic mixture is optically inactive.**

A and B are diastereomers of C.

a. The bp's of **A** and **B** are the same. The bp's of **A** and **C** are different.
b. Pure **A:** optically active
   Pure **B:** optically active
   Pure **C:** optically inactive
   Equal mixture of **A** and **B:** optically inactive
   Equal mixture of **A** and **C:** optically active
c. There would be two fractions: one containing **A** and **B** (optically inactive), and one containing **C** (optically inactive).

5.35

R isomer
celery ketone

S isomer
licorice

5.36

a, b.

three stereogenic centers

5.37

A   identical   B

C   D

a. **A** and **B,** same *R,S* assignment, identical
b. **A** and **C,** opposite *R,S* assignment, enantiomers
c. **A** and **D,** one stereogenic center different, diastereomers
d. **C** and **D,** one stereogenic center different, diastereomers

5.38 Use the definitions from Answer 5.2.

a.   and

one up, one down   both up
trans   **cis**
Both compounds are
1,2-dimethylcyclohexane.
one cis, one trans = **stereoisomers**

b.   and

same molecular formula C_7H_{14}
different connectivity
constitutional isomers

c.   and

same molecular formula $C_{10}H_{16}O$
different connectivity
constitutional isomers

d.   and

$C_9H_{16}O$ $C_8H_{14}O$
different molecular formula
not isomers

5.39 Use the definitions from Answer 5.3.

5.40

5.41 A plane of symmetry cuts the molecule into **two identical halves.**

5.42 Use the directions from Answer 5.6 to locate the stereogenic centers.

a.

d.

f.

b.

c.

e.

5.43 Stereogenic centers are circled.

a.

amoxicillin

b.

norethindrone

c.

heroin

5.44

a.

amphetamine

b.

ketoprofen

5.45 Assign priority based on the rules in Answer 5.11.

a. –CD₃, –CH₃

D higher mass than H
higher priority

c. –CH₂Cl, –CH₂CH₂CH₂Br

C bonded to Cl
higher priority

b. –CH(CH₃)₂, –CH₂OH

C bonded to O
higher priority

d. –CH₂NH₂, –NHCH₃

higher atomic number
higher priority

5.46 Assign priority based on the rules in Answer 5.11.

a. –F > –OH > –NH$_2$ > –CH$_3$

b. –(CH$_2$)$_3$CH$_3$ > –CH$_2$CH$_2$CH$_3$ > –CH$_2$CH$_3$ > –CH$_3$

c. –NH$_2$ > –CH$_2$NHCH$_3$ > –CH$_2$NH$_2$ > –CH$_3$

d. –COOH > –CHO > –CH$_2$OH > –H

e. –Cl > –SH > –OH > –CH$_3$

f. –C≡CH > –CH=CH$_2$ > –CH(CH$_3$)$_2$ > –CH$_2$CH$_3$

5.47 Use the rules in Answer 5.13 to assign *R* or *S* to each stereogenic center.

a.
counterclockwise
S isomer

b.
clockwise, but H in front
S isomer

c.
It looks like an *S* isomer, but we must reverse the answer, *S* to *R*.
R isomer

d.
R, R

e.
S

f.

g.

h.

5.48

a.
re-draw

b.
re-draw

5.49

5.50
citalopram
S isomer

5.51

a. (*R*)-3-methylhexane

c. (3*R*,5*S*,6*R*)-5-ethyl-3,6-dimethylnonane

b. (4*R*,5*S*)-4,5-diethyloctane

d. (3*S*,6*S*)-6-isopropyl-3-methyldecane

5.52

a.

(*S*)-3-methylhexane

b.

(4*R*,6*R*)-4-ethyl-6-methyldecane

c.

(3*R*,5*S*,6*R*)-5-isobutyl-3,6-dimethylnonane

5.53

paclitaxel

5.54

a.

2 stereogenic centers
2^2 = 4 possible stereoisomers

b.

0 stereogenic centers

c.

4 stereogenic centers
2^4 = 16 possible stereoisomers

5.55

a. 1*R*,2*S*

ephedrine

b. 1*S*,2*S*

pseudoephedrine

c. Ephedrine and pseudoephedrine
are diastereomers. One stereogenic
center is the same; one is different.

d.

enantiomer of (–)-ephedrine

diastereomer of (–)-ephedrine

5.56

a.

A   **B**

Pair of enantiomers: **A** and **B**.
Pairs of diastereomers: **A** and **C**, **B** and **C**.

C

identical

meso compound

b.

A   **B**   **C**   **D**

Pairs of enantiomers: **A** and **B**, **C** and **D**.
Pairs of diastereomers: **A** and **C**, **A** and **D**, **B** and **C**, **B** and **D**.

c.

A

identical

meso compound

B   **C**

Pair of enantiomers: **B** and **C**.
Pairs of diastereomers: **A** and **B**, **A** and **C**.

d.

A

identical

B

identical

Pair of diastereomers: **A** and **B**.
Meso compounds: **A** and **B**.

5.57

a.

enantiomer   diastereomer   or   diastereomer

b.

enantiomer   diastereomer

c.

enantiomer   diastereomer   or   diastereomer

5.58

achiral   achiral   chiral   chiral   achiral   achiral

5.59 Explain each statement.

a. All molecules have a mirror image, but only chiral molecules have enantiomers. **A** is not chiral, and therefore, does not have an enantiomer.

b. **B** has one stereogenic center, and therefore, has an enantiomer. Only compounds with two or more stereogenic centers have diastereomers.

c. **C** is chiral and has two stereogenic centers, and therefore, has both an enantiomer and a diastereomer.

d. rotate → **D** has two stereogenic centers, but is a meso compound. Therefore, it has a diastereomer, but no enantiomer because it is achiral.

plane of symmetry

e. **E** has two stereogenic centers, but is a meso compound. Therefore, it has a diastereomer, but no enantiomer because it is achiral.

plane of symmetry

5.60

5.61 Re-draw each Newman projection and determine the *R,S* configuration. Then determine how the molecules are related.

a. **A** and **B** are identical.
b. **A** and **C** are enantiomers.

c. **A** and **D** are diastereomers.
d. **C** and **D** are diastereomers.

5.62

A (trans, *R*) and **B** (cis, *R*) are diastereomers.
A (trans, *R*) and **C** (trans, *R*) are identical molecules.
A (trans, *R*) and **D** (trans, *S*) are enantiomers.
A (trans, *R*) and **E** are constitutional isomers.

5.63

a.

enantiomers

b.
one different configuration
diastereomers

c.
mirror images
not superimposable
enantiomers

d.
enantiomers

e.
2S,3S   2S,3S
identical

f.
1,4-trans   1,4-cis
diastereomers

5.64

a. **A** and **B** are constitutional isomers.
   A and **C** are constitutional isomers.
   B and **C** are diastereomers (cis and trans).
   C and **D** are enantiomers.

b.

plane of symmetry

A   B   C   D

A has two   **achiral**   **chiral**   **chiral**
planes of symmetry.

achiral

mirror images and not
superimposable
enantiomers

c. Alone, **C** and **D** would be optically active.
d. **A** and **B** have a plane of symmetry.
e. **A** and **B** have different boiling points.
   B and **C** have different boiling points.
   C and **D** have the same boiling point.
f. **B** is a meso compound.
g. An equal mixture of **C** and **D** is optically inactive because it is a racemic mixture.
   An equal mixture of **B** and **C** would be optically active.

5.65

quinine

$$ee = \frac{[\alpha] \text{ mixture}}{[\alpha] \text{ pure enantiomer}} \times 100\%$$

quinine = **A**
quinine's enantiomer = **B**

a.

$$\frac{-50}{-165} \times 100\% = 30\% \ ee$$

$$\frac{-83}{-165} \times 100\% = 50\% \ ee$$

$$\frac{-120}{-165} \times 100\% = 73\% \ ee$$

b. 30% *ee* = 30% excess one compound (**A**)
remaining 70% = mixture of 2 compounds (35% each **A** and **B**)
Amount of **A** = 30 + 35 = **65%**
Amount of **B** = **35%**

50% *ee* = 50% excess one compound (**A**)
remaining 50% = mixture of 2 compounds (25% each **A** and **B**)
Amount of **A** = 50 + 25 = **75%**
Amount of **B** = **25%**

73% *ee* = 73% excess of one compound (**A**)
remaining 27% = mixture of 2 compounds (13.5% each **A** and **B**)
Amount of **A** = 73 + 13.5 = **86.5%**
Amount of **B** = **13.5%**

c. $[\alpha] = +165$
d. 80% – 20% = 60% *ee*

e. $60\% = \dfrac{[\alpha] \text{ mixture}}{-165} \times 100\%$

$[\alpha]$ mixture = –99

5.66

amygdalin

a. The 11 stereogenic centers are circled. Maximum number of stereoisomers = 2^{11} = 2048
b. Enantiomers of mandelic acid:

c. 60% – 40% = 20% *ee*
20% = $[\alpha]$ mixture/–154 x 100%
$[\alpha]$ mixture = –31

d. $ee = \dfrac{+50}{+154} \times 100\% = 32\% \ ee$

$[\alpha]$ for (*S*)-mandelic acid = +154

32% excess of the *S* enantiomer
68% of racemic *R* and *S* = 34% *S* and 34% *R*

S enantiomer: 32% + 34% = 66%
R enantiomer = 34%

5.67

artemisinin

mefloquine

a. Each stereogenic center is circled.
b. The stereogenic centers in mefloquine are labeled.
c. Artemisinin has seven stereogenic centers.
   $2^n = 2^7 = 128$ possible stereoisomers
d. One N atom in mefloquine is sp^2 and one is sp^3.
e. Two molecules of artemisinin cannot intermolecularly H-bond because there are no O–H or N–H bonds.

f.

5.68

a. Each stereogenic center is circled.

saquinavir
Trade name Invirase

c. diastereomer

b. enantiomer

d. constitutional isomer

new amine

new aldehyde

5.69 Allenes contain an *sp* hybridized carbon atom doubly bonded to two other carbons. This makes the double bonds of an allene perpendicular to each other. When each end of the allene has two like substituents, the allene contains two planes of symmetry and it is achiral. When each end of the allene has two different groups, the allene has no plane of symmetry and it becomes chiral.

These two substituents are at 90° to these two substituents.

Allene **A** contains two planes of symmetry,
making it **achiral.**

no plane of symmetry

chiral

allene → **mycomycin**

re-draw

The substituents on each end of the allene in mycomycin are
different. Therefore, mycomycin is **chiral.**

5.70

six stereogenic centers

5.71

discodermolide

a. The 13 tetrahedral stereogenic centers are circled.

b. Because there is restricted rotation around a C–C double bond, groups on the end of the double bond cannot interconvert. Whenever the substituents on each end of the double bond are different from each other, the double bond is a stereogenic site. Thus, the following two double bonds are isomers:

These compounds are isomers.

There are three stereogenic sites due to the double bonds in discodermolide, labeled with arrows.

c. The maximum number of stereoisomers for discodermolide must include the 13 tetrahedral stereogenic centers and the three double bonds.
Maximum number of stereoisomers = 2^{16} = 65,536.

5.72 When the spiro compound has a plane of symmetry, it is achiral.

a.

achiral

b.

chiral

c.

achiral

d.

chiral

5.73

These salts are **diastereomers,** and they are now separable by physical methods because they have different physical properties.

Chapter 6: Understanding Organic Reactions

Chapter Review

Writing organic reactions (6.1)

- Use curved arrows to show the movement of electrons. Full-headed arrows are used for electron pairs and half-headed arrows are used for single electrons.

- Reagents can be drawn either on the left side of an equation or over an arrow. Catalysts are drawn over or under an arrow.

Types of reactions (6.2)

[1] Substitution	Z = H or a heteroatom
[2] Elimination	Two σ bonds are broken. A π bond is formed.
[3] Addition	A π bond is broken. Two σ bonds are formed.

Important trends

Values compared	Trend
Bond dissociation energy and **bond strength**	The *higher* the bond dissociation energy, the *stronger* the bond (6.4).

Increasing size of the halogen →

CH₃—F   CH₃—Cl   CH₃—Br   CH₃—I

ΔH^o = 456 kJ/mol   351 kJ/mol   293 kJ/mol   234 kJ/mol

← Increasing bond strength |

E_a and **reaction rate**	The *larger* the energy of activation, the *slower* the reaction (6.9A).
E_a and **rate constant**	The *higher* the energy of activation, the *smaller* the rate constant (6.9B).

Equilibrium always favors the species *lower* in energy.

Reactive intermediates (6.3)

- Breaking bonds generates reactive intermediates.
- Homolysis generates radicals with unpaired electrons.
- Heterolysis generates ions.

Reactive intermediate	General structure	Reactive feature	Reactivity
radical		unpaired electron	electrophilic
carbocation		positive charge; only six electrons around C	electrophilic
carbanion		net negative charge; lone electron pair on C	nucleophilic

Energy diagrams (6.7, 6.8)

Conditions favoring product formation (6.5, 6.6)

Variable	Value	Meaning
K_{eq}	$K_{eq} > 1$	More product than starting material is present at equilibrium.
$\Delta G°$	$\Delta G° < 0$	The energy of the products is **lower** than the energy of the reactants.
$\Delta H°$	$\Delta H° < 0$	The bonds in the products are **stronger** than the bonds in the reactants.
$\Delta S°$	$\Delta S° > 0$	The product is **more disordered** than the reactant.

Equations (6.5, 6.6)

Factors affecting reaction rate (6.9)

Factor	Effect
energy of activation	higher E_a → slower reaction
concentration	higher concentration → faster reaction
temperature	higher temperature → faster reaction

Practice Test on Chapter Review

1. Label each statement as TRUE (T) or FALSE (F) for a reaction with $K_{eq} = 0.5$ and $E_a = 18$ kJ/mol. Ignore entropy considerations.

    a. The reaction is faster than a reaction with $K_{eq} = 8$ and $E_a = 18$ kJ/mol.
    b. The reaction is faster than a reaction with $K_{eq} = 0.5$ and $E_a = 12$ kJ/mol.
    c. $\Delta G°$ for the reaction is a positive value.
    d. The starting materials are lower in energy than the products of the reaction.
    e. The reaction is exothermic.

2.a. Which of the following statements is true about an endothermic reaction, ignoring entropy considerations?

    1. The bonds in the products are stronger than the bonds in the starting materials.
    2. $K_{eq} < 1$.
    3. A catalyst speeds up the rate of the reaction and gives a larger amount of product.
    4. Statements (1) and (2) are both true.
    5. Statements (1), (2), and (3) are all true.

  b. Which of the following statements is true about a reaction with $K_{eq} = 10^3$ and $E_a = 2.5$ kJ/mol? Ignore entropy considerations.

    1. The reaction is faster than a reaction with $E_a = 4$ kJ/mol.
    2. The starting materials are higher in energy than the products of the reaction.
    3. $\Delta G°$ is positive.
    4. Statements (1) and (2) are both true.
    5. Statements (1), (2), and (3) are all true.

3.a. Draw the transition state for the following reaction.

  b. Draw the transition state for the following one-step elimination reaction.

Answers to Practice Test

1. a. F
   b. F
   c. T
   d. T
   e. F

2. a. 2
   b. 4

3. a.

$$\left[\begin{array}{c} \overset{\delta+}{\diagup}\diagdown \\ \quad \overset{H}{\underset{H}{\diagup}}\text{--O:}\delta+ \end{array}\right]^{\ddagger}$$

b.

$$\left[\begin{array}{c} \diagup\diagup \overset{\ddot{\text{Br}}:\,\delta-}{\diagup} \\ \overset{H}{\diagdown}\underset{\delta-}{\ddot{\text{O}}\text{CH}_3} \end{array}\right]^{\ddagger}$$

Answers to Problems

6.1 [1] In a **substitution reaction,** one group replaces another.
[2] In an **elimination reaction,** elements of the starting material are lost and a π bond is formed.
[3] In an **addition reaction,** elements are added to the starting material.

a.

Br replaces OH =
substitution reaction

b.

addition of 2 H's
addition reaction

c.

Cl replaces H =
substitution reaction

d.

elements lost
(H + OH)

π bond formed

elimination reaction

6.2 The elements of Cl and OH are added to a π bond, so the reaction is an addition.

A   B

6.3 **Heterolysis** means one atom gets both of the electrons when a bond is broken. A carbocation is a C with a positive charge, and a carbanion is a C with a negative charge.

a.
heterolysis
Electrons go to the more
electronegative atom, O.

carbocation   :ÖH

b.
heterolysis
Electrons go to the more
electronegative atom, Br.

carbocation   :Br:⁻

c.
heterolysis
Electrons go to the more
electronegative atom, C.

carbanion   Li⁺

6.4 Use **full-headed arrows** to show the movement of electron pairs, and **half-headed arrows** to show the movement of single electrons.

a.

b.

6.5

6.6 Increasing number of electrons between atoms = increasing bond strength = increasing bond dissociation energy = decreasing bond length.
Increasing size of an atom = increasing bond length = decreasing bond strength.

a.

higher bond dissociation energy   S is larger than O. longer, weaker bond

b.

higher bond dissociation energy   single bond fewer electrons

6.7 **To determine $\Delta H°$ for a reaction:**
[1] Add the bond dissociation energies for all bonds *broken* in the equation (+ values).
[2] Add the bond dissociation energies for all of the bonds *formed* in the equation (– values).
[3] *Add the energies together* to get the $\Delta H°$ for the reaction.
A positive $\Delta H°$ means the reaction is *endothermic*. **A negative $\Delta H°$** means the reaction is *exothermic*.

a. $CH_3CH_2-Br + H_2O \longrightarrow CH_3CH_2-OH + HBr$

[1] Bonds broken

	$\Delta H°$ (kJ/mol)
CH_3CH_2-Br	+ 285
$H-OH$	+ 498
Total	+ 783 kJ/mol

[2] Bonds formed

	$\Delta H°$ (kJ/mol)
CH_3CH_2-OH	– 393
$H-Br$	– 368
Total	– 761 kJ/mol

[3] Overall $\Delta H°$ =

sum in Step [1] + sum in Step [2]

+ 783 kJ/mol
– 761 kJ/mol

ANSWER: + 22 kJ/mol
endothermic

b. $CH_4 + Cl_2 \longrightarrow CH_3Cl + HCl$

[1] Bonds broken			[2] Bonds formed			[3] Overall $\Delta H° =$
	$\Delta H°$ (kJ/mol)			$\Delta H°$ (kJ/mol)		sum in Step [1]
CH_3-H	+ 435		CH_3-Cl	− 351		+
$Cl-Cl$	+ 242		$H-Cl$	− 431		sum in Step [2]
Total	+ 677 kJ/mol		Total	− 782 kJ/mol		+ 677 kJ/mol
						− 782 kJ/mol
						ANSWER: − 105 kJ/mol
						exothermic

6.8 Use the directions from Answer 6.7. In determining the number of bonds broken or formed, you must take into account the coefficients needed to balance an equation.

a. $CH_4 + 2 O_2 \longrightarrow CO_2 + 2 H_2O$

[1] Bonds broken			[2] Bonds formed			[3] Overall $\Delta H° =$
	$\Delta H°$ (kJ/mol)			$\Delta H°$ (kJ/mol)		sum in Step [1]
CH_3-H	+ 435 x 4 =	+ 1740	$OC-O$ − 535 x 2 =	− 1070		+
$O-O$	+ 497 x 2 =	+ 994	$HO-H$ − 498 x 4 =	− 1992		sum in Step [2]
Total		+ 2734 kJ/mol	Total	− 3062 kJ/mol		+ 2734 kJ/mol
						− 3062 kJ/mol
						ANSWER: − 328 kJ/mol

b. $2 CH_3CH_3 + 7 O_2 \longrightarrow 4 CO_2 + 6 H_2O$

[1] Bonds broken			[2] Bonds formed			[3] Overall $\Delta H° =$
	$\Delta H°$ (kJ/mol)			$\Delta H°$ (kJ/mol)		sum in Step [1]
CH_3CH_2-H	+ 410 x 12 =	+ 4920	$OC-O$ − 535 x 8 =	− 4280		+
$O-O$	+ 497 x 7 =	+ 3479	$HO-H$ − 498 x 12 =	− 5976		sum in Step [2]
$C-C$	+ 368 x 2 =	+736	Total	− 10256 kJ/mol		+ 9135 kJ/mol
Total		+ 9135 kJ/mol				− 10256 kJ/mol
						ANSWER: − 1121 kJ/mol

6.9 Use the following relationships to answer the questions:
If $K_{eq} = 1$, then $\Delta G° = 0$; if $K_{eq} > 1$, then $\Delta G° < 0$; if $K_{eq} < 1$, then $\Delta G° > 0$.

a. A negative value of $\Delta G°$ means the equilibrium favors the product and $K_{eq} > 1$. Therefore, $K_{eq} = 1000$ is the answer.

b. A lower value of $\Delta G°$ means a larger value of K_{eq}, and the products are more favored. $K_{eq} = 10^{-2}$ is larger than $K_{eq} = 10^{-5}$, so $\Delta G°$ is lower.

6.10 Use the relationships from Answer 6.9.

    a. $K_{eq} = 5.5$. $K_{eq} > 1$ means that the equilibrium favors the **product.**
    b. $\Delta G° = 40$ kJ/mol. A positive $\Delta G°$ means the equilibrium favors the **starting material.**

6.11 When the *product* is lower in energy than the *starting material,* the equilibrium favors the *product.* When the *starting material* is lower in energy than the *product,* the equilibrium favors the *starting material.*

    a. $\Delta G°$ **is positive,** so the equilibrium favors the starting material. Therefore the *starting material is lower in energy than the product.*
    b. $K_{eq} > 1$, so the equilibrium favors the product. Therefore the *product is lower in energy than the starting material.*
    c. $\Delta G°$ **is negative,** so the equilibrium favors the product. Therefore the *product is lower in energy than the starting material.*
    d. $K_{eq} < 1$, so the equilibrium favors the starting material. Therefore *the starting material is lower in energy than the product.*

6.12

    a. $K_{eq} > 1$, so the **product** (the conformation on the right) is favored at equilibrium.
    b. $\Delta G°$ for this process must be **negative,** because the product is favored.
    c. $\Delta G°$ is somewhere between 0 and −6 kJ/mol.

6.13 A positive $\Delta H°$ favors the starting material. A negative $\Delta H°$ favors the product.
    a. $\Delta H°$ is positive (80 kJ/mol). The starting material is favored.
    b. $\Delta H°$ is negative (−40 kJ/mol). The product is favored.

6.14
    a. **False.** The reaction is endothermic.
    b. **True.** This assumes that $\Delta G°$ is approximately equal to $\Delta H°$.
    c. **False.** $K_{eq} < 1$.
    d. **True.**
    e. **False.** The starting material is favored at equilibrium.

6.15
    a. **True.**
    b. **False.** $\Delta G°$ for the reaction is negative.
    c. **True.**
    d. **False.** The bonds in the product are stronger than the bonds in the starting material.
    e. **True.**

6.16

6.17 A transition state is drawn with dashed lines to indicate the partially broken and partially formed bonds. Any atom that gains or loses a charge contains a partial charge in the transition state.

6.18

a. Reaction **A–C** is exothermic. Reaction **A–B** is endothermic.
b. Reaction **A–C** is faster.
c. Reaction **A–C** generates a lower-energy product.
d. See labels.
e. See labels.
f. See labels.

6.19

a. Two steps, because there are two energy barriers.
b. See labels.
c. See labels.
d. One reactive intermediate is formed (see label).
e. The first step is rate-determining, because its transition state is at higher energy.
f. The overall reaction is endothermic, because the energy of the products is higher than the energy of the reactants.

6.20

relative energies: C < A < B
B ⟶ C is rate-determining.

Energy

Reaction coordinate

6.21 E_a, **concentration, and temperature affect reaction rate.** $\Delta H°$, $\Delta G°$, and K_{eq} do not affect reaction rate.

    a. E_a = **4 kJ/mol** corresponds to a faster reaction rate.
    b. A temperature of **25 °C** will have a faster reaction rate, because a higher temperature corresponds to a faster reaction.
    c. **No change:** K_{eq} does not affect reaction rate.
    d. **No change:** $\Delta H°$ does not affect reaction rate.

6.22

    a. **False.** The reaction occurs at the same rate as a reaction with K_{eq} = 8 and E_a = 80 kJ/mol.
    b. **False.** The reaction is slower than a reaction with K_{eq} = 0.8 and E_a = 40 kJ/mol.
    c. **True.**
    d. **True.**
    e. **False.** The reaction is endothermic.

6.23 All reactants in the rate equation determine the rate of the reaction.

[1] rate = $k[CH_3CH_2Br][^-OH]$

    a. Tripling the concentration of CH_3CH_2Br only → **The rate is tripled.**
    b. Tripling the concentration of ^-OH only → **The rate is tripled.**
    c. Tripling the concentration of both $CH_3CH_2CH_2Br$ and ^-OH → **The rate increases by a factor of 9 (3 × 3 = 9).**

[2] rate = $k[(CH_3)_3COH]$

    a. Doubling the concentration of $(CH_3)_3COH$ → **The rate is doubled.**
    b. Increasing the concentration of $(CH_3)_3COH$ by a factor of 10 → **The rate increases by a factor of 10.**

6.24 The rate equation is determined by the rate-determining step.

a.   Br + ^-OH ⟶ + H_2O + Br^- one step
rate = $k[CH_3CH_2CH_2Br][^-OH]$

b.   Br   slow → ^-OH fast → + H_2O
+ Br^- two steps
The slow step determines the rate equation.
rate = $k[(CH_3)_3CBr]$

6.25 A catalyst is not used up or changed in the reaction. It only speeds up the reaction rate.

a.
OH and H are added to the starting material.

$CH_2=CH_2$ $\xrightarrow[\text{H}_2\text{SO}_4]{\text{H}_2\text{O}}$ CH_3CH_2OH

H_2SO_4 is not used up = **catalyst.**

b.
I^- not used up = **catalyst.**

CH_3Cl $\xrightarrow[^-OH]{I^-}$ CH_3OH

^-OH substitutes for Cl^-.

6.26

a. [structure] $\longrightarrow$ [radical] + $\cdot H$

radical

b. [structure with OH] $\longrightarrow$ $CH_3\overset{+}{C}H_2$ + ^-OH

carbocation

6.27

propane

$CH_3-CH_2CH_3 \longrightarrow \cdot CH_3 + \cdot CH_2CH_3$

$\Delta H^o = 356$ kJ/mol
This bond is formed from two sp^3 hybridized C's.

propene

$CH_3-CH=CH_2 \longrightarrow \cdot CH_3 + \cdot CH=CH_2$

$\Delta H^o = 385$ kJ/mol
This bond is formed from one sp^2 and one sp^3 hybridized C. The higher percent s-character in one C makes a stronger bond; thus, the bond dissociation energy is higher.

6.28 Use the directions from Answer 6.1.

a.
HO OH
[cyclohexane structure] $\longrightarrow$ [cyclohexanone structure]

elements lost
(H + OH)

π bond formed

elimination reaction

b.
[acetone structure] $\longrightarrow$ [structure]
H OH

addition of 2 H's
addition reaction

6.29 Use the rules in Answer 6.4 to draw the arrows.

a. [structure] $\longrightarrow$ [structure] + $:\overset{..}{\underset{..}{Cl}}:^-$

b. [structure] + $:\overset{..}{\underset{..}{Br}}-\overset{..}{\underset{..}{Br}}:$ $\longrightarrow$ [structure] + $:\overset{..}{\underset{..}{Br}}\cdot$

c. $CH_3CH_2-\overset{..}{\underset{..}{Br}}: + {}^-:\overset{..}{O}H \longrightarrow CH_3CH_2\overset{..}{O}H + :\overset{..}{\underset{..}{Br}}:^-$

d. [structure]$-H + {}^-:\overset{..}{O}H \longrightarrow$ [structure] + $H_2\overset{..}{O}:$

6.30

6.31

a.

A   B   C

b. The conversion of **A** to **C** is an addition because a π bond is broken and an O atom is added.

c.

6.32 Draw the curved arrows to identify the product **X.**

A   B   X

6.33 Follow the curved arrows to identify the intermediate **Y.**

C   D   Y

6.34 Use the rules from Answer 6.6.

$I-CCl_3$ $Br-CCl_3$ $Cl-CCl_3$

largest halogen   **intermediate**   smallest halogen
weakest bond   **bond strength**   **strongest bond**

6.35 Use the directions from Answer 6.7.

a. ·OH + CH₄ ⟶ ·CH₃ + H₂O

[1] Bonds broken		[2] Bonds formed		[3] Overall ΔH° =
	ΔH° (kJ/mol)		ΔH° (kJ/mol)	+ 435 kJ/mol
				– 498 kJ/mol
CH₃–H	+ 435 kJ/mol	H–OH	– 498 kJ/mol	ANSWER: – 63 kJ/mol

b. CH₃–OH + HBr ⟶ CH₃–Br + H₂O

[1] Bonds broken		[2] Bonds formed		[3] Overall ΔH° =
	ΔH° (kJ/mol)		ΔH° (kJ/mol)	
CH₃–OH	+ 389	CH₃–Br	– 293	+ 757 kJ/mol
H–Br	+ 368	H–OH	– 498	– 791 kJ/mol
Total	+ 757 kJ/mol	Total	– 791 kJ/mol	ANSWER: – 34 kJ/mol

6.36

hybrid:

6.37 The more stable radical is formed by a reaction with a smaller ΔH°.

This C–H bond is stronger.   A ΔH° = 410 kJ/mol = less stable radical

This C–H bond is weaker.   B ΔH° = 397 kJ/mol = more stable radical

Because the bond dissociation for cleavage of the C–H bond to form radical **A** is higher, more energy must be added to form it. This makes **A** higher in energy and therefore less stable than **B**.

6.38 Use the rules from Answer 6.11.
a. K_{eq} = 0.5. K_{eq} is less than one, so the **starting material** is favored.
b. ΔG° = –100 kJ/mol. ΔG° is less than 0, so the **product** is favored.
c. ΔH° = 8.0 kJ/mol. ΔH° is positive, so the **starting material** is favored.
d. K_{eq} = 16. K_{eq} is greater than one, so the **product** is favored.
e. ΔG° = 2.0 kJ/mol. ΔG° is greater than zero, so the **starting material** is favored.
f. ΔH° = 200 kJ/mol. ΔH° is positive, so the **starting material** is favored.
g. ΔS° = 8 J/(K·mol). ΔS° is greater than zero, so the **product** is more disordered and favored.
h. ΔS° = –8 J/(K·mol). ΔS° is less than zero, so the **starting material** is more disordered and favored.

6.39

    a. A negative $\Delta G°$ must have $K_{eq} > 1$. $K_{eq} = 10^2$.

    b. K_{eq} = [products]/[reactants] = [1]/[5] = 0.2 = K_{eq}. $\Delta G°$ is positive.

    c. A negative $\Delta G°$ has $K_{eq} > 1$, and a positive $\Delta G°$ has $K_{eq} < 1$. $\Delta G° = -8$ kJ/mol will have a larger K_{eq}.

6.40

R	K_{eq}
$-CH_2CH_3$	23
$-C(CH_3)_3$	4000

    a. The equatorial conformation is present in the larger amount at equilibrium, because the K_{eq} is greater than 1.

    b. The cyclohexane with the $-C(CH_3)_3$ group will have the greater amount of equatorial conformation at equilibrium, because this group has the higher K_{eq}.

    c. The cyclohexane with the $-CH_2CH_3$ group will have the greater amount of axial conformation at equilibrium, because this group has the lower K_{eq}.

    d. The cyclohexane with the $-C(CH_3)_3$ group will have the more negative $\Delta G°$, because it has the larger K_{eq}.

    e. The larger the R group, the more favored the equatorial conformation.

6.41 Reactions resulting in an increase in entropy are favored. When a single molecule forms two molecules, there is an increase in entropy.

    a.

increased number of molecules
$\Delta S°$ is positive.
products favored

    b. $CH_3CO_2CH_3$ + H_2O ⟶ CH_3CO_2H + CH_3OH

no change in the number of molecules
neither favored

6.42 Use the directions in Answer 6.17 to draw the transition state. Nonbonded electron pairs are drawn in at reacting sites.

6.43

a.

Reaction coordinate

· one step **A → B**
· exothermic because **B** lower than **A**

b.

Reaction coordinate $\Delta H^\circ_{overall}$

· two steps
· **A** lowest energy
· **B** highest energy
· $E_{a(A\text{-}B)}$ **is rate-determining,** because the transition state for Step [1] is higher in energy.

6.44

a.

b.   ·Cl + CH₄ ⟶ ·CH₃ + HCl

[1] **Bonds broken**		[2] **Bonds formed**		[3] **Overall** $\Delta H^\circ =$
	ΔH° (kJ/mol)		ΔH° (kJ/mol)	+ 435 kJ/mol
CH₃—H	+ 435 kJ/mol	H—Cl	– 431 kJ/mol	– 431 kJ/mol
				ANSWER: + 4 kJ/mol

c.

Reaction coordinate

d. E_a for the reverse reaction is the difference in energy between the products and the transition state, 12 kJ/mol.

Reaction coordinate

6.45

Reaction coordinate

a. **B, D,** and **F** are transition states.
b. **C** and **E** are reactive intermediates.
c. The overall reaction has **three steps.**
d. **A–C** is endothermic.
   C–E is exothermic.
   E–G is exothermic.
e. The overall reaction is exothermic.

6.46

Because pK_a (CH$_3$CO$_2$H) = 4.8 and pK_a [(CH$_3$)$_3$COH] = 18, the weaker acid is formed as product, and equilibrium favors the products. Thus, $\Delta H°$ is negative, and the products are lower in energy than the starting materials.

6.47

a. Step [1] breaks one π bond and the H–Cl bond, and one C–H bond is formed. $\Delta H°$ for this step should be positive, because more bonds are broken than formed.

b. Step [2] forms one bond. $\Delta H°$ for this step should be negative, because one bond is formed and none is broken.

c. Step [1] is rate-determining, because it is more difficult.

d. Transition state for Step [1]:   Transition state for Step [2]:

e.

6.48 E_a, concentration, catalysts, rate constant, and temperature affect reaction rate so (c), (d), (e), (g), and (h) affect rate.

6.49

  a. **rate = k[CH$_3$Br][NaCN]**
  b. Double [CH$_3$Br] = **rate doubles.**
  c. Halve [NaCN] = **rate halved.**
  d. Increase both [CH$_3$Br] and [NaCN] by factor of 5 = [5][5] = **rate increases by a factor of 25.**

6.50

  a.

  b. Only the slow step is included in the rate equation: **Rate = k[CH$_3$O$^-$][CH$_3$COCl]**
  c. CH$_3$O$^-$ is in the rate equation. Increasing its concentration by 10 times would increase the rate by **10 times.**
  d. When both reactant concentrations are increased by 10 times, the rate increases by **100 times (10 × 10 = 100).**
  e. This is a **substitution reaction** (OCH$_3$ substitutes for Cl).

6.51

  a. **True:** Increasing temperature increases reaction rate.
  b. **True:** If a reaction is fast, then it has a large rate constant.
  c. **False: Corrected**—There is no relationship between $\Delta G°$ and reaction rate.
  d. **False: Corrected**—When the E_a is large, *the rate constant is small.*
  e. **False: Corrected**—There is no relationship between K_{eq} and reaction rate.
  f. **False: Corrected**—Increasing the concentration of a reactant increases the rate of a reaction *only if the reactant appears in the rate equation.*

6.52

  a.

  b. Two π bonds in **A** are broken and one π bond in **B** is broken. Two new σ bonds in **C** are formed (in bold), as well as a new π bond.
  c. The reaction should be exothermic because more energy is released in forming two new C–C σ bonds than is required to break two C–C π bonds.
  d. Entropy favors the reactants for two reasons. There are two molecules of reactant and only one product. The reactants are both acyclic and the product has a ring with fewer degrees of freedom.
  e. The Diels–Alder reaction is an addition reaction because π bonds are broken and new σ bonds are formed.

6.53

a. The first mechanism has one step: **Rate = $k[(CH_3)_3CI][^-OH]$**

b. The second mechanism has two steps, but only the first step would be in the rate equation, because it is slow and therefore rate-determining: **Rate = $k[(CH_3)_3CI]$**

c. Possibility [1] is second order; possibility [2] is first order.

d. These rate equations can be used to show which mechanism is plausible by changing the concentration of ^-OH. If this affects the rate, then possibility [1] is reasonable. If it does not affect the rate, then possibility [2] is reasonable.

e.

$A = (CH_3)_3CI + {}^-OH$
$B = (CH_3)_2C=CH_2 + I^- + H_2O$

f.

6.54 The difference in both the acidity and the bond dissociation energy of CH_3CH_3 versus $HC{\equiv}CH$ is due to the same factor: percent *s*-character. The difference results because one process is based on homolysis and one is based on heterolysis.

Bond dissociation energy:

$$CH_3CH_2{-}H \qquad\qquad HC{\equiv}C{-}H$$

sp³ hybridized   *sp* hybridized
25% *s*-character   50% *s*-character
Higher percent *s*-character makes
this bond shorter and stronger.

Acidity: To compare acidity, we must compare the stability of the conjugate bases:

$CH_3\overset{-}{C}H_2$ $HC \equiv C^{-}$

sp³ hybridized   sp hybridized
25% s-character   50% s-character
Now a higher percent s-character
stabilizes the conjugate base making the
starting acid more acidic.

6.55 a. Re-draw **A** to see more clearly how cyclization occurs.

b.

6.56

a.

b.

c. C–H$_a$ is weaker than C–H$_b$ because the carbon radical formed when the C–H$_a$ bond is broken is highly resonance stabilized. This means the bond dissociation energy for C–H$_a$ is lower.

6.57 In Reaction [1], the number of molecules of reactants and products stays the same, so entropy is not a factor. In Reaction [2], a single molecule of starting material forms two molecules of products, so entropy increases. This makes $\Delta G°$ more favorable, thus increasing K_{eq}.

6.58

ethyl acetate

+ H$_2$O $K_{eq} = 4$

To increase the yield of ethyl acetate, H$_2$O can be removed from the reaction mixture, or there can be a large excess of one of the starting materials.

6.59

a.

phenol

resonance stabilized
less energy for homolysis

ethanol   no resonance stabilization

Less energy is required for cleavage of C$_6$H$_5$O–H because homolysis forms the more stable radical.

b.

Csp^2–O
higher % s-character
shorter bond

Csp^3–O
lower % s-character
longer bond

Chapter 7 Alkyl Halides and Nucleophilic Substitution

Chapter Review

General facts about alkyl halides

- Alkyl halides contain a halogen atom X bonded to an sp^3 hybridized carbon (7.1).
- Alkyl halides are named as halo alkanes, with the halogen as a substituent (7.2).
- Alkyl halides have a polar C–X bond, so they exhibit dipole–dipole interactions but are incapable of intermolecular hydrogen bonding (7.3).
- The polar C–X bond containing an electrophilic carbon makes alkyl halides reactive towards nucleophiles and bases (7.5).

The central theme (7.6)

- Nucleophilic substitution is one of the two main reactions of alkyl halides. A nucleophile replaces a leaving group on an sp^3 hybridized carbon.

- One σ bond is broken and one σ bond is formed.
- There are two possible mechanisms: S$_N$1 and S$_N$2.

S$_N$1 and S$_N$2 mechanisms compared

	S$_N$2 mechanism	S$_N$1 mechanism
[1] Mechanism	• One step (7.11B)	• Two steps (7.12B)
[2] Alkyl halide	• Order of reactivity: CH_3X > RCH_2X > R_2CHX > R_3CX (7.11D)	• Order of reactivity: R_3CX > R_2CHX > RCH_2X > CH_3X (7.12D)
[3] Rate equation	• rate = $k[RX][:Nu^-]$ • second-order kinetics (7.11A)	• rate = $k[RX]$ • first-order kinetics (7.12A)
[4] Stereochemistry	• backside attack of the nucleophile (7.11C) • inversion of configuration at a stereogenic center	• trigonal planar carbocation intermediate (7.12C) • racemization at a stereogenic center
[5] Nucleophile	• favored by stronger nucleophiles (7.15B)	• favored by weaker nucleophiles (7.15B)
[6] Leaving group	• better leaving group → faster reaction (7.15C)	• better leaving group → faster reaction (7.15C)
[7] Solvent	• favored by polar aprotic solvents (7.15D)	• favored by polar protic solvents (7.15D)

Important trends

- The best leaving group is the weakest base. Leaving group ability increases left-to-right across a row and down a column of the periodic table (7.7).

- Nucleophilicity decreases across a row of the periodic table (7.8A).

- Nucleophilicity decreases down a column of the periodic table in polar aprotic solvents (7.8C).

- Nucleophilicity increases down a column of the periodic table in polar protic solvents (7.8C).

- The stability of a carbocation increases as the number of R groups bonded to the positively charged carbon increases (7.13).

Important principles

Principle	Example
• Electron-donating groups (such as R groups) stabilize a positive charge (7.13A).	• 3° Carbocations (R_3C^+) are more stable than 2° carbocations (R_2CH^+), which are more stable than 1° carbocations (RCH_2^+).
• Steric hindrance decreases nucleophilicity but not basicity (7.8B).	• $(CH_3)_3CO^-$ is a stronger base but a weaker nucleophile than $CH_3CH_2O^-$.
• Hammond postulate: In an endothermic reaction, the more stable product is formed faster. In an exothermic reaction, this fact is not necessarily true (7.14).	• S_N1 reactions are faster when more stable (more substituted) carbocations are formed, because the rate-determining step is endothermic.
• Planar, sp^2 hybridized atoms react with reagents from both sides of the plane (7.12C).	• A trigonal planar carbocation reacts with nucleophiles from both sides of the plane.

Practice Test on Chapter Review

1. Give the IUPAC name for the following compound, including the appropriate R,S prefix.

2.a. Which of the following carbocations is the most stable?

b. Which of the following anions is the best leaving group?

    1. CH_3^- 2. ^-OH 3. H^- 4. $^-NH_2$ 5. Cl^-

c. Which species is the strongest nucleophile in polar protic solvents?

    1. F^- 2. ^-OH 3. Cl^- 4. H_2O 5. ^-SH

d. Which of the following statements is true about the given reaction?

   1. The reaction follows second-order kinetics.
   2. The rate of the reaction increases when the solvent is changed from CH_3CH_2OH to DMSO.
   3. The rate of the reaction increases when the leaving group is changes from Br to F.
   4. Statements (1) and (2) are both true.
   5. Statements (1), (2), and (3) are all true.

3. Rank the following compounds in order of *increasing* reactivity in an S_N1 reaction. Rank the *least reactive* compound as **1**, the *most reactive* compound as **4**, and the compounds of intermediate reactivity as **2** and **3**.

     A   B   C   D

4. Consider the following two nucleophilic substitution reactions, labeled Reaction [1] and Reaction [2]. (Only the starting materials are drawn.) Then answer True (T) or False (F) to each of the following statements.

    Reaction [1] $(CH_3CH_2)_3CBr$ + CH_3OH ⟶

    Reaction [2] $CH_3CH_2CH_2Br$ + $^-OCH_3$ ⟶

   a. The rate equation for Reaction [1] is rate $= k[(CH_3CH_2)_3CBr][CH_3OH]$.
   b. Changing the leaving group from Br^- to Cl^- decreases the rate of both reactions.
   c. Changing the solvent from CH_3OH to $(CH_3)_2S=O$ increases the rate of Reaction [2].
   d. Doubling the concentration of both $CH_3CH_2CH_2Br$ and $^-OCH_3$ in Reaction [2] doubles the rate of the reaction.
   e. If entropy is ignored and K_{eq} for Reaction [1] is < 1, then the reaction is exothermic.
   f. If entropy is ignored and $\Delta H°$ is negative for Reaction [1], then the bonds in the product are stronger than the bonds in the starting materials.
   g. The energy diagram for Reaction [2] exhibits only one energy barrier.

5. Draw the organic products formed in the following reactions. **Use wedges and dashed wedges to show stereochemistry in compounds with stereogenic centers.**

a.

                $^-C\equiv C-H$

b.

                CH_3OH
          (Consider substitution only.)

c.

                NaOH

d.

Answers to Practice Test

1. (S)-7-chloro-3-ethyldecane

3. **A**–4
   B–1
   C–3
   D–2

4. a. F
   b. T
   c. T
   d. F
   e. F
   f. T
   g. T

5. a.

   b.

   +

   c.

   d.

2. a. 4
   b. 5
   c. 5
   d. 4

Answers to Problems

7.1 Classify the alkyl halide as 1°, 2°, or 3° **by counting the number of carbons bonded directly to the carbon bonded to the halogen.**

a, b.

2°, neither   2°, neither
3°, neither
     telfairine

3°, neither
1°, neither
3°, allylic
vinyl
halomon
2°, neither

7.2 To name a compound with the IUPAC system:

[1] **Name the parent** chain by finding the longest carbon chain.

[2] **Number the chain** so the first substituent gets the lower number. Then **name and number all substituents,** giving like substituents a prefix (di, tri, etc.). **To name the halogen substituent, change the *-ine* ending to *-o*.**

[3] **Combine all parts,** alphabetizing substituents, and ignoring all prefixes except iso.

a. [1]

5 carbon alkane = **pentane**

[2]   **2-methyl**

2   Cl ← **3-chloro**

[3]   **3-chloro-2-methylpentane**

b.

[1] Br — 7 carbon alkane = **heptane**

[2] **2-bromo** Br — 2 — **5,5-dimethyl**

[3] **2-bromo-5,5-dimethylheptane**

c.

[1] 6 carbon cycloalkane = **cyclohexane** — Br

[2] **2-methyl** — **1-bromo** — Br — 1

[3] **1-bromo-2-methylcyclohexane**

d.

[1] 10 carbon alkane = **decane** — F

[2] **2** **6** **7** — **7-isopropyl** — **2-fluoro** — **6-ethyl**

[3] **6-ethyl-2-fluoro-7-isopropyldecane**

7.3 To work backwards from a name to a structure:

[1] Find the parent name and draw that number of carbons. Use the suffix to identify the functional group **(-ane = alkane).**

[2] Arbitrarily number the carbons in the chain. Add the substituents to the appropriate carbon.

a. 3-chloro-2-methyl**hexane**

[1]   6 carbon alkane

1 2 3 4 5 6

[2]   **methyl at C2**

Cl ← **chloro at C3**

b. 4-ethyl-5-iodo-2,2-dimethyl**octane**

[1]   8 carbon alkane

1 2 3 4 5 6 7 8

[2]   **ethyl at C4**

2 methyls at C2   I ← **iodo at C5**

c. *cis*-1,3-dichloro**cyclopentane**

[1]   5 carbon cycloalkane

[2]   chloro groups at C1 and C3, both on the same side

Cl   Cl
C1   **C3**

d. 1,1,3-tribromo**cyclohexane**

[1]   6 carbon cycloalkane

[2]   3 Br groups

Br
C3 →   ↓ Br
Br

C1

e. ***sec*-butyl** bromide

[1]   4 carbon alkyl group   [2]

Br ← **bromide**

f. 6-ethyl-3-iodo-3,5-dimethylnonane

[1]   9 carbon chain   [2]

7.4   a. Because an sp^2 hybridized C has a higher percent *s*-character than an sp^3 hybridized C, it holds electron density closer to C. This pulls a little more electron density towards C, away from Cl, and thus a C_{sp^2}–Cl bond is less polar than a C_{sp^3}–Cl bond.

b.

lowest boiling point sp^3 C–Cl bond   larger halogen, sp^3 C–Br bond
                                 intermediate   **highest boiling point**
                                 boiling point

7.5   a. Since chondrocole A has 10 C's and only one functional group capable of hydrogen bonding to water (an ether), it is insoluble in H_2O. Because it is organic, it is soluble in CH_2Cl_2.

b.

Three stereogenic centers are labeled with (*).

c.

enantiomer   constitutional isomer

7.6   To draw the products of a nucleophilic substitution reaction:
[1] **Find the sp^3 hybridized electrophilic carbon** with a leaving group.
[2] **Find the nucleophile** with lone pairs or electrons in π bonds.
[3] **Substitute the nucleophile for the leaving group** on the electrophilic carbon.

a.   +   ⁻OCH₂CH₃   ⟶   +   Br⁻

Br   OCH₂CH₃
↑   ↑
leaving group   nonbonded e⁻ pairs
                        nucleophile

b.

leaving group   nonbonded e⁻ pairs
nucleophile

c.

leaving group   nonbonded e⁻ pairs
nucleophile

d.

leaving group   nonbonded e⁻ pairs
nucleophile

7.7 Use the steps from Answer 7.6 and then draw the proton transfer reaction.

a.

nucleophile

leaving group

b.

leaving group   **nucleophile**

7.8 Draw the structure of CPC using the steps from Answer 7.6.

nucleophile   substitution   **leaving group**

CPC

7.9

A   ticlopidine

These atoms come from the nucleophile.

7.10 Compare the leaving groups based on these trends:
- Better leaving groups are weaker bases.
- A neutral leaving group is always better than its conjugate base.

a. Cl⁻, I⁻

farther down a column
of the periodic table
less basic
better leaving group

b. NH₃, NH₂⁻

neutral compound
less basic
better leaving group

c. H₂O, H₂S

farther down a column
of the periodic table
less basic
better leaving group

7.11 Good leaving groups include Cl⁻, Br⁻, I⁻, and H₂O.

a.

Br⁻ is a **good
leaving group.**

b.

no good leaving group
⁻OH is too strong a base.

c.

H₂O is a **good
leaving group.**

d.

no good leaving group
H⁻ is too strong a base.

7.12 To decide whether the equilibrium favors the starting material or the products, **compare the nucleophile and the leaving group.** The reaction proceeds towards the weaker base.

a.

nucleophile
better leaving group
weaker base

pK_a (HBr) = –9

leaving group

pK_a (NH₃) = 38

| Reaction favors
starting material. |

b.

nucleophile

pK_a (HCN) = 9.1

leaving group
better leaving group
weaker base

pK_a (HI) = –10

| Reaction favors product. |

7.13 Use these three rules to find the stronger nucleophile in each pair:

[1] Comparing two nucleophiles having the *same attacking atom*, **the stronger base is a stronger nucleophile.**

[2] **Negatively charged nucleophiles** are always **stronger than their conjugate acids.**

[3] **Nucleophilicity decreases left to right across a row of the periodic table,** when comparing species of similar charge.

a. NH₃, NH₂⁻

A negatively charged
nucleophile is stronger
than its conjugate acid.
stronger nucleophile

b. CH₃NH₂, CH₃OH

Across a row of the periodic
table, nucleophilicity decreases
with species of the same charge.
stronger nucleophile

c.

same attacking atom
(O)
stronger base
stronger nucleophile

7.14 *Polar protic solvents* are capable of hydrogen bonding, and therefore must contain a **H bonded to an electronegative O or N.** *Polar aprotic solvents* are incapable of hydrogen bonding, and therefore do not contain any O–H or N–H bonds.

a. contains 2 O–H bonds
polar protic

b. no O–H bonds
polar aprotic

c. no O–H bonds
polar aprotic

7.15 • In *polar protic solvents,* **the trend in nucleophilicity is opposite to the trend in basicity** down a column of the periodic table, so nucleophilicity increases.
• In *polar aprotic solvents,* **the trend is identical to basicity,** so nucleophilicity decreases down a column.

a. Br⁻ and Cl⁻ in polar protic solvent

farther down the column
**more nucleophilic
in protic solvent**

c. HS⁻ and F⁻ in polar protic solvent

farther down the column
and left in the row
**more nucleophilic
in protic solvent**

In polar protic solvents:
nucleophilicity increases

O   F
S   nucleophilicity
increases

b. ⁻OH and Cl⁻ in polar aprotic solvent

farther up the column
and to the left in the row
more basic
more nucleophilic

In polar aprotic solvents:
nucleophilicity increases

O   F
Cl   nucleophilicity
increases

7.16 The stronger base is the stronger nucleophile, except in polar protic solvents, where nucleophilicity increases down a column. For other rules, see Answers 7.13 and 7.15.

a.
H_2O
no charge
weakest nucleophile

⁻OH
negatively charged
intermediate nucleophile

⁻NH_2
negatively charged
farther left in periodic table
strongest nucleophile

b.
Br⁻
Basicity decreases down a
column in polar aprotic solvents.
weakest nucleophile

F⁻
Basicity decreases
across a row.
intermediate nucleophile

⁻OH
strongest nucleophile

c.
H_2O
weakest nucleophile

CH_3COO^-
weaker base than ⁻OH
intermediate nucleophile

⁻OH
strongest nucleophile

7.17 To determine what nucleophile is needed to carry out each reaction, look at the product to see what has replaced the leaving group.

a.
SH replaces Br.
HS⁻ is needed.

c.
$OCOCH_3$ replaces Br.
CH_3COO^- is needed.

b.
OCH_2CH_3 replaces Br.
$CH_3CH_2O^-$ is needed.

d.
C≡CH replaces Br.
HC≡C⁻ is needed.

7.18 The general rate equation for an S_N2 reaction is rate = k[RX][:Nu$^-$].

    a. [RX] is tripled, and [:Nu$^-$] stays the same: **rate triples.**
    b. Both [RX] and [:Nu$^-$] are tripled: **rate increases by a factor of 9 (3 × 3 = 9).**
    c. [RX] is halved, and [:Nu$^-$] stays the same: **rate halved.**
    d. [RX] is halved, and [:Nu$^-$] is doubled: **rate stays the same (1/2 × 2 = 1).**

7.19 All S_N2 reactions have one step. The transition state in an S_N2 reaction has **dashed bonds to both the leaving group and the nucleophile,** and must contain partial charges.

7.20 To draw the products of S_N2 reactions, **replace the leaving group by the nucleophile, and then draw the stereochemistry with *inversion* at the stereogenic center.**

7.21 *Increasing* the number of R groups *increases* crowding of the transition state and *decreases* the rate of an S_N2 reaction.

a.

2° alkyl halide   1° alkyl halide
               faster reaction

b.

2° alkyl halide   3° alkyl halide
faster reaction

7.22

7.23 In a first-order reaction, **the rate changes with any change in [RX].** The rate is independent of any change in [:Nu⁻].
a. [RX] is tripled, and [:Nu⁻] stays the same: **rate triples.**
b. Both [RX] and [:Nu⁻] are tripled: **rate triples.**
c. [RX] is halved, and [:Nu⁻] stays the same: **rate is halved.**
d. [RX] is halved, and [:Nu⁻] is doubled: **rate is halved.**

7.24 In S$_N$1 reactions, racemization always occurs at a stereogenic center. Draw two products, with the two possible configurations at the stereogenic center.

7.25 Carbocations are classified by the number of R groups bonded to the carbon: 0 R groups = methyl, 1 R group = 1°, 2 R groups = 2°, and 3 R groups = 3°.

a.
2 R groups
2° carbocation

b.
1 R group
1° carbocation

c.
3 R groups
3° carbocation

d.
2 R groups
2° carbocation

7.26 For carbocations: Increasing number of R groups = Increasing stability.

1° carbocation
least stable

2° carbocation
**intermediate
stability**

3° carbocation
most stable

7.27 For carbocations: Increasing number of R groups = Increasing stability.

1° carbocation
least stable

2° carbocation
**intermediate
stability**

3° carbocation
most stable

7.28 The rate of an S$_N$1 reaction increases with increasing alkyl substitution.

a.

3° alkyl halide
faster S$_N$1 reaction

1° alkyl halide
slower S$_N$1 reaction

b.

3° alkyl halide
faster S$_N$1 reaction

2° alkyl halide
slower S$_N$1 reaction

7.29 • For **methyl and 1° alkyl halides,** only S$_N$2 will occur.
• For **2° alkyl halides,** S$_N$1 and S$_N$2 will occur.
• For **3° alkyl halides,** only S$_N$1 will occur.

a.

2° alkyl halide
S$_N$1 and S$_N$2

b.

1° alkyl halide
S$_N$2

c.

2° alkyl halide
S$_N$1 and S$_N$2

d.

3° alkyl halide
S$_N$1

7.30 • Draw the product of nucleophilic substitution for each reaction.
• For **methyl and 1° alkyl halides,** only S$_N$2 will occur.
• For **2° alkyl halides,** S$_N$1 and S$_N$2 will occur and other factors determine which mechanism operates.
• For **3° alkyl halides,** only S$_N$1 will occur.

7.31 First decide whether the reaction will proceed via an S$_N$1 or S$_N$2 mechanism. Then draw the products with stereochemistry.

a.

2° alkyl halide
S$_N$1 and S$_N$2

Weak nucleophile
favors S$_N$1.

enantiomers

+ HBr   **S$_N$1 = racemization at the stereogenic C**

b.

1° alkyl halide
S$_N$2 only

+ Cl$^-$ **S$_N$2 = inversion at the stereogenic C**

7.32 Compounds with better leaving groups react faster. Weaker bases are better leaving groups.

a.   Cl   or   I

weaker base
better leaving group

b.   Br   or   I

weaker base
better leaving group

c.   OH   or $\overset{+}{O}H_2$

weaker base
better leaving group

d.   OH   or

weaker base
better leaving group

7.33 • **Polar protic solvents** favor the S$_N$1 mechanism by solvating the intermediate carbocation and halide.
  • **Polar aprotic solvents** favor the S$_N$2 mechanism by making the nucleophile stronger.

a. CH_3CH_2OH	b. CH_3CN	c. CH_3COOH	d. $CH_3CH_2OCH_2CH_3$
polar protic solvent	polar aprotic solvent	polar protic solvent	polar aprotic solvent
contains an O–H bond	no O–H or N–H bond	contains an O–H bond	no O–H or N–H bond
favors S$_N$1	**favors S$_N$2**	**favors S$_N$1**	**favors S$_N$2**

7.34 Compare the solvents in the reactions below. **For the solvent to increase the reaction rate of an S_N1 reaction, the solvent must be *polar protic*. For the solvent to increase the reaction rate of an S_N2 reaction, the solvent must be *polar aprotic*.**

a.

$$\text{(structure)} + CH_3OH \xrightarrow[\text{DMSO}]{\overset{CH_3OH}{\text{or}}} \text{(structure)} + HCl$$

Cl

$3° \, RX - S_N1$ reaction

CH_3OH
Polar protic solvent
increases the rate of an
S_N1 reaction.

b.

$$\text{(structure)} Br + {}^-OH \xrightarrow[\text{DMF}]{\overset{H_2O}{\text{or}}} \text{(structure)} OH + Br^-$$

$1° \, RX - S_N2$ reaction

$DMF \, [HCON(CH_3)_2]$
Polar aprotic solvent
increases the rate of an
S_N2 reaction.

c.

$$\text{(structure)} + CH_3O^- \xrightarrow[\text{HMPA}]{\overset{CH_3OH}{\text{or}}} \text{(structure)} + Cl^-$$

H Cl   H OCH₃

$2° \, RX$ strong nucleophile

S_N2 reaction

$HMPA \, [(CH_3)_2N]_3P{=}O$
Polar aprotic solvent
increases the rate of an
S_N2 reaction.

7.35 To predict whether the reaction follows an S_N1 or S_N2 mechanism:
[1] **Classify RX as a methyl, 1°, 2°, or 3° halide.** (Methyl, 1° = S_N2; 3° = S_N1; 2° = either.)
[2] **Classify the nucleophile as strong or weak.** (Strong favors S_N2; weak favors S_N1.)
[3] **Classify the solvent as polar protic or polar aprotic.** (Polar protic favors S_N1; polar aprotic favors S_N2.)

a.

$$\text{(structure)} Br + \text{(structure)} O^- \longrightarrow \text{(structure)} O + Br^- \qquad S_N2 \text{ reaction}$$

1° alkyl halide
S_N2

b.

$$\text{(structure)} Br + N_3^- \longrightarrow \text{(structure)} N_3 + Br^-$$

2° alkyl halide   Strong nucleophile
S_N1 or S_N2   **favors S_N2.**

S_N2 reaction = ***inversion*** at the stereogenic center
The leaving group was "up."
The nucleophile attacks from below.

c.

$$\text{(structure)} I + CH_3OH \longrightarrow \text{(structure)} OCH_3 + HI \qquad S_N1 \text{ reaction}$$

3° alkyl halide   Weak nucleophile
S_N1   **favors S_N1.**

d.

$$\text{(structure)} Cl + H_2O \longrightarrow \text{(structure)} OH + \text{(structure)} HO + HCl$$

3° alkyl halide   Weak nucleophile
S_N1   **favors S_N1.**

S_N1 reaction
forms **two enantiomers.**

7.36

$$\text{(structure A)} \quad \overset{CH_3 {-} \overset{+}{SR_2}}{\longrightarrow} \quad \text{(structure)} \quad \xrightarrow{\substack{\text{loss of} \\ \text{a proton}}} \quad \text{(structure nicotine)}$$

A   + SR₂   nicotine

7.37 Vinyl carbocations are even less stable than 1° carbocations.

vinyl carbocation
least stable

1° carbocation
**intermediate
stability**

2° carbocation
most stable

7.38 Convert each ball-and-stick model to a skeletal or condensed structure and draw the reactants.

a.

carbon
framework

nucleophile

b.

carbon
framework

nucleophile

c.

nucleophile

carbon
framework

d.

carbon
framework

nucleophile

7.39

$CH_3\ddot{O}{:}^-$ + Cl—CH₂CH₃ ⟶ $CH_3\ddot{O}CH_2CH_3$ $CH_3CH_2\ddot{O}{:}^-$ + Cl—CH₃ ⟶ $CH_3\ddot{O}CH_2CH_3$

7.40 Use the directions from Answer 7.2 to name the compounds.

a.

7 C chain = heptane
bromo at C2
ethyl at C5
one stereogenic center–*R*
(*R*)-2-bromo-5-ethylheptane

b.

5 C ring = cyclopentane
chloro at C1
isopropyl at C3
trans isomer—one substituent up, one down
(1*R*,3*R*)-*trans*-1-chloro-3-isopropylcyclopentane

7.41

a.   inversion (equatorial to axial)

polar aprotic solvent
S_N2 reaction

S_N2 reaction   Large *tert*-butyl group is in more roomy equatorial position.

b.   inversion (axial to equatorial)

polar aprotic solvent
S_N2 reaction

7.42 Use the directions from Answer 7.2 to name the compounds.

a. [1]   [2] 3,3-dimethyl / 4-methyl / 1-fluoro   [3] **1-fluoro-3,3,4-trimethylpentane**

5 carbon alkane = **pentane**

b. [1]   [2] 3-ethyl / 2-methyl / 1-iodo   [3] **3-ethyl-1-iodo-2-methylhexane**

6 carbon alkane = **hexane**

c. [1]   [2] 5,5-dimethyl / 1,3-dichloro   [3] **1,3-dichloro-5,5-dimethylhexane**

6 carbon alkane = **hexane**

d. [1]   [2] 1-bromo / 3-iodo   [3] ***cis*-1-bromo-3-iodocyclopentane**

5 carbon cycloalkane =
cyclopentane

e. [1]   [2] 6-bromo / 2-chloro / 6-methyl   [3] **6-bromo-2-chloro-6-methyloctane**

8 carbon alkane = **octane**

f. [1]

6 carbon alkane = hexane
(Indicate the *R,S*
designation also)

[2]

4,4-dimethyl   **(*R*)-2-iodo**

[3] (*R*)-2-iodo-4,4-dimethylhexane

Clockwise
R

7.43 To work backwards to a structure, use the directions in Answer 7.3.

a. 3-bromo-4-ethyl**heptane**

4-ethyl
3-bromo

c. 1-bromo-4-ethyl-3-fluoro**octane**

4-ethyl
1-bromo   3-fluoro

e. (1*R*,2*R*)-*trans*-1-bromo-2-chloro**cyclohexane**

1*R*
Br ◄— 1-bromo
Cl ◄— 2-chloro
2*R*

b. 1,1-dichloro-2-methyl**cyclohexane**

Cl ◄
Cl ◄   1,1-dichloro
2-methyl

d. (*S*)-3-iodo-2-methyl**nonane**

2-methyl
3*S*   3-iodo

f. (*R*)-4,4,5-trichloro-3,3-dimethyl**decane**

5*R*
Cl   H   3,3-dimethyl
5   4   3
Cl   Cl
4,4,5-trichloro

7.44

1-chloro
1°
1-chloropentane

2°
3-chloro   Cl
3-chloropentane

3-methyl
Cl 1   3
1°
1-chloro
1-chloro-3-methylbutane

1-chloro
1°
2   1
Cl
1-chloro-2,2-dimethylpropane

2-chloro
3°
2-methyl   Cl
2
2-chloro-2-methylbutane

Two stereoisomers

2-chloro   Cl
2°
2-chloropentane
[* denotes stereogenic center]

2   3
Cl   H 4
1
Clockwise
"4" in *back* =
R

2   3
Cl   H 4
1
Clockwise
"4" in *front* =
S

Two stereoisomers

3-methyl
2°
2-chloro   Cl
2-chloro-3-methylbutane
[* denotes stereogenic center]

3   2
4 H   Cl
1
Counterclockwise
"4" in *front* =
R

3   2
4 H   Cl
1
Counterclockwise
"4" in *back* =
S

1-chloro
1°

2-methyl →

Two stereoisomers

3   4
H

3   4
H

1-chloro-2-methylbutane

[* denotes stereogenic center]

Clockwise
"4" in *front* =
S

Clockwise
"4" in *back* =
R

7.45

a. Br   or   Br

larger surface area =
stronger intermolecular forces =
higher boiling point

b. I   or   Br

larger halide = more polarizable =
higher boiling point

7.46 Use the steps from Answer 7.6 and then draw the proton transfer reaction, when necessary.

a.
Cl
leaving group

O⁻
nucleophile

+   Cl⁻

b.
I
leaving group

+ Na⁺ ⁻CN
nucleophile

CN   + NaI

c.
I
leaving group

+ $H_2\ddot{O}$:
nucleophile

OH   +   HI

d.
Cl
leaving group

+ $\ddot{O}H$
nucleophile

O   +   HCl

e.
Br
leaving group

+ Na⁺ ⁻OCH₃
nucleophile

OCH₃

+   NaBr

f.
Cl
leaving group

+ $\ddot{S}$
nucleophile

S⁺   +   Cl⁻

7.47 A good leaving group is a weak base.

a.
bad leaving group
⁻OH is a strong base.

b.
Cl⁻ good leaving group
weak base

c.
This has only C–C
and C–H bonds.
no good leaving group

d.
good leaving group
H_2O is a weak base.

7.48 Use the rules from Answer 7.10.

a. increasing leaving group ability: Cl⁻ < Br⁻ < I⁻

most basic   least basic
worst leaving   **best leaving**
group   **group**

b. increasing leaving group ability: NH_3 < H_2O < H_2S

most basic   least basic
worst leaving   **best leaving**
group   **group**

7.49 Compare the nucleophile and the leaving group in each reaction. The reaction will occur if it proceeds towards the weaker base. Remember that the stronger the acid (lower pK_a), the weaker the conjugate base.

a.

NH_2

+ I⁻ ✗ →

I

+ ⁻NH_2

| Reaction will not occur. |

weaker base
pK_a (HI) = –10

stronger base
pK_a (NH_3) = 38

b.

I + ⁻O–CH_3 →

O + I⁻

| Reaction will occur. |

stronger base
pK_a (CH_3OH) = 15.5

weaker base
pK_a (HI) = –10

7.50 Use the directions in Answer 7.13.

a. • In a **polar protic solvent** (CH_3OH), nucleophilicity *increases down a column* of the periodic table, so: $CH_3CH_2S^-$ is more nucleophilic than $CH_3CH_2O^-$.
• For two species with the same attacking atom, the more basic is the more nucleophilic, so $CH_3CH_2O^-$ is more nucleophilic than CH_3COO^-.

CH_3COO^- < $CH_3CH_2O^-$ < $CH_3CH_2S^-$

b. Compare the nucleophilicity of N, S, and O. In a polar aprotic solvent (acetone), nucleophilicity parallels basicity.

CH_3SH < CH_3OH < CH_3NH_2

c. In a **polar aprotic solvent** (acetone), nucleophilicity parallels basicity. Across a row and down a column of the periodic table nucleophilicity decreases.

Cl⁻ < F⁻ < ⁻OH

d. Nucleophilicity decreases across a row so ⁻SH is more nucleophilic than Cl⁻.
In a **polar protic solvent** (CH_3OH), nucleophilicity increases down a column, so Cl⁻ is more nucleophilic than F⁻.

F⁻ < Cl⁻ < ⁻SH

7.51 *Polar protic solvents* **are capable of hydrogen bonding,** so they must contain a H bonded to an electronegative O or N. *Polar aprotic solvents* **are incapable of hydrogen bonding,** so they do not contain any O–H or N–H bonds.

a. $(CH_3)_2CHOH$

contains O–H bond
protic

c. CH_2Cl_2

no O–H or N–H bond
aprotic

e. $N(CH_3)_3$

no O–H or N–H bond
aprotic

b. CH_3NO_2

no O–H or N–H bond
aprotic

d. NH_3

contains N–H bond
protic

f. $HCONH_2$

contains an N–H bond
protic

7.52

The amine N is more nucleophilic
because the electron pair is
localized on the N.

The amide N is less nucleophilic
because the electron pair is
delocalized by resonance.

7.53

1° alkyl halide
S$_N$2 reaction

a. Mechanism:

b. Energy diagram:

c. Transition state:

d. Rate equation: one-step reaction with both nucleophile and alkyl halide in the only step:
rate = k[R–Br][$^-$CN]

e. [1] The leaving group is changed from Br$^-$ to I$^-$:

Leaving group becomes less basic → a better leaving group → faster reaction.

[2] The solvent is changed from acetone to CH$_3$CH$_2$OH:

Solvent changed to polar protic → decreases reaction rate.

[3] The alkyl halide is changed from CH$_3$(CH$_2$)$_4$Br to CH$_3$CH$_2$CH$_2$CH(Br)CH$_3$:

Changed from 1° to 2° alkyl halide → the alkyl halide gets more crowded and the reaction rate decreases.

[4] The concentration of $^-$CN is increased by a factor of 5.

Reaction rate will increase by a factor of 5.

[5] The concentration of both the alkyl halide and $^-$CN are increased by a factor of 5:

Reaction rate will increase by a factor of 25 (5 × 5 = 25).

7.54 a. CH₃CH₂Br reacts faster than CH₃CH₂Cl because Br⁻ is a better leaving group than Cl⁻.

b. The S_N2 reaction with NaOH is faster than the S_N2 reaction with NaOCOCH₃ because ⁻OH is a stronger nucleophile than ⁻OCOCH₃.

c. The S_N2 reaction is faster in the polar aprotic solvent DMSO because the nucleophile ⁻OCH₃ is stronger.

7.55 All S_N2 reactions proceed with backside attack of the nucleophile. When nucleophilic attack occurs at a stereogenic center, inversion of configuration occurs.

7.56 For carbocations: **Increasing number of R groups = Increasing stability.**

a.

1° carbocation
least stable   2° carbocation
**intermediate
stablity**   3° carbocation
most stable

b.

1° carbocation
least stable   2° carbocation
**intermediate
stablity**   3° carbocation
most stable

7.57 Both **A** and **B** are resonance stabilized, but the N atom in **B** is more basic and therefore more willing to donate its electron pair.

A

B

more basic
N atom

This resonance form stabilizes the carbocation more than the equivalent resonance structure for **A**. Thus, **B** is more stable than **A**.

7.58

a. Mechanism:
S_N1 only

b. Energy diagram:

c. Transition states:

d. Rate equation: **rate = k[(CH₃)₂CICH₂CH₃]**

e. [1] Leaving group changed from I⁻ to Cl⁻: **rate decreases** because I⁻ is a better leaving group.
   [2] Solvent changed from H₂O (polar protic) to DMF (polar aprotic):
       rate decreases because polar protic solvent favors S_N1.
   [3] Alkyl halide changed from 3° to 2°: **rate decreases** because 2° carbocations are less stable.
   [4] [H₂O] increased by factor of five: **no change in rate** because H₂O is not in rate equation.
   [5] [R–X] and [H₂O] increased by factor of five: **rate increases** by a factor of five. (Only the concentration of R–X affects the rate.)

7.59 a. S_N1 reaction of (CH₃)₃CI is faster than S_N1 reaction with (CH₃)₃CCl because I⁻ is a better leaving group than Cl⁻.
   b. S_N1 reaction is faster with the 3° alkyl halide (CH₃)₃CBr than with the 1° alkyl halide (CH₃)₂CHCH₂Br.
   c. S_N1 reaction is faster with the polar protic solvent H₂O rather than the aprotic solvent DMSO.

7.60

a.

b.

7.61 The 1° alkyl halide is also allylic, so it forms a resonance-stabilized carbocation. Increasing the stability of the carbocation by resonance increases the rate of the S_N1 reaction.

resonance-stabilized carbocation

Use each resonance structure individually to continue the mechanism:

7.62

A   B   C

Vinyl halides like **A** do not react by either an S$_N$1 or S$_N$2 mechanism. With S$_N$2, the rate is 3° < 2° < 1°, and with S$_N$1, the rate is 1° < 2° < 3°.
a.   S$_N$2 reactivity: **A < C < B**
b.   S$_N$1 reactivity: **A < B < C**

7.63

a.

1° alkyl halide
S$_N$2 only

b.

2° alkyl halide   strong nucleophile
S$_N$1 and S$_N$2   polar aprotic solvent
Both favor S$_N$2.

reaction at a stereogenic center
inversion of configuration

c.

3° alkyl halide
S$_N$1 only

d.

2° alkyl halide   Weak nucleophile
S$_N$1 and S$_N$2   **favors S$_N$1.**

reaction at a stereogenic center
racemization of product

e.

2° alkyl halide
S_N1 and S_N2

strong nucleophile
polar aprotic solvent
Both favor S_N2.

reaction at a stereogenic center
inversion of configuration

f.

2° alkyl halide
S_N1 and S_N2

Weak nucleophile
favors S_N1.

two products – **diastereomers**
Nucleophile attacks
from above and below.

7.64 The reaction follows an S_N2 mechanism.

7.65 An S_N1 mechanism means the reaction occurs in a stepwise fashion by way of a carbocation.

7.66

diphenhydramine

7.67 First decide whether the reaction will proceed via an S$_N$1 or S$_N$2 mechanism (Answer 7.38), and then draw the mechanism.

3° alkyl halide
S$_N$1 only

can attack from
above or below

+ HBr

7.68

nucleophile

+ H$_2$O

+ Br$^-$

C$_7$H$_{10}$O$_2$

leaving group

7.69

+ HCO$_3^-$

NaHCO$_3$ + NaBr +

nicotine

7.70

a. Two diastereomers (**C** and **D**) are formed as products from the two enantiomers of **A**.

b.

Since both stereogenic centers in **D** have the *S* configuration and the corresponding stereogenic centers in quinapril are also *S*, **D** is needed to synthesize the drug.

Both have the *S* configuration

quinapril

7.71

7.72 In the first reaction, substitution occurs at the stereogenic center. Because an achiral, planar carbocation is formed, the nucleophile can attack from either side, thus generating a racemic mixture.

3° alkyl halide

(*R*)-6-bromo-2,6-dimethylnonane

$CH_3\overset{..}{O}H$

S_N1

achiral, planar carbocation

two steps

racemic mixture
optically inactive

In the second reaction, the starting material contains a stereogenic center, but the nucleophile does not attack at that carbon. Because a bond to the stereogenic center is not broken, the configuration is retained and a chiral product is formed.

(R)-2-bromo-2,5-dimethylnonane

Reaction does not occur at the stereogenic center.

optically active

7.73

a. The nucleophile has replaced the leaving group. Missing reagent:

b. The nucleophile has replaced the leaving group. Missing reagent: $^-C\equiv CH$

c. The nucleophile has replaced the halide. Starting material:

d. The nucleophile has replaced the halide. Starting material:

The leaving group must have the opposite orientation to the position of the nucleophile in the product.

7.74 To devise a synthesis, look for the carbon framework and the functional group in the product. **The carbon framework is from the alkyl halide and the functional group is from the nucleophile.**

a. carbon framework, functional group

b. carbon framework, functional group

c. carbon framework, functional group

d.

carbon framework | functional group

2° halide

Na⁺ ⁻O~~~

This path is preferred.
The strong nucleophile favors an S$_N$2 reaction so an unhindered 1° alkyl halide reacts faster.

or

functional group | carbon framework

1° halide

e.

carbon framework | functional group

~~Cl

Na⁺ ⁻O

7.75

B
very crowded 3° halide

Na⁺ :ÖCH₃
C

E

preferred method
The strong nucleophile favors S$_N$2 reaction, so the alkyl halide should be unhindered for a faster reaction.

:Ö:⁻ Na⁺
D

CH₃I
A

ÖCH₃
E

unhindered methyl halide

7.76

H–C≡C–H   →(NaH)   H–C≡C⁻
                      A
                      + H₂

B

↓ NaH

+ H₂

C

addition of H₂
(1 equiv)

muscalure

D

7.77

a.

:Ö:–H   [1] Na⁺H:⁻   :Ö:⁻   [2] CH₃–Br: :Ö:–CH₃   + Na⁺ Br⁻

(Chapter 9)

+ H₂

b.

(Chapter 11)

c.

(Chapter 23)

7.78

quinuclidine   triethylamine

| This electron pair is more hindered by the three CH₂CH₃ groups. |

The three alkyl groups are "tied back" in a ring, making the electron pair more available.

These bulky groups around the N cause steric hindrance and this decreases nucleophilicity.

This electron pair on quinuclidine is much more available than the one on triethylamine.

less steric hindrance
more nucleophilic

7.79

[1]   + H₂

CH₃—Br

[2] →   minor product

+   NaBr

CH₃—Br

[2] →   major product

7.80

a.   base →   intramolecular S_N2 →

b.   base →   intramolecular S_N2 →

c.

base → 3° alkyl halide harder reaction → intramolecular S_N2

d.

base → 3° alkyl halide harder reaction → intramolecular S_N2

7.81

Cl bonded to sp^2 C cannot undergo S_N1.

Cl bonded to sp^3 C
no resonance stabilization possible for the carbocation formed here

Cl bonded to sp^3 C
Resonance-stabilized carbocation forms.
best for S_N1

$CH_3\ddot{O}H$ (1 equiv) →

re-draw

$CH_3\ddot{O}H$ →

+ $:\ddot{C}l:^-$

↓

+ HCl

K

7.82

a.

(S)-1-phenylpropan-1-ol
$[\alpha] = -48$

(R)-1-phenylpropan-1-ol
$[\alpha] = +48$

$$ee = \frac{[\alpha] \text{ mixture}}{[\alpha] \text{ pure enantiomer}} \times 100\%$$

$$= \frac{+5.0}{+48} \times 100\% = 10.\% \text{ excess of } R \text{ isomer}$$

90% racemic mixture = 45% R and 45% S
Total R isomer = 45 + 10 = 55% R isomer

b. The *R* product is the product of inversion and it predominates.

c. The weak nucleophile favors an S$_N$1 reaction, which occurs by way of an intermediate carbocation. Perhaps there is more inversion than retention because H$_2$O attacks the intermediate carbocation while the Br$^-$ leaving group is still in the vicinity of the carbocation. The Br$^-$ would then shield one side of the carbocation and backside attack would be slightly favored.

Chapter 8 Alkyl Halides and Elimination Reactions

Chapter Review

A comparison between nucleophilic substitution and β-elimination

Nucleophilic substitution—A nucleophile attacks a carbon atom (7.6).

β-Elimination—A base attacks a proton (8.1).

Similarities	Differences
• In both reactions RX acts as an electrophile, reacting with an electron-rich reagent. • Both reactions require a **good leaving group X:⁻** willing to accept the electron density in the C–X bond.	• In substitution, a nucleophile attacks a single carbon atom. • In elimination, a Brønsted–Lowry base removes a proton to form a π bond, and two carbons are involved in the reaction.

The importance of the base in E2 and E1 reactions (8.9)

The strength of the base determines the mechanism of elimination.
- Strong bases favor E2 reactions.
- Weak bases favor E1 reactions.

E1 and E2 mechanisms compared

	E2 mechanism	E1 mechanism
[1] Mechanism	• one step (8.4B)	• two steps (8.6B)
[2] Alkyl halide	• rate: $R_3CX > R_2CHX >$ RCH_2X (8.4C)	• rate: $R_3CX > R_2CHX >$ RCH_2X (8.6C)
[3] Rate equation	• rate = $k[RX][B:]$ • second-order kinetics (8.4A)	• rate = $k[RX]$ • first-order kinetics (8.6A)
[4] Stereochemistry	• anti periplanar arrangement of H and X (8.8)	• trigonal planar carbocation intermediate (8.6B)
[5] Base	• favored by strong bases (8.4B)	• favored by weak bases (8.6C)
[6] Leaving group	• better leaving group → faster reaction (8.4B)	• better leaving group → faster reaction (Table 8.4)
[7] Solvent	• favored by polar aprotic solvents (8.4B)	• favored by polar protic solvents (Table 8.4)
[8] Product	• more substituted alkene favored (Zaitsev rule, 8.5)	• more substituted alkene favored (Zaitsev rule, 8.6C)

Summary chart on the four mechanisms: S_N1, S_N2, E1, and E2 (8.11)

Alkyl halide type	Conditions	Mechanism
1° RCH_2X	strong nucleophile	S_N2
	strong bulky base	E2
2° R_2CHX	strong base and nucleophile	S_N2 + E2
	strong bulky base	E2
	weak base and nucleophile	S_N1 + E1
3° R_3CX	weak base and nucleophile	S_N1 + E1
	strong base	E2

Zaitsev rule

- β-Elimination affords the more stable product having the more substituted double bond.
- Zaitsev products predominate in E2 reactions except when a cyclohexane ring prevents trans diaxial arrangement.

Practice Test on Chapter Review

1. Which of the following is true about an E1 reaction?
   1. The reaction is faster with better leaving groups.
   2. The reaction is fastest with 3° alkyl halides.
   3. The reaction is faster with stronger bases.
   4. Statements (1) and (2) are true.
   5. Statements (1), (2), and (3) are all true.

2. Consider the S_N2 and E1 reaction mechanisms. What effect on the rate of the reaction is observed when each of the following changes is made? Fill in each box of the table with one of the following phrases: **increases, decreases,** or **remains the same.**

Change	S_N2 mechanism	E1 mechanism
a. The alkyl halide is changed from $(CH_3)_3CBr$ to $CH_3CH_2CH_2CH_2Br$.		
b. The solvent is changed from $(CH_3)_2CO$ to CH_3CH_2OH.		
c. The nucleophile/base is changed from ^-OH to H_2O.		
d. The alkyl halide is changed from CH_3CH_2Cl to CH_3CH_2I.		
e. The concentration of the base/nucleophile is increased by a factor of five.		

3. Rank the following compounds in order of *increasing* reactivity in an **E2 elimination** reaction. Rank the *most reactive* compound as **3,** the *least reactive* compound as **1,** and the compound of intermediate reactivity as **2.**

4. Draw the organic products formed in the following reactions.

5.a. Fill in the appropriate alkyl halide needed to synthesize the following compound as a single product using the given reagents.

C $\xrightarrow{\text{K}^+ \ ^-\text{OC(CH}_3)_3}$

b. What starting material is needed for the following reaction? The starting material must yield product cleanly, in one step without any other organic side products.

D $\xrightarrow{\text{K}^+ \ ^-\text{OC(CH}_3)_3}$ $CH_3CH_2CH=CHCH_3$
(cis and trans mixture)

6. Draw all products formed in the following reaction.

$\xrightarrow{\text{CH}_3\text{OH}}$

Answers to Practice Test

1. 4

2. S_N2 E1
   a. increases   decreases
   b. decreases   increases
   c. decreases   same
   d. increases   increases
   e. increases   same

3. A–2
   B–1
   C–3

4.

a.

b.

c.

d.

5.

a.

C

b.
   Br
   (or Cl or I)
   D

6.

Answers to Problems

8.1 • The carbon bonded to the leaving group is the **α carbon.** Any carbon bonded to it is a **β carbon.**
  • **To draw the products of an elimination reaction:** Remove the leaving group from the α carbon and a H from the β carbon and form a π bond.

a.

b.

$+ \quad CH_3CH=C(CH_3)CH_2CH_3$

c.

8.2 **Alkenes are classified by the number of carbon atoms bonded to the double bond.** A monosubstituted alkene has one carbon atom bonded to the double bond, a disubstituted alkene has two carbon atoms bonded to the double bond, etc.

a.

4 C's bonded to C=C
tetrasubstituted

2 C's bonded to each C=C
disubstituted

3 C's bonded to each C=C
trisubstituted

vitamin A

b.

vitamin D$_3$

3 C's bonded to each C=C
trisubstituted

2 C's bonded to the C=C
disubstituted

8.3 To have stereoisomers at a C=C, the two groups on each end of the double bond must be different from each other.

a.

two CH$_3$ groups
no stereoisomers
possible

b.

two CH$_3$ groups
no stereoisomers
possible

two different groups on both ends
stereoisomers possible

c.

two different groups
on both ends
stereoisomers
possible

two identical groups on one end
no stereoisomers
possible

8.4

a.
2 CH₃'s on → one end

Only this C=C exhibits stereoisomerism.

2 H's on one end ←

b. A diastereomer has a different 3-D arrangement of groups but the carbon skeleton and the double bonds must stay in the original positions.

diastereomer   different arrangement of groups around this double bond

8.5 Two definitions:
   • **Constitutional isomers** differ in the connectivity of the atoms.
   • **Stereoisomers** differ only in the 3-D arrangement of the atoms in space.

a.

and

different connectivity of atoms
constitutional isomers

c.

cis   and   trans

different arrangement of atoms in space
stereoisomers

b.

trans   and   trans

identical

d.

and

different connectivity of atoms
constitutional isomers

8.6 Two rules to predict the relative stability of alkenes:
   [1] Trans alkenes are generally more stable than cis alkenes.
   [2] The stability of an alkene increases as the number of R groups on the C=C increases.

a.

monosubstituted   or   disubstituted
more stable

c.

or

trisubstituted
more stable   disubstituted

b.

cis   or   trans
more stable

8.7

Alkene **A** is more stable than alkene **B** because the double bond in **A** is in a six-membered ring. The double bond in **B** is in a four-membered ring, which has considerable angle strain due to the small ring size.

8.8 In an E2 mechanism, four bonds are involved in the single step. Use curved arrows to show these simultaneous actions:
   [1] The base attacks a hydrogen on a β carbon.
   [2] A π bond forms.
   [3] The leaving group comes off.

8.9 In both cases, the rate of elimination decreases.

8.10 As the number of R groups on the carbon with the leaving group increases, the rate of an E2 reaction increases.

8.11 Use the following characteristics of an E2 reaction to answer the questions:
   [1] E2 reactions are second order and one step.
   [2] More substituted halides react faster.
   [3] Reactions with strong bases or better leaving groups are faster.
   [4] Reactions with polar aprotic solvents are faster.

Rate equation: rate = k[RX][Base]
   a. tripling the concentration of the alkyl halide = **rate triples**
   b. halving the concentration of the base = **rate is halved**
   c. changing the solvent from CH_3OH to DMSO = **rate increases** (Polar aprotic solvent is better for E2.)
   d. changing the leaving group from I^- to Br^- = **rate decreases** (I^- is a better leaving group.)
   e. changing the base from ^-OH to H_2O = **rate decreases** (weaker base)
   f. changing the alkyl halide from CH_3CH_2Br to $(CH_3)_2CHBr$ = **rate increases** (More substituted halide reacts faster.)

8.12 According to the Zaitsev rule, the major product in a β-elimination reaction has the *more* substituted double bond.

a.

loss of H and Br

trisubstituted
major product

+

(+ stereoisomer)
disubstituted
minor product

b.

loss of H and Br

trisubstituted
minor product

+

tetrasubstituted
major product

+

disubstituted
minor product

c.

loss of H and Cl

monosubstituted
minor product

+

(+ stereoisomer)
disubstituted
major product

d.

loss of H and Cl

trisubstituted
ONLY product

8.13 An E1 mechanism has two steps:
   [1] The leaving group comes off, creating a carbocation.
   [2] A base pulls off a proton from a β carbon, and a π bond forms.

$+ CH_3\ddot{O}H$ $\xrightarrow{[1]}$ $+ CH_3\ddot{O}H + Cl^-$ $\xrightarrow{[2]}$ $+ CH_3\overset{+}{O}H_2 + Cl^-$

transition state [1]:

$\left[\begin{array}{c} \delta+ \\ :\ddot{Cl}: \ \delta- \end{array} \right]^{\ddagger}$

transition state [2]:

$\left[\begin{array}{c} \delta+ \quad H \\ O \\ H \quad \underset{\delta+}{} \ CH_3 \end{array} \right]^{\ddagger}$

8.14 According to the Zaitsev rule, the major product in a β-elimination reaction has the *more* substituted double bond.

a.

(+ stereoisomer)
trisubstituted
major product

disubstituted

b.

tetrasubstituted
major product

disubstituted

trisubstituted

8.15 Use the following characteristics of an **E1 reaction** to answer the questions:
  [1] E1 reactions are first order and two steps.
  [2] More substituted halides react faster.
  [3] Weaker bases are preferred.
  [4] Reactions with better leaving groups are faster.
  [5] Reactions in polar protic solvents are faster.

Rate equation: rate = k[RX]. The base doesn't affect rate.
  a. doubling the concentration of the alkyl halide = **rate doubles**
  b. doubling the concentration of the base = **no change** (The base is not in the rate equation.)
  c. changing the alkyl halide from $(CH_3)_3CBr$ to $CH_3CH_2CH_2Br$ = **rate decreases** (More substituted halides react faster.)
  d. changing the leaving group from Cl^- to Br^- = **rate increases** (better leaving group)
  e. changing the solvent from DMSO to CH_3OH = **rate increases** (Polar protic solvent favors E1.)

8.16 Both S_N1 and E1 reactions occur by forming a carbocation. To draw the products:
  [1] **For the S_N1 reaction,** substitute the nucleophile for the leaving group.
  [2] **For the E1 reaction,** remove a proton from a β carbon and create a new π bond.

a.

leaving group

nucleophile and base

S_N1 product

E1 products

b.

leaving group

nucleophile and base

S_N1 product

E1 products

8.17 The E2 elimination reactions will occur in the anti periplanar orientation as drawn. To draw the product of elimination, maintain the orientation of the remaining groups around the C=C.

a.

The two benzene rings are anti in this conformation (one wedge, one dashed wedge).

The two benzene rings remain on opposite sides of the newly formed C=C. This makes them **trans.**

diastereomers

b.

The two benzene rings are gauche in this conformation (both drawn on dashed wedges, behind the plane).

The two benzene rings remain on the same side of the newly formed C=C. This makes them **cis.**

8.18 Note: The Zaitsev products predominate in E2 elimination *except* when substituents on a cyclohexane ring prevent a **trans diaxial** arrangement of H and X.

a.

two conformations

axial H's

A

Use this conformation. It has Cl axial and two axial H's.

B

β₁

β₂   A

two different axial H's

–OH

[loss of H(β₂) + Cl]

+

[loss of H(β₁) + Cl]

re-draw

re-draw

disubstituted

trisubstituted
major product

b.

two
conformations

A

CH₃
Cl
H
H
B CH(CH₃)₂

Use this conformation.
It has Cl axial and
one axial H.

CH₃
β₂
Cl
H
H
H
β₁
H
B
CH(CH₃)₂

only one axial H
on a β carbon

⁻OH

CH₃
H
H
H
CH(CH₃)₂

[loss of H(β₁) + Cl]

=

disubstituted
only product

8.19 Draw the chair conformations of *cis*-1-chloro-2-methylcyclohexane and its trans isomer. For E2 elimination reactions to occur, **there must be a H and X trans diaxial to each other.**

Two conformations of the cis isomer:

CH₃   Cl
H
H
H

CH₃
H
H
Cl
H
H

A
reacting conformation (axial Cl)

This reacting conformation has only one group axial,
making it more stable and present in a higher
concentration than **B**. This makes a **faster
elimination reaction** with the cis isomer.

Two conformations of the trans isomer:

H
H
H
Cl
CH₃

Cl
H
H
H
CH₃

B
reacting conformation (axial Cl)

This conformation is less stable than **A**,
because both CH₃ and Cl are axial.
**This slows the rate of elimination from
the trans isomer.**

8.20 **E2 reactions are favored by strong negatively charged bases** and occur with 1°, 2°, and 3° halides, with 3° being the most reactive.
E1 reactions are favored by weaker neutral bases and do not occur with 1° halides because they would have to form highly unstable carbocations.

a.

Cl   +   ⁻OCH₃ ⟶

strong negatively
charged base
E2

c.

Cl   +   CH₃OH ⟶

weak neutral
base
E1

b.

I   +   H₂O ⟶

weak neutral
base
E1

d.

Br   +   ⁻OC(CH₃)₃ ⟶

strong negatively
charged base
E2

8.21 Draw the alkynes that result from removal of two equivalents of HX.

a. (cyclohexyl-CHCl-CHCl-CH₂CH₃) $\xrightarrow{\ ^-NH_2\ }$ (cyclohexyl-C≡C-CH₂CH₃)

c. (CH₃)₂C(Br)(Br)CH₂CH₃ $\xrightarrow{\ ^-NH_2\ }$ (CH≡C-CH₂CH₃) + (CH₃-C≡C-CH₃)

b. (CH₃CH₂CH₂-CHCl₂) $\xrightarrow[\text{DMSO}]{\text{KOC(CH}_3)_3}$ (CH₃CH₂-C≡CH)

d. (Ph-CHBr-CHBr-Ph) $\xrightarrow{\ ^-NH_2\ }$ (Ph-C≡C-Ph)

8.22

a. (CH₃CH₂CH₂CH₂CH₂CH₂CH₂Cl) $\xrightarrow[\substack{\text{strong bulky base} \\ \textbf{E2}}]{\text{K}^+\ {}^-\text{OC(CH}_3)_3}$ (CH₃CH₂CH₂CH₂CH=CH₂)

1° halide
S_N2 or E2

b. (CH₃CH₂CHCl-CH₃) $\xrightarrow[\substack{\text{strong base} \\ \textbf{S}_N\textbf{2 and E2}}]{\ ^-\text{OH}\ }$ (CH₃CH₂CH(OH)CH₃) + (CH₃CH=CHCH₃) + (CH₂=CHCH₂CH₃)

2° halide
any mechanism

S_N2 product   (+ stereoisomer)   monosubstituted
disubstituted   minor E2 product
major E2 product

c. (cyclohexane with I and ethyl) $\xrightarrow[\substack{\text{weak base} \\ \textbf{S}_N\textbf{1 and E1}}]{\text{CH}_3\text{CH}_2\text{OH}}$ (cyclohexane-OCH₂CH₃ with ethyl) + (cyclohexylidene-ethyl) + (cyclohexene-ethyl)

3° halide
no S_N2

S_N1 product   E1 product   E1 product

d. (CH₃CH₂CH₂CH₂C(CH₃)₂Cl) $\xrightarrow[\substack{\text{strong base} \\ \textbf{E2}}]{\text{CH}_3\text{CH}_2\text{O}^-}$ (CH₃CH₂CH=C(CH₃)CH₃) + (CH₃CH₂CH₂C(=CH₂)CH₃)

3° halide
no S_N2

major E2 product   **minor E2 product**

8.23

3° halide
no S_N2   weak base
S_N1 and E1

(cyclohexane with Br and CH₃) $\xrightarrow[\substack{\textbf{overall} \\ \textbf{reaction}}]{\text{CH}_3\text{OH}}$ (cyclohexane-ÖCH₃ with CH₃) + (dimethylcyclohexene) + HBr

The steps:

(cyclohexyl cation with CH₃) $\xrightarrow[\textbf{S}_N\textbf{1}]{\text{CH}_3\ddot{\text{O}}\text{H}}$ (cyclohexane-O⁺(CH₃)(H) with :Br:⁻)

+ Br⁻

or

(cyclohexyl cation with H and CH₃)   :B̈r:⁻ $\xrightarrow{\textbf{E1}}$

8.24 More substituted alkenes are more stable. Trans alkenes are generally more stable than cis alkenes. Order of stability:

B less stable   <   C   <   A most stable

8.25 The trans isomer (**E**) reacts faster. During elimination, Br must be axial to give trans diaxial elimination. In the trans isomer, the more stable conformation has the bulky *tert*-butyl group in the more roomy equatorial position and Br in the axial position. In the cis isomer (**D**), elimination can occur only when both the *tert*-butyl and Br groups are axial, a conformation that is not energetically favorable.

cis
D
cis-1-bromo-3-*tert*-butylcyclohexane

This conformation must react, but it contains two axial groups.

trans
E
trans-1-bromo-3-*tert*-butylcyclohexane

preferred conformation

8.26 Translate each model to a structure and arrange H and Br to be anti periplanar.

H and Br 180° away from each other

H and Br 180° away from each other

8.27

a.

b.

(+ stereoisomer)   +   (+ stereoisomer)

c.

(+ stereoisomer)   +

d.

8.28 To give only one product in an elimination reaction, **the starting alkyl halide must have only one type of β carbon with H's.**

a.

c.   Two β carbons are identical.

b.

d.   Two β carbons are identical.

8.29 To have stereoisomers, the two groups on each end of the double bond must be different from each other.

a.

2 H's on one end

2 CH₃'s on one end

two different groups at each end **can have stereoisomers**

b.

two different groups at each end **can have stereoisomers**

2 H's on one end

c.

two different groups on each end of each double bond **All C=C's can have stereoisomerism.**

8.30 Use the definitions in Answer 8.5.

a. and

different connectivity
constitutional isomers

c. and

trans   trans   trans   trans

identical

b. and

stereoisomers

d. and

stereoisomers

8.31 Use the rules from Answer 8.6 to rank the alkenes.

B

monosubstituted

least stable

A

disubstituted

**intermediate
stability**

C

trisubstituted

most stable

8.32 **A larger negative value for $\Delta H°$ means the reaction is more exothermic.**   Because both but-1-ene and *cis*-but-2-ene form the same product (butane), these data show that but-1-ene was higher in energy to begin with, **because more energy is released in the hydrogenation reaction.**

but-1-ene   +   H$_2$ $\longrightarrow$

$\Delta H° = -127$ kJ/mol

cis-but-2-ene   +   H$_2$ $\longrightarrow$

$\Delta H° = -120$ kJ/mol

but-1-ene

cis-but-2-ene

Energy

larger $\Delta H°$ for
but-1-ene
higher in energy

butane

smaller $\Delta H°$ for *cis*-but-2-ene
lower in energy, more stable

8.33

a.

$(CH_3)_3CO^-$

(+ stereoisomer)
(loss of β$_2$ H)
major product
disubstituted

+

(loss of β$_1$ H)
monosubstituted

b.

DBU

only product

c.

(+ stereoisomer)
(loss of β_1 H)
major product
tetrasubstituted

(+ stereoisomer)
(loss of β_2 H)
trisubstituted

(loss of β_3 H)
disubstituted

d.

(loss of β_2 H)
major product
trisubstituted

(loss of β_1 H)
disubstituted

8.34 To give only one alkene as the product of elimination, the alkyl halide must have either:
• only one β carbon with a hydrogen atom, or
• all identical β carbons, so the resulting elimination products are identical

a.

b.

c.

8.35

a. Mechanism:

by-products

b. Rate = k[R–Br][$^-$OC(CH$_3$)$_3$]
   [1] Solvent changed to DMF (polar aprotic) = **rate increases**
   [2] [$^-$OC(CH$_3$)$_3$] decreased = **rate decreases**
   [3] Base changed to $^-$OH = **rate decreases** (weaker base)
   [4] Halide changed to 2° = **rate increases** (More substituted RX reacts faster.)
   [5] Leaving group changed to I$^-$ = **rate increases** (better leaving group)

8.36

1-chloro-1-methyl-
cyclopropane

The dehydrohalogenation of an alkyl halide usually forms the more stable alkene. In this case, **A** is more stable than **B** even though **A** contains a disubstituted C=C whereas **B** contains a trisubstituted C=C. The double bond in **B** is part of a three-membered ring, and is less stable than **A** because of severe angle strain around both C's of the double bond.

8.37

a.

trans isomer more stable
major product

b.

trans isomer more stable
major product

8.38

a.

tetrasubstituted   trisubstituted   disubstituted
major product

b.

trisubstituted
This isomer is more stable—
large groups farther away.
major product

trisubstituted

disubstituted

c.

disubstituted   trisubstituted
major product

8.39 Use the rules from Answer 8.20.

a.

Br

2° halide

$^-$OCH$_3$

strong base
E2

(+ cis isomer)

b.

Br

2° halide

CH$_3$OH

weak base
E1

(+ cis isomer)

c.

1° halide

I

$^-$OC(CH$_3$)$_3$

strong base
E2

d.

3° halide

H_2O
weak base
E1

e.

2° halide

¯OH
strong base
E2

f.

2° halide

¯OH
strong base
E2

8.40 The order of reactivity is the same for both E2 and E1: 1° < 2° < 3°.

2° halide

3° halide

3° halide +
better leaving group

⟶

Increasing reactivity in E1 and E2

8.41

a.

1-chloro-1-methylcyclohexane
3° RX reacts facter in E2
(strong base)

1-chloro-3-methylcyclohexane

b.

2° RX

3° RX reacts faster in E1
(weak base)

c. The mechanism is E2 because ¯OH is a strong base, so the reaction is faster in a polar aprotic
solvent (DMSO).

8.42

bromocyclodecane   *cis*-cyclodecene

In a ten-membered ring, the cis isomer is more stable and,
therefore, the preferred elimination product. The trans
isomer is less stable because strain is introduced when two
ends of the double bond are connected in a trans
arrangement in this medium-sized ring.

8.43

	but-1-ene	but-2-ene
	from loss of β_1 H	from loss of β_2 H
Na$^+$ $^-$OCH$_2$CH$_3$	19%	81%
K$^+$ $^-$OC(CH$_3$)$_3$	33%	67%

The H's on the CH$_2$ group of the β_2 carbon are more sterically hindered than the H's on the CH$_3$ group of the β_1 carbon. Because K$^+$$^-$OC(CH$_3$)$_3$ is a much bulkier base than Na$^+$$^-OCH_2CH_3$, it is easier to remove the more accessible H on β_1, giving K$^+$$^-$OC(CH$_3$)$_3$ a higher percentage of but-1-ene than Na$^+$$^-OCH_2CH_3$.

8.44 H and Br must be anti during the E2 elimination. Rotate if necessary to make them anti; then eliminate.

8.45

b.

8.46 To react by E2, the Br must be axial and this can only happen in the trans isomer when the large *tert*-butyl group is also axial, an energetically unfavorable conformation.

trans isomer
two equatorial groups
more stable conformation

two axial groups
highly destabilized

The cis isomer has an axial Br in its more stable conformation that keeps the large *tert*-butyl group equatorial. As a result, the cis isomer reacts faster.

cis isomer
more stable conformation
axial Br

8.47

a.

C2   C3
2-chloro-3-methylpentane

H and Cl are arranged anti in each stereoisomer, for anti periplanar elimination.

A   enantiomers   B   C   enantiomers   D

–HCl   –HCl   –HCl   –HCl

identical   identical

b. Two different alkenes are formed as products.

c. The products are diastereomers: Two enantiomers (**A** and **B**) give identical products. **A** and **B** are diastereomers of **C** and **D**. Each pair of enantiomers gives a single alkene. Thus, diastereomers give diastereomeric products.

8.48 To undergo an E2 reaction the Cl must be axial and the conformation with an axial Cl is highly unstable. Thus, no E2 reaction can occur because the needed conformation is too destabilized.

A

more stable conformation
All substituents are equatorial.

highly destabilized
All substituents are axial.

8.49 The alkyl chloride must have a Cl and H anti periplanar with all the substituents having the same arrangement that they have in the alkene product.

or

8.50

a.

b.

c.

d.

8.51

a.

b.

c.

8.52

2,3-dibromobutane

sp sp

A

$CH_3CH=C=CH_2$

sp

B

C

8.53 Use the "Summary chart on the four mechanisms: S_N1, S_N2, E1, or E2" on p. 8–2 to answer the questions.

a. Both S_N1 and E1 involve carbocation intermediates.
b. Both S_N1 and E1 have two steps.
c. S_N1, S_N2, E1, and E2 all have increased reaction rates with better leaving groups.
d. Both S_N2 and E2 have increased rates when changing from CH_3OH (a protic solvent) to $(CH_3)_2SO$ (DMSO—an aprotic solvent).
e. In S_N1 and E1 reactions, the rate depends on only the alkyl halide concentration.
f. Both S_N2 and E2 are concerted reactions.
g. CH_3CH_2Br and NaOH react by an S_N2 mechanism.
h. Racemization occurs in S_N1 reactions.
i. In S_N1, E1, and E2 mechanisms, 3° alkyl halides react faster than 1° or 2° halides.
j. E2 and S_N2 reactions follow second-order rate equations.

8.54

a.

1° halide
S_N2 or E2

$^-OC(CH_3)_3$
sterically hindered base
E2

b.

1° halide
S_N2 or E2

$^-OCH_2CH_3$
strong nucleophile
S_N2

c.

dihalide

$^-NH_2$
(2 equiv)
strong base

d.

1° halide
S_N2 or E2

DBU
sterically
hindered
base
E2

e.

2° halide
S_N1, S_N2, E1, E2

$^-OC(CH_3)_3$
sterically
hindered
base
E2

major product

+

f.

3° halide
no S_N2

CH_3CH_2OH
weak base

OCH$_2$CH$_3$

S_N1 product

+

+

E1 products

g.

diahlide

2 NaNH$_2$

h.

3° halide
no S_N2

H$_2$O
weak base

OH

S_N1 product

+

(+ stereoisomer)
E1 product

+

E1 product

8.55 [1] NaOCOCH$_3$ is a good nucleophile and weak base, and substitution is favored. [3] KOC(CH$_3$)$_3$
is a strong, bulky base that reacts by E2 elimination when there is a β hydrogen in the alkyl halide.

a.

Cl [1] NaOCOCH$_3$

OCOCH$_3$

b.

Cl

[1] NaOCOCH$_3$

OCOCH$_3$

[2] NaOCH$_3$

OCH$_3$

[2] NaOCH$_3$

OCH$_3$ +

S_N2

E2

[3] KOC(CH$_3$)$_3$

[3] KOC(CH$_3$)$_3$

c.

[1] NaOCOCH₃

[2] NaOCH₃

[3] KOC(CH₃)₃

8.56

a.

⁻OH

strong base
S_N2 and E2

2° halide
S_N1, S_N2, E1, E2

S_N2 product
inversion at
stereogenic center

+ major E2 product + minor E2 product + minor E2 product

b.

H₂O

weak base
S_N1 and E1

2° halide
S_N1, S_N2, E1, E2

S_N1 products

major E1 product + minor E1 product + minor E1 product

c.

CH₃OH

weak base
S_N1 and E1

3° halide
no S_N2

S_N1 products

E1 product

d.

KOH

strong base
S_N2 and E2

2° halide
S_N1, S_N2, E1, E2

S_N2 product
inversion at
stereogenic center

+ E2 product

(trans diaxial elimination of D, Br)

8.57

a.

CH₃OH

weak base
S_N1 and E1

S_N1 + S_N1 + E1 + E1

+ E1

b.

KOH

strong base
E2

+ +

8.58

a.

3° halide   strong bulky base   **E2**

major product
more substituted alkene

No substitution occurs with a strong bulky base and a 3° RX. The C with the leaving group is too crowded for an S_N2 substitution to occur. Elimination occurs instead by an E2 mechanism.

b.

1° halide   strong nucleophile S_N2

All elimination reactions are slow with 1° halides.
The strong nucleophile reacts by an S_N2 mechanism instead.

c.

3° halide   strong base   **E2**   ← minor product only

More substituted
alkene is favored.

d.

2° halide   good nucleophile, weak base S_N2 **favored**   minor product only

major product

The 2° halide can react by an E2 or S_N2 reaction with a negatively charged nucleophile or base. Since I⁻ is a weak base, substitution by an S_N2 mechanism is favored.

8.59

3° halide, weak base:
S_N1 **and E1**

a.

overall
reaction

+ HCl

The steps:

S_N1 + HCl   Any base (such as CH_3CH_2OH or Cl⁻) can be used to remove a proton to form an alkene. If Cl⁻ is used, HCl is formed as a reaction by-product. If CH_3CH_2OH is used, $(CH_3CH_2OH_2)^+$ is formed instead.

or

E1   + HCl

or

E1   + HCl

b.

3° halide
strong base
E2

overall
reaction

Each product:

one step

or

one step

8.60 Draw the products of each reaction with the 1° alkyl halide.

a.

NaOCH₂CH₃

strong
nucleophile
S_N2

c.

DBU

sterically
hindered base
E2

b.

KCN

strong
nucleophile
S_N2

8.61

H₂O:

H₂O:
above

+ Br⁻

re-draw

H₂O:
below

+ HBr

+ HBr

+ HBr

8.62

good nucleophile

CH$_3$COO$^-$ is a good nucleophile and a weak base, so it favors substitution by S$_N$2.

strong base

(only)

20%   80%

The strong base gives both S$_N$2 and E2 products, but because the 2° RX is somewhat hindered to substitution, the E2 product is favored.

8.63

CH$_3$OH

+   +   +   +   HCl

3° halide
weak base
S$_N$1 and E1

8.64 E2 elimination needs a leaving group and a hydrogen in the **trans diaxial** position.

Two different conformations:

This conformation has Cl's axial, but no H's axial.

This conformation has no Cl's axial.

For elimination to occur, a cyclohexane must have a H and Cl in the trans diaxial arrangement. Neither conformation of this isomer has both atoms—H and Cl—axial; thus, this isomer only slowly loses HCl by elimination.

8.65

H and Br are *anti periplanar.* Elimination can occur.

CH₃O⁻ → -HBr → major product

H (in the ring) and Br are **NOT** *anti periplanar.* Elimination can**not** occur using this H. Instead elimination must occur with the H on the CH₃ group.

Elimination can occur here.

CH₃O⁻ → -HBr

Elimination cannot occur in the ring because the required anti periplanar geometry is not present.

8.66

leaving group

DBN
overall reaction

A sequence of two reactions forms the final product: E2 elimination opens the five-membered ring. Then the sulfur nucleophile displaces the Cl⁻ leaving group to form the six-membered ring.

E2

S_N2

8.67

a.

Δ →

SeOC₆H₅ and H are on the same side of the ring.
syn elimination

b.

rotate →

Zn →

Both Br atoms are on the opposite sides of the C–C bond.
anti elimination

8.68 One equivalent of NaNH₂ removes one mole of HBr in an anti periplanar fashion from each dibromide. Two modes of elimination are possible for each compound.

a.

H and Br are anti periplanar on C1 and C2 as drawn.

A

loss of H from C1
loss of Br from C2
C

loss of H from C2
loss of Br from C1
D

B

Rotate to make H and Br anti periplanar.

Two different rotations are needed.

Rotate to make H and Br anti periplanar.

loss of H from C1
loss of Br from C2
F

loss of H from C2
loss of Br from C1
E

b. **C** and **F** are diastereomers.
D and **E** are diastereomers.
C and **D** are constitutional isomers.
E and **F** are constitutional isomers.

Chapter 9 Alcohols, Ethers, and Related Compounds

Chapter Review

General facts about ROH, ROR, and epoxides

- All three compounds contain an O atom that is sp^3 hybridized and tetrahedral (9.2).

- All three compounds have polar C–O bonds, but only alcohols have an O–H bond for intermolecular hydrogen bonding (9.4).

- Alcohols and ethers do not contain a good leaving group. Nucleophilic substitution can occur only after the OH (or OR) group is converted to a better leaving group (9.7A).

- Epoxides have a leaving group located in a strained three-membered ring, making them reactive to strong nucleophiles and acids HZ that contain a nucleophilic atom Z (9.16).

A new reaction of carbocations (9.9)

- Less stable carbocations rearrange to more stable carbocations by shift of a hydrogen atom or an alkyl group. Besides rearrangement, carbocations also react with nucleophiles (7.12) and bases (8.6).

Preparation of alcohols, ethers, and epoxides (9.6)

[1] Preparation of alcohols

$$R-X \ + \ ^-OH \longrightarrow R-OH \ + \ X^-$$

- The mechanism is S_N2.
- The reaction works best for CH_3X and 1° RX.

[2] Preparation of alkoxides (a Brønsted–Lowry acid–base reaction)

$$R-O-H \ + \ Na^+H^- \longrightarrow R-O^- \ Na^+ \ + \ H_2$$

alkoxide

[3] Preparation of ethers (Williamson ether synthesis)

$$R-X \ + \ ^-OR' \longrightarrow R-OR' \ + \ X^-$$

- The mechanism is S_N2.
- The reaction works best for CH_3X and 1° RX.

[4] Preparation of epoxides (intramolecular S_N2 reaction)

- A two-step reaction sequence:
  [1] Removal of a proton with base forms an alkoxide.
  [2] Intramolecular **S_N2** reaction forms the epoxide.

Reactions of alcohols

[1] Dehydration to form alkenes

a. Using strong acid (9.8, 9.9)

- Order of reactivity: $R_3COH > R_2CHOH > RCH_2OH$.
- The mechanism for 2° and 3° ROH is E1; carbocations are intermediates and rearrangements occur.
- The mechanism for 1° ROH is E2.
- The Zaitsev rule is followed.

b. Using $POCl_3$ and pyridine (9.10)

- The mechanism is E2.
- No carbocation rearrangements occur.

[2] Reaction with HX to form RX (9.11)

R—OH + H—X $\longrightarrow$ R—X + H_2O

- Order of reactivity: $R_3COH > R_2CHOH > RCH_2OH$.
- The mechanism for 2° and 3° ROH is S_N1; carbocations are intermediates and rearrangements occur.
- The mechanism for CH_3OH and 1° ROH is S_N2.

[3] Reaction with other reagents to form RX (9.12)

R—OH + $SOCl_2$ $\xrightarrow{\text{pyridine}}$ R—Cl

R—OH + PBr_3 $\longrightarrow$ R—Br

- Reactions occur with CH_3OH and 1° and 2° ROH.
- The reactions follow an S_N2 mechanism.

[4] Reaction with tosyl chloride to form alkyl tosylates (9.13A)

R—OH + Cl—S(=O)(=O)—⟨C₆H₄⟩—CH₃ $\xrightarrow{\text{pyridine}}$ R—O—S(=O)(=O)—⟨C₆H₄⟩—CH₃

R—OTs

- The C–O bond is not broken, so the configuration at a stereogenic center is retained.

Reactions of alkyl tosylates

Alkyl tosylates undergo either substitution or elimination depending on the reagent (9.13B).

- Substitution is carried out with strong :Nu⁻, so the mechanism is S_N2.

- Elimination is carried out with strong bases, so the mechanism is E2.

Reactions of ethers

Only one reaction is useful: Cleavage with strong acids (9.14)

R—O—R' + H—X (2 equiv) (X = Br or I) $\longrightarrow$ R—X + R'—X + H_2O

- With 2° and 3° R groups, the mechanism is S_N1.
- With CH_3 and 1° R groups, the mechanism is S_N2.

Reactions involving thiols and sulfides (9.15)

[1] Preparation of thiols

$$R-X \quad + \quad {}^-SH \longrightarrow \boxed{R-SH} \quad + \quad X^-$$

- The mechanism is S_N2.
- The reaction works best for CH_3X and 1° RX.

[2] Oxidation and reduction involving thiols

a. Oxidation of thiols to disulfides

$$R-SH \xrightarrow{\text{Br}_2 \text{ or } \text{I}_2} \boxed{RS-SR}$$

b. Reduction of disulfides to thiols

$$RS-SR \xrightarrow[\text{HCl}]{\text{Zn}} \boxed{R-SH}$$

[3] Preparation of sulfides

$$R-X \quad + \quad {}^-SR' \longrightarrow \boxed{R-SR'} \quad + \quad X^-$$

- The mechanism is S_N2.
- The reaction works best for CH_3X and 1° RX.

[4] Reaction of sulfides to form sulfonium ions

$$R'_2S \quad + \quad R-X \longrightarrow \boxed{R'_2\overset{+}{S}-R} \quad + \quad X^-$$

- The mechanism is S_N2.
- The reaction works best for CH_3X and 1° RX.

Reactions of epoxides

Epoxide rings are opened with nucleophiles :Nu⁻ and acids HZ (9.16).

- The reaction occurs with backside attack, resulting in trans or anti products.
- With :Nu⁻, the mechanism is S_N2, and nucleophilic attack occurs at the *less* substituted C.
- With HZ, the mechanism is between S_N1 and S_N2, and attack of Z⁻ occurs at the *more* substituted C.

Practice Test on Chapter Review

1. Give the IUPAC name for each of the following compounds.

a.

b.

2. Draw the organic products formed in each reaction. Draw all stereogenic centers using wedges and dashed wedges.

a. $\xrightarrow[\text{(2 equiv)}]{\text{HI}}$

b. $\xrightarrow[\text{[2] CH}_3\text{CH}_2\text{Br}]{\text{[1] NaH}}$

c. $\xrightarrow[\text{[2] NaOCH}_3]{\text{[1] TsCl, pyr}}$

d. $\xrightarrow{\text{HBr}}$

e. $\xrightarrow[\text{[2] H}_2\text{O}]{\text{[1] NaCN}}$

f. $\xrightarrow{\text{H}_2\text{SO}_4}$

3. What starting material is needed for the following reaction?

$\xrightarrow[\text{[2] NaOCH}_3]{\text{[1] TsCl, pyridine}}$

4. What alkoxide and alkyl halide are needed to make the following ether?

Answers to Practice Test

1.a. 3-ethoxy-2-methylhexane

b. 4-ethyl-7-methyl-octan-3-ol

2.
a.

b.

c.

d.

e.

f.
(+ stereoisomer)

3.

4.

Answers to Problems

9.1 **Alcohols** are classified as 1°, 2°, or 3°, depending on the number of carbon atoms bonded to the carbon with the OH group. Five ether oxygens are circled.

9.2 To name an alcohol:

[1] **Find the longest chain that has the OH group as a substituent.** Name the molecule as a derivative of that number of carbons by changing the *-e* ending of the alkane to the suffix *-ol.*

[2] **Number the carbon chain to give the OH group the lower number.** When the OH group is bonded to a ring, the ring is numbered beginning with the OH group, and the "1" is usually omitted.

[3] Apply the other rules of nomenclature to complete the name.

a. [1] 5 carbons = **pentanol** [2] [3] **3,3-dimethylpentan-1-ol**

b. [1] 6 carbon ring = **cyclohexanol** [2] 2-methyl, 1 [3] *cis*-**2-methylcyclohexanol**

c. [1] 9 carbons = **nonanol** [2] 6-methyl, 5-ethyl, 3 [3] **5-ethyl-6-methylnonan-3-ol**

9.3 To work backwards from a name to a structure:
[1] Find the parent name and draw its structure.
[2] Add the substituents to the long chain.

a. 7,7-dimethyl**octan-4-ol**

c. 2-*tert*-butyl-3-methyl**cyclohexanol**

b. 5-methyl-4-propyl**heptan-3-ol**

d. *trans*-**cyclohexane-1,2-diol**

or

9.4 **To name simple ethers:**
[1] Name both alkyl groups bonded to the oxygen.
[2] Arrange these names alphabetically and add the word ***ether.*** For symmetrical ethers, name the alkyl group and add the prefix ***di.***

To name ethers using the IUPAC system:
[1] Find the two alkyl groups bonded to the ether oxygen. The smaller chain becomes the substituent, named as an alkoxy group.
[2] Number the chain to give the lower number to the first substituent.

a. **common name:**

methyl   butyl

butyl methyl ether

IUPAC name:

← 4 C's, butane

substituent:
methoxy

1-methoxybutane

b. **common name:**

cyclohexyl

methyl

cyclohexyl methyl ether

IUPAC name:

← methoxy
   substituent

6 C's, cyclohexane
methoxycyclohexane

c. **common name:**

propyl   propyl

dipropyl ether

IUPAC name:

propoxy   propane

1-propoxypropane

9.5   Three ways to name epoxides:

[1] Epoxides are named as derivatives of oxirane, the simplest epoxide.

[2] Epoxides can be named by considering the oxygen as a substituent called an **epoxy** group, bonded to a hydrocarbon chain or ring. Use two numbers to designate which two atoms the oxygen is bonded to.

[3] Epoxides can be named as **alkene oxides** by mentally replacing the epoxide oxygen by a double bond. Name the alkene (Chapter 10) and add the word *oxide*.

a.
Three possibilities:
[1] **methyloxirane**
[2] **1,2-epoxypropane**
[3] **propene oxide**

c.
Three possibilities:
[1] *cis*-**2-methyl-3-propyloxirane**
[2] *cis*-**2,3-epoxyhexane**
[3] *cis*-**hex-2-ene oxide**

b.
1-methyl
epoxy group

Two possibilities:
[1] 6 carbons = cyclohexane
1,2-epoxy-1-methylcyclohexane
[2] **1-methylcyclohexene oxide**

9.6   Two rules for boiling point:

[1] **The stronger the intermolecular forces the higher the bp.**

[2] **Bp increases as the extent of the hydrogen bonding increases.** For alcohols with the same number of carbon atoms: 3° ROH < 2° ROH < 1° ROH.

a.

VDW
lowest bp

VDW
DD
intermediate bp

VDW
DD
hydrogen
bonding
highest bp

b.

3° ROH
lowest bp

2° ROH
intermediate bp

1° ROH
highest bp

9.7   Strong nucleophiles (like ⁻CN) favor S$_N$2 reactions. The use of crown ethers in nonpolar solvents increases the nucleophilicity of the anion, and this increases the rate of the S$_N$2 reaction. The nucleophile does not appear in the rate equation for the S$_N$1 reaction. Nonpolar solvents cannot solvate carbocations, so this disfavors S$_N$1 reactions as well.

9.8   Draw the products of substitution in the following reactions by substituting OH or OR for X in the starting material.

a.   Br + ⁻OH ⟶   OH + Br⁻ **alcohol**

b.   Cl + ⁻OCH$_3$ ⟶   OCH$_3$ + Cl⁻ **unsymmetrical ether**

c.   I + O⁻ ⟶   O + I⁻ **unsymmetrical ether**

d.   Br + ⁻OCH$_2$CH$_3$ ⟶   OCH$_2$CH$_3$ + Br⁻ **unsymmetrical ether**

9.9 Two possible routes to **X** are shown. Path [2] with a 1° alkyl halide is preferred. Path [1] cannot occur because the leaving group would be bonded to an sp^2 hybridized C, making it an unreactive aryl halide.

aryl halide

1° alkyl halide

9.10 NaH and NaNH₂ are strong bases that will remove a proton from an alcohol, creating a nucleophile.

a.

b.

c.

d.

$C_6H_{10}O$

9.11 Dehydration follows the Zaitsev rule, so the more stable, more substituted alkene is the major product.

a.

TsOH

(+ cis isomer)

b.

TsOH

(+ stereoisomer)

trisubstituted
major product

disubstituted
minor product

+ H_2O

c.

OH

TsOH

trisubstituted
major product

disubstituted
minor product

+ H_2O

9.12 The rate of dehydration increases as the number of R groups increases.

1° alcohol
slowest reaction

2° alcohol
**intermediate
reactivity**

3° alcohol
fastest reaction

9.13

transition state [1]:

$$\left[\begin{array}{c} \overset{\delta+}{\ddot{O}H} \\ \overset{\delta-}{H} ---OSO_3H \end{array} \right]^{\ddagger}$$

transition state [2]:

$$\left[\begin{array}{c} \overset{\delta+}{\ddot{O}H_2} \\ H--OSO_3H \\ \delta- \end{array} \right]^{\ddagger}$$

9.14

H H HSO_4^-

β

rearranged 3° carbocation

+ H_2SO_4

| This alkene is also formed in addition to **Y** from the rearranged carbocation. |

The initially formed 2° carbocation gives two alkenes:

HSO_4^- H   H

or

H

H HSO_4^-

9.15

a.

2° carbocation → rearrangement 1,2-H shift → 3° carbocation **more stable**

b.

2° carbocation → rearrangement 1,2-H shift → 3° carbocation **more stable**

c.

2° carbocation → rearrangement 1,2-methyl shift → 3° carbocation **more stable**

9.16

H_2O overall reaction

:ÖH + $+$ + HCl

The steps:

+ Cl⁻

and

2° carbocation

Rearrangement of H forms a more stable carbocation.

3° carbocation

9.17

a.

OH → HCl → Cl + H_2O

b.

OH → HI → I + H_2O

c.

OH → HBr → Br + H_2O

9.18 • **CH₃OH and 1° alcohols** follow an S_N2 mechanism, which results in inversion of configuration.
• **Secondary (2°) and 3° alcohols** follow an S_N1 mechanism, which results in racemization at a stereogenic center.

a.

OH, D → HI → I, D   1° alcohol, so **inversion of configuration**

b.

3° alcohol, so Br⁻ attacks from above and below. The product is achiral.

achiral starting material   achiral product

c.

3° alcohol = **racemization**

9.19

a.

c.

(product formed after a 1,2-H shift)

b.

(product formed after a 1,2-CH₃ shift)

9.20 Substitution reactions of alcohols using SOCl₂ proceed by an S_N2 mechanism. Therefore, there is **inversion of configuration** at a stereogenic center.

Reactions using SOCl₂ proceed by an S_N2 mechanism = **inversion of configuration.**

9.21 Substitution reactions of alcohols using PBr₃ proceed by an S_N2 mechanism. Therefore, there is inversion of configuration at a stereogenic center.

Reactions using PBr₃ proceed by an S_N2 mechanism = **inversion of configuration.**

9.22 Stereochemistry for conversion of ROH to RX by reagent:
[1] **HX**—with 1°, S_N2, so inversion of configuration; with 2° and 3°, S_N1, so racemization.
[2] **SOCl₂**—S_N2, so inversion of configuration.
[3] **PBr₃**—S_N2, so inversion of configuration.

a.

c.

S_N2 = **inversion**

b.

3° alcohol, S_N1 = **racemization**

9.23 To do a two-step synthesis with this starting material:

[1] Convert the OH group into a good leaving group (by using either PBr₃ or SOCl₂).

[2] Add the nucleophile for the S_N2 reaction.

9.24

a.

b.

9.25

a.

b.

c.

(Substitution is favored over elimination.)

9.26

9.27 These reagents can be classified as follows:

[1] SOCl₂, PBr₃, HCl, and HBr replace OH with X by a substitution reaction.

[2] Tosyl chloride (TsCl) makes OH a better leaving group by converting it to OTs.

[3] Strong acids (H₂SO₄) and POCl₃ (pyridine) result in elimination by dehydration.

9.28

9.29 Ether cleavage can occur by either an S$_N$1 or S$_N$2 mechanism, but neither mechanism can occur when the ether O atom is bonded to an aromatic ring. An S$_N$1 reaction would require formation of a highly unstable carbocation on a benzene ring, a process that does not occur. An S$_N$2 reaction would require backside attack through the plane of the aromatic ring, which is also not possible. Thus, cleavage of the Ph–OCH$_3$ bond does not occur.

9.30

4-ethyl-2-methylhexane-1-thiol

4,6-dimethyloctane-2-thiol

9.31

grapefruit mercaptan

backside attack

9.32

a. 1-ethylthiobutane
or
butyl ethyl sulfide

b. 2-methyl-1-methylthiocyclopentane

9.33

a.

b. + Cl⁻

9.34 Two rules for the reaction of an epoxide:
[1] Nucleophiles attack from the **back side** of the epoxide.
[2] Negatively charged nucleophiles attack at the **less substituted carbon.**

a.   [1] $CH_3CH_2O^-$ →   OH   OCH$_2$CH$_3$
     [2] H_2O

Attack here:
less substituted C
backside attack

b.   [1] $H-C\equiv C^-$ →
     [2] H_2O

Attack here:
less substituted C
backside attack

9.35 In both isomers, ⁻OH attacks from the back side at either C–O bond.

cis-2,3-dimethyloxirane

enantiomers

identical

trans-2,3-dimethyloxirane

Rotate around the C–C bond to
see the plane of symmetry.

meso compound

9.36 Remember the difference between negatively charged nucleophiles and neutral nucleophiles:
- **Negatively charged nucleophiles attack first,** followed by protonation, and the nucleophile attacks at the *less* **substituted carbon.**
- **Neutral nucleophiles have protonation first,** followed by nucleophilic attack at the *more* **substituted carbon.**

Trans or anti products are always formed, regardless of the nucleophile.

a.

neutral nucleophile:
attack at **more**
substituted C

b.

[1] ⁻CN

[2] H₂O

negatively charged
nucleophile:
attack at **less**
substituted C

c.

CH₃CH₂OH

H₂SO₄

neutral nucleophile:
attack at **more**
substituted C

d.

[1] CH₃O⁻

[2] CH₃OH

negatively charged
nucleophile:
attack at **less**
substituted C

9.37

a.

6 C ring ⟶ cyclohexanol
2 CH₃'s at C3
3,3-dimethylcyclohexanol

b.

ethyl isobutyl ether
or
1-ethoxy-2-methylpropane

c.

1,2-epoxy-1-ethylcyclopentane
or
1-ethylcyclopentene oxide

9.38

a. (1R,2R)-2-isobutylcyclopentanol

b. 2° alcohol

A

c. stereoisomer

(1R,2S)-2-isobutylcyclopentanol

d. constitutional isomer

(1S,3S)-3-isobutylcyclopentanol

e. constitutional isomer with an ether

butoxycyclopentane

f.

[1]

HO A → NaH → ⁻O

[2] HO → H₂SO₄ → +

[3] HO → POCl₃ / pyridine → +

[4] HO → HCl → Cl

[5] HO → SOCl₂ / pyridine → Cl

[6] HO → TsCl / pyridine → TsO

9.39

a. HO, H → HBr → Br, H (**S**) + H, Br (**R**)   2° alcohol
S_N1 = racemization

b. HO, H → PBr₃ → H, Br (**R**)   PBr₃ follows
S_N2 = inversion.

c. HO, H → HCl → Cl, H (**S**) + H, Cl (**R**)   2° alcohol
S_N1 = racemization

d. HO, H → SOCl₂ / pyridine → H, Cl (**R**)   SOCl₂ follows
S_N2 = inversion.

9.40 Use the directions from Answer 9.2.

a.

[1] OH
7 carbons = **heptanol**

[2] ← 4-ethyl
3
OH
6-methyl

[3] **4-ethyl-6-methylheptan-3-ol**

b.

[1] OH
8 carbons = **octanol**

[2] ← 5-methyl
4-ethyl
3 OH

[3] **4-ethyl-5-methyloctan-3-ol**

c.

[1] HO⟶[structure]⟵OH
cyclohexanediol

[2] HO⟶[structure, positions 4, 3, 2, 1]⟵OH
(labels: 4, 3, 2, 1)

[3] **2-*sec*-butylcyclohexane-1,4-diol**

d.

[1] [structure] OH / OH OH
7 carbons = heptanetriol

[2] [structure with labels 4, 3, 2] OH / OH OH
↑ **5-methyl**

[3] **5-methylheptane-2,3,4-triol**

e.

[1] HO⟍[structure]⟋
5 carbons = cyclopentanol

[2] HO⟍[structure, label 1]⟋
↙ **3-isopropyl**

[3] ***trans*-3-isopropylcyclopentanol**

9.41 Use the rules from Answers 9.4 and 9.5.

a. [structure]
dicyclohexyl ether

b. **4,4-dimethyl**
↑
longest chain =
heptane

↑
substituent =
3-ethoxy

3-ethoxy-4,4-dimethylheptane

c.
1,2-epoxy-2-methylhexane
or **2-butyl-2-methyloxirane**
or **2-methylhexene oxide**

d. (labels 8, 4, 3, 1)
4-ethylthio-3-methyloctane

e. [structure with labels 2, 4, 1] —SH
2,2,4-trimethylcyclopentanethiol

9.42 Use the directions from Answer 9.3.

a. *trans*-2-methyl**cyclohexanol**

b. 2,3,3-trimethyl**butan-2-ol**

c. 6-*sec*-butyl-7,7-diethyl**decan-4-ol**

d. 3-chloro**propane-1,2-diol**

e. 1,2-epoxy-1,3,3-trimethyl**cyclohexane**

f. 1-ethoxy-3-ethyl**heptane**

g. (2*R*,3*S*)-3-isopropyl**hexan-2-ol**

h. (*S*)-2-ethoxy-1,1-dimethyl**cyclopentane**

i. 4-ethyl**heptane-3-thiol**

j. 1-isopropylthio-2-methyl**cyclohexane**

9.43 Melting points depend on intermolecular forces and symmetry. $(CH_3)_2CHCH_2OH$ has a lower melting point than $CH_3CH_2CH_2CH_2OH$ because branching decreases surface area and makes $(CH_3)_2CHCH_2OH$ less symmetrical, so it packs less well. Although $(CH_3)_3COH$ has the most branching and least surface area, it is the most symmetrical, so it packs best in a crystalline lattice, giving it the highest melting point.

−108 °C	−90 °C	26 °C
lowest melting point	intermediate melting point	highest melting point

9.44 Stronger intermolecular forces increase boiling point. All of the compounds can hydrogen bond, but both diols have more opportunity for hydrogen bonding because they have two OH groups, making their bp's higher than the bp of butan-1-ol. Propane-1,2-diol can also intramolecularly hydrogen bond. Intramolecular hydrogen bonding decreases the amount of intermolecular hydrogen bonding, so the bp of propane-1,2-diol is somewhat lower.

Increasing boiling point

butan-1-ol	propane-1,2-diol	propane-1,3-diol
118 °C	187 °C	215 °C

9.45

a. [structure] OH $\xrightarrow{H_2SO_4}$ [structure]

b. [structure] OH $\xrightarrow{NaH}$ [structure] O⁻ Na⁺

c. [structure] OH $\xrightarrow[ZnCl_2]{HCl}$ [structure] Cl

d. [structure] OH $\xrightarrow{HBr}$ [structure] Br

e. [structure] OH $\xrightarrow[\text{pyridine}]{SOCl_2}$ [structure] Cl

f. [structure] OH $\xrightarrow{PBr_3}$ [structure] Br

g. [structure] OH $\xrightarrow[\text{pyridine}]{TsCl}$ [structure] OTs

h. [structure] OH $\xrightarrow{[1]\ NaH}$ [structure] O⁻ Na⁺ $\xrightarrow{[2]\ \text{Br}}$ [structure] O

i. [structure] OH $\xrightarrow{[1]\ TsCl}$ [structure] OTs $\xrightarrow{[2]\ NaSH}$ [structure] SH

j. [structure] OH $\xrightarrow[\text{pyridine}]{POCl_3}$ [structure]

9.46 Dehydration follows the Zaitsev rule, so the more stable, more substituted alkene is the major product.

a. [structure] OH $\xrightarrow{TsOH}$ [structure] + [structure]

tetrasubstituted   disubstituted
major product

b. [structure] OH $\xrightarrow{TsOH}$ [structure] + [structure]

c. [structure] OH $\xrightarrow{TsOH}$ [structure] + [structure]

trisubstituted   disubstituted
major product

d. [structure] OH $\xrightarrow{TsOH}$ [structure] + [structure]

tetrasubstituted   disubstituted
major product

two products formed
by carbocation rearrangement

9.47 OTs is a good leaving group and will easily be replaced by a nucleophile. Draw the products by substituting the nucleophile in the reagent for OTs in the starting material.

a. [structure] OTs $\xrightarrow[S_N2]{CH_3SH}$ [structure] S + HOTs

b. [structure] OTs $\xrightarrow[S_N2]{NaOCH_2CH_3}$ [structure] O + Na⁺ ⁻OTs

c. [structure] OTs $\xrightarrow[S_N2]{NaOH}$ [structure] OH + Na⁺ ⁻OTs

d. [structure] OTs $\xrightarrow[E2]{K^+\ {}^-OC(CH_3)_3}$ [structure] + (CH₃)₃COH + K⁺ ⁻OTs

9.48

a.

HBr

2° Alcohol will undergo S_N1.
racemization

b.

HCl
ZnCl₂

1° Alcohol will undergo S_N2.
inversion

c.

SOCl₂
pyridine

SOCl₂ always implies S_N2.
inversion

d.

TsCl
pyridine

KI
S_N2
inversion

Configuration is maintained.
C–O bond is not broken.

9.49

TsO

K₂CO₃

inversion
of configuration

9.50

(R)-hexan-2-ol

(a) NaH → A =

(b) TsCl pyridine → C =

(c) PBr₃ → E =

CH₃I → B =

CH₃O⁻ → D =

CH₃O⁻ → F =

Routes (a) and (c) give identical products, labeled **B** and **F**.

9.51

a.

H₂SO₄
(–H₂O)

A

2° carbocation

1,2-H shift

3° carbocation

major product

b.

POCl₃, pyridine
E2

A H

H and OH must be trans in the E2 reaction.

The major products are different because the mechanisms are different. With H_2SO_4, the reaction proceeds by an E1 mechanism involving a carbocation rearrangement. With $POCl_3$, the mechanism is E2 and the H and OH must be trans diaxial. No H on the C with the CH_3 group is trans to the OH group, so only one product forms.

9.52 Acid-catalyzed dehydration follows an E1 mechanism for 2° and 3° ROH with an added step to make a good leaving group. The three steps are:

[1] Protonate the oxygen to make a good leaving group.
[2] Break the C–O bond to form a carbocation.
[3] Remove a β hydrogen to form the π bond.

9.53 With $POCl_3$ (pyridine), elimination occurs by an E2 mechanism. Because only one carbon has a β hydrogen, only one product is formed. With H_2SO_4, the mechanism of elimination is E1. A 2° carbocation rearranges to a 3° carbocation, which has three pathways for elimination.

9.54 To draw the mechanism:

[1] Protonate the oxygen to make a good leaving group.

[2] Break the C–O bond to form a carbocation.

[3] Look for possible rearrangements to make a more stable carbocation.

[4] Remove a β hydrogen to form the π bond.

Dark and light circles are meant to show where the carbons in the starting material appear in the product.

9.55

3-methylbutan-2-ol

The 2° alcohol reacts by an S_N1 mechanism to form a carbocation that rearranges.

2-methylpropan-1-ol

The 1° alcohol reacts with HBr by an S_N2 mechanism.
no carbocation intermediate = no rearrangement possible

9.56

A

H—Cl

two resonance structures
for the carbocation

+ H$_2$O

:Cl:⁻

B

:Cl:⁻

C

9.57

H—OSO$_3$H

+ HSO$_4$⁻

+ H$_2$O

1,2-shift

+ HSO$_4$⁻

+ H$_2$SO$_4$

9.58

H—OSO$_3$H

+ HSO$_4$⁻

+ H$_2$O

HSO$_4$⁻

+ H$_2$SO$_4$

9.59

a.

O + Br

2° halide

Br + ⁻O

1° halide

less hindered RX
preferred path

b.

O⁻ + Br

1° halide

less hindered RX
preferred path

Br + ⁻O

2° halide

c.

1° halide   **1° halide**

Neither path preferred.

9.60 A tertiary halide is too hindered and an aryl halide is too unreactive to undergo a Williamson ether synthesis.

Two possible sets of starting materials:

aryl halide
unreactive in S_N2

3° alkyl halide
**too sterically
hindered for S_N2**

9.61

a.

c.

b.

9.62

a.

overall reaction

The steps:

b.

9.63

+ CF$_3$CO$_2^-$ CF$_3$CO$_2^-$ + CF$_3$CO$_2$H

9.64

a. HBr

b. H$_2$O / H$_2$SO$_4$

c. [1] CH$_3$CH$_2$O$^-$ [2] H$_2$O

d. [1] HC≡C$^-$ [2] H$_2$O

e. [1] $^-$OH [2] H$_2$O

f. [1] CH$_3$S$^-$ [2] H$_2$O

9.65

a. CH$_3$CH$_2$OH / H$_2$SO$_4$

b. [1] CH$_3$CH$_2$O$^-$ Na$^+$ [2] H$_2$O

c. HBr

d. [1] NaCN [2] H$_2$O

9.66

a. Na$^+$H:$^-$ — C$_4$H$_8$O

The 2 CH$_3$ groups are anti in the starting material, making them trans in the product.

b. Na$^+$H:$^-$ — C$_4$H$_8$O

The 2 CH$_3$ groups are gauche in the starting material, making them cis in the product.

c. rotate — Na$^+$H:$^-$ — backside attack — C$_4$H$_8$O

9.67 First, use the names to draw the structures of the starting material and both products. Because the product has two OH groups, one OH must come from the epoxide oxygen, and one must come from the nucleophile, either ⁻OH or H_2O.

With ⁻OH, the nucleophile attacks at the less substituted end of the epoxide to form the *R* isomer of the product.

(*R*)-2-ethyl-2-methyloxirane

[1] Na⁺ ⁻OH
[2] H_2O

HO
OH
← from the nucleophile

(*R*)-2-methylbutane-1,2-diol

With H_2O and H_2SO_4, H_2O attacks at the more substituted end of the epoxide, from the back side, and the *S* isomer is formed.

(*R*)-2-ethyl-2-methyloxirane

H_2O
H_2SO_4

HO
OH ← from the nucleophile

(*S*)-2-methylbutane-1,2-diol

9.68

+ HB⁺

9.69

a.

KOC(CH₃)₃

Bulky base favors E2.

b.

HBr

Keep the stereochemistry at the stereogenic center [*] the same here because no bond to it is broken.

c.

Br₂

d.

KSH

e.

PBr₃

S$_N$2 inversion

f.

TsCl
pyridine

CH₃CO₂⁻

g.

HBr

h.

[1] NaOCH$_3$

[2] H$_2$O

i.

NaH

ii.

ii.

j.

HI
(2 equiv)

+ I–CH$_3$ + H$_2$O

k.

l.

9.70

[1] 2 CH$_3$SO$_2$Cl
pyridine

[2] Na$_2$S

S + CH$_3$SO$_2^-$

[1] 2 CH$_3$SO$_2$Cl
pyridine

S$_N$2

S$_N$2

+ CH$_3$SO$_2^-$

9.71

a.

HCl
or
SOCl$_2$, pyridine

b.

H$_2$SO$_4$

c.

[1] Na$^+$ H$^-$

[2] CH$_3$Cl

d.

[1] TsCl, pyridine

[2] ⁻CN

Make OH a good
leaving group (use TsCl);
then add ⁻CN.

9.72

9.73

9.74 With the cis isomer, ⁻OH acts as a nucleophile to displace Br⁻ from the back side, forming a trans diol (**A**). With the trans isomer, the two functional groups are arranged in a manner that allows an intramolecular S_N2. ⁻OH removes a proton to form an alkoxide, which can then displace Br⁻ by intramolecular backside attack to afford an ether (**B**). Such a reaction is not possible with the cis isomer because the nucleophile and leaving group are on the same side.

9.75 If the base is not bulky, it can react as a nucleophile and open the epoxide ring. The bulky base cannot act as a nucleophile, and will only remove the proton.

9.76 First form the 2° carbocation. Then lose a proton to form each product.

9.77

9.78

a.

X

1,2-shift

+ H₃O⁺

1,2-shift

Y

b. Other elimination products can form from carbocations **X** and **Y.**

X

+ H₃O⁺

X

+ H₃O⁺

Y

+ H₃O⁺

9.79

a.

pinacol

1,2-CH₃
shift

H_2SO_4 +

pinacolone

b. Two different carbocations can form. The carbocation with the (+) charge adjacent to the benzene
rings (**A**) is more stable, so it is preferred.

9.80

9.81 The conversion of **X** to **Y** requires two operations. **X** contains both a nucleophile (NH_2) and a leaving group (OSO_2CH_3), so an intramolecular S_N2 reaction forms an aziridine. Because the aziridine is strained, the amine nucleophile (CH_2=$CHCH_2NH_2$) opens the ring by backside attack, resulting in the trans stereochemistry of the two N's on the six-membered ring.

Chapter 10 Alkenes

Chapter Review

General facts about alkenes

- Alkenes contain a carbon–carbon double bond consisting of a stronger σ bond and a weaker π bond. Each carbon is *sp²* hybridized and trigonal planar (10.1).
- Alkenes are named using the suffix *-ene* (10.3).
- Alkenes with different groups on each end of the double bond exist as a pair of diastereomers, identified by the prefixes *E* and *Z* (10.3B).

- Alkenes have weak intermolecular forces, giving them low mp's and bp's, and making them water insoluble. A cis alkene is more polar than a trans alkene, giving it a slightly higher boiling point (10.4).

- A π bond is electron rich and much weaker than a σ bond, so alkenes undergo addition reactions with electrophiles (10.8).

Stereochemistry of alkene addition reactions (10.8)

A reagent XY adds to a double bond in one of three different ways:

- **Syn addition**—X and Y add from the same side.

  - Syn addition occurs in **hydroboration.**

- **Anti addition**—X and Y add from opposite sides.

  - Anti addition occurs in **halogenation** and **halohydrin formation.**

- **Both syn and anti addition** occur when carbocations are intermediates.

$$\ce{C=C} \xrightarrow[\text{or}]{\underset{H_2O,\ H^+}{H-X}} \ \overset{H\ \ X(OH)}{\underset{}{C-C}} \quad \text{and} \quad \overset{H}{\underset{X(OH)}{C-C}}$$

- Syn and anti addition occur in **hydrohalogenation** and **hydration.**

Addition reactions of alkenes

[1] **Hydrohalogenation**—Addition of HX (X = Cl, Br, I) (10.9–10.11)

$$R\diagup\diagdown + H-X \longrightarrow \boxed{\underset{\text{alkyl halide}}{R\overset{X}{\diagup}}}$$

- The mechanism has two steps.
- Carbocations are formed as intermediates.
- Carbocation rearrangements are possible.
- Markovnikov's rule is followed. H bonds to the less substituted C to form the more stable carbocation.
- Syn and anti addition occur.

[2] **Hydration** and related reactions—Addition of H_2O or ROH (10.12)

$$R\diagup\diagdown + H-OH \xrightarrow{H_2SO_4} \boxed{\underset{\text{alcohol}}{R\overset{OH}{\diagup}}}$$

$$R\diagup\diagdown + H-OR \xrightarrow{H_2SO_4} \boxed{\underset{\text{ether}}{R\overset{OR}{\diagup}}}$$

For both reactions:

- The mechanism has three steps.
- Carbocations are formed as intermediates.
- Carbocation rearrangements are possible.
- Markovnikov's rule is followed. H bonds to the less substituted C to form the more stable carbocation.
- Syn and anti addition occur.

[3] **Halogenation**—Addition of X_2 (X = Cl or Br) (10.13–10.14)

$$R\diagup\diagdown + X-X \longrightarrow \boxed{\underset{\text{vicinal dihalide}}{R\overset{X}{\diagup}\diagdown X}}$$

- The mechanism has two steps.
- Bridged halonium ions are formed as intermediates.
- No rearrangements occur.
- Anti addition occurs.

[4] **Halohydrin formation**—Addition of OH and X (X = Cl, Br) (10.15)

$$R\diagup\diagdown + X-X \xrightarrow{H_2O} \boxed{\underset{\text{halohydrin}}{R\overset{OH}{\diagup}\diagdown X}}$$

- The mechanism has three steps.
- Bridged halonium ions are formed as intermediates.
- No rearrangements occur.
- X bonds to the less substituted C.
- Anti addition occurs.
- NBS in DMSO and H_2O adds Br and OH in the same fashion.

[5] **Hydroboration–oxidation**—Addition of H_2O (10.16)

- Hydroboration has a one-step mechanism.
- No rearrangements occur.
- OH bonds to the less substituted C.
- Syn addition of H_2O results.

Practice Test on Chapter Review

1. Give the IUPAC name for the following compounds.

a.

b.

2. Draw the organic products formed in the following reactions. Draw all stereogenic centers using wedges and dashes.

a.   HCl

b.   [1] BH_3
      [2] H_2O_2, ⁻OH

c. Br_2

d. H_2O
      H_2SO_4

3. Fill in the table with the stereochemistry observed in the reaction of an alkene with each reagent. Choose from syn, anti, or both syn and anti.

Reagent	Stereochemistry
a. [1] 9-BBN; [2] H_2O_2, ⁻OH	
b. H_2O, H_2SO_4	
c. Cl_2, H_2O	
d. HI	
e. Br_2	

4. a. In which of the following reactions are carbocation rearrangements observed?
   1. hydrohalogenation
   2. halohydrin formation
   3. hydroboration–oxidation
   4. Carbocation rearrangements are observed in reactions (1) and (2).
   5. Carbocation rearrangements are observed in reactions (1), (2), and (3).

   b. Which of the following products are formed when HCl is added to 3-methylpent-1-ene?
   1. 2-chloro-2-methylpentane
   2. 3-chloro-3-methylpentane
   3. 1-chloro-3-methylpentane
   4. Both (1) and (2) are formed.
   5. Products (1), (2), and (3) are all formed.

Answers to Practice Test

1. a. (E)-4-ethyl-2,5-
      dimethylnon-3-ene
   b. (E)-4-isopropyl-2-
      methyloct-3-ene

2.

3. a. syn
   b. both
   c. anti
   d. both
   e. anti

4. a. 1
   b. 2

Answers to Problems

10.1

Six alkenes of molecular formula C_5H_{10}:

trans   cis

diastereomers

10.2 To determine the number of degrees of unsaturation:
   [1] Calculate the maximum number of H's ($2n + 2$).
   [2] Subtract the actual number of H's from the maximum number.
   [3] Divide by two.

a. C_6H_6
  [1] maximum number of H's = $2n + 2 = 2(6) + 2 = 14$
  [2] subtract actual from maximum = $14 - 6 = 8$
  [3] divide by two = $8/2 =$ **4 degrees of unsaturation**

b. C_8H_{18}
  [1] maximum number of H's = $2n + 2 = 2(8) + 2 = 18$
  [2] subtract actual from maximum = $18 - 18 = 0$
  [3] divide by two = $0/2 =$ **0 degrees of unsaturation**

c. C_7H_8O
  Ignore the O.
  [1] maximum number of H's = $2n + 2 = 2(7) + 2 = 16$
  [2] subtract actual from maximum = $16 - 8 = 8$
  [3] divide by two = $8/2 =$ **4 degrees of unsaturation**

d. $C_7H_{11}Br$
  Because of Br, add one more H ($11 + 1$ H = 12 H's).
  [1] maximum number of H's = $2n + 2 = 2(7) + 2 = 16$
  [2] subtract actual from maximum = $16 - 12 = 4$
  [3] divide by two = $4/2 =$ **2 degrees of unsaturation**

e. C_5H_9N
  Because of N, subtract one H ($9 - 1$ H = 8 H's).
  [1] maximum number of H's = $2n + 2 = 2(5) + 2 = 12$
  [2] subtract actual from maximum = $12 - 8 = 4$
  [3] divide by two = $4/2 =$ **2 degrees of unsaturation**

10.3 Use the directions from Answer 10.2. Ignore O in calculating degrees of unsaturation. Add one more H for each halogen. Subtract one H for each N.

a. $C_{19}H_{21}N_3O$ is equivalent to $C_{19}H_{18}$ when calculating degrees of unsaturation.
For 19 C's, the maximum number of H's = $2n + 2 = 2(19) + 2 = 40$ H's
Subtract actual from maximum = $40 - 18 = 22$ H's fewer
Divide by two = $22/2 = 11$ degrees of unsaturation

b. $C_{17}H_{16}F_6N_2O$ is equivalent to $C_{17}H_{20}$ when calculating degrees of unsaturation.
For 17 C's, the maximum number of H's = $2n + 2 = 2(17) + 2 = 36$ H's
Subtract actual from maximum = $36 - 20 = 16$ H's fewer
Divide by two = $16/2 = 8$ degrees of unsaturation

10.4 To name an alkene:

[1] Find the longest chain that contains the double bond. Change the ending from *-ane* to *-ene.*

[2] Number the chain to give the double bond the lower number. The alkene is named by the first number.

[3] Apply all other rules of nomenclature.

To name a cycloalkene:

[1] When a double bond is located in a ring, it is always located between C1 and C2. Omit the "1" in the name. Change the ending from *-ane* to *-ene.*

[2] Number the ring clockwise or counterclockwise to give the first substituent the lower number.

[3] Apply all other rules of nomenclature.

a. [1] 5 C chain with double bond **pentene** [2] **1** ← 3-methyl **pent-1-ene** [3] **3-methylpent-1-ene**

b. [1] 7 C chain with double bond **heptene** [2] **3** **hept-3-ene** 3-ethyl [3] **3-ethylhept-3-ene**

c. [1] 5 C chain with double bond **pentene**

[2] **2-ethyl** **4-methyl** **1 pent-1-ene**

[3] **2-ethyl-4-methylpent-1-ene**

d. [1] 5 C ring with a double bond **cyclopentene**

[2] **1** **4** **3** **3,4-dimethyl**

[3] **3,4-dimethylcyclopentene**

e. [1] 6 C ring with a double bond **cyclohexene**

[2] **1-methyl** **5-tert-butyl** **1**

[3] **5-tert-butyl-1-methylcyclohexene**

10.5 Use the rules from Answer 10.4 to name the compounds. Enols are named to give the OH the lower number. Compounds with two C=C's are named with the suffix *-adiene.*

a. [1] 6 C chain with double bond **hexene**

[2] **3** **1** OH **4-ethyl**

[3] **4-ethylhex-3-en-1-ol**

b. [1] OH 8 C chain with double bond **octene**

[2] **1** **4** OH **7** **5-ethyl** **6-methyl**

[3] **5-ethyl-6-methyloct-7-en-4-ol**

c. [1] 7 C chain with two double bonds **heptadiene**

[2] **5** **2** **2-methyl** **6-methyl**

[3] **2,6-dimethylhepta-2,5-diene**

10.6 To label an alkene as *E* or *Z*:

[1] **Assign priorities** to the two substituents *on each end* using the rules for *R,S* nomenclature.

[2] **Assign *E* or *Z*** depending on the location of the two higher priority groups.

- The *E* prefix is used when the two higher priority groups are on **opposite sides** of the double bond.
- The *Z* prefix is used when the two higher priority groups are on the **same side** of the double bond.

a.

higher priority →

Two higher priority groups are
on opposite sides: *E* isomer.

Br ← **higher priority**

c.

higher priority → OCH₃

higher priority → C₆H₅

← **higher priority**

higher priority

kavain

In both double bonds, the two higher priority
groups are on opposite sides: *E* isomers.

b.

higher priority →

← **higher priority**

Two higher priority groups are
on the same side: *Z* isomer.

10.7

skeletal structure of 11-*cis*-retinal

E   E   Z

E

Z

CHO

10.8 To work backwards from a name to a structure:

[1] Find the parent name and functional group and draw, remembering that the double bond is
between C1 and C2 for cycloalkenes.

[2] Add the substituents to the appropriate carbons.

a. (*Z*)-4-ethylhept-3-ene

7 carbons

The higher priority groups are
on the same side = *Z*.

← 4-ethyl

The double bond is
between C3 and C4.

c. (*Z*)-2-bromo-1-iodohex-1-ene

6 carbons

The double bond is
between C1 and C2.

I

Br

The higher priority groups are
on the same side = *Z*.

b. (*E*)-3,5,6-trimethyloct-2-ene

8 carbons

The double bond is
between C2 and C3.

The higher priority groups
are on opposite sides = *E*.

3,5,6-trimethyl

10.9

(2*Z*,6*E*)-3-ethyl-7-methyldeca-2,6-dien-1-ol

10 C chain with OH at C1
2 C=C's beginning at C2, C6

10

E 6

Z 1

3

OH

10.10 To rank the isomers by increasing boiling point:
Look for polarity differences: *small net dipoles* make an alkene more polar, giving it a higher boiling point than an alkene with *no net dipole*. Cis isomers have a higher boiling point than their trans isomers.

All dipoles cancel.
smallest surface area
no net dipole
lowest bp

Two dipoles cancel.
no net dipole
trans isomer
intermediate bp

Two dipoles reinforce.
net dipole
cis isomer
highest bp

10.11 Increasing number of double bonds = decreasing melting point.

stearic acid
no double bonds
highest melting point

stearidonic acid

4 double bonds
lowest melting point

linolenic acid

3 double bonds
intermediate melting point

10.12

a. (H₂SO₄)

b. (NaOCH₂CH₃)
(+ cis isomer)

10.13 To draw the products of an addition reaction:
[1] Locate the two bonds that will be broken in the reaction. Always break the π bond.
[2] Draw the product by forming two new σ bonds.

a. HCl → two new σ bonds

b. HCl → two new σ bonds

c. HCl → two new σ bonds

10.14 **Addition reactions of HX occur in two steps:**
[1] The double bond attacks the H atom of HX to form a carbocation.
[2] X⁻ attacks the carbocation to form a C–X bond.

10.15 Addition to alkenes follows Markovnikov's rule: When HX adds to an unsymmetrical alkene, the H bonds to the C that has more H's to begin with.

10.16 To determine which alkene will react faster, draw the carbocation that forms in the rate-determining step. The more stable, more substituted the carbocation, the lower the E_a to form it and the faster the reaction.

10.17 Look for rearrangements of a carbocation intermediate to explain these results.

10.18 Addition of HX to alkenes involves the formation of carbocation intermediates. Rearrangement of the carbocation will occur if it forms a more stable carbocation.

a.

3° carbocation
no rearrangement

b.

2° carbocation   2° carbocation

Rearrangement would not further
stabilize either carbocation.
no rearrangement

c.

2° carbocation

Rearrangement would not further
stabilize the carbocation.

2° carbocation
rearrangement

1,2-H shift

3° carbocation
more stable

Br

Br

10.19 To draw the products, remember that addition of HX proceeds via a carbocation intermediate.

a.

HBr

Br
new stereogenic center

Br   Br

+

enantiomers

b.

Addition of H⁺ (from HCl) from above and
below by Markovnikov's rule forms
an achiral 3° carbocation.

Cl⁻ attacks
from above
and below.

H H

achiral, trigonal planar
3° carbocation

Cl   +   Cl

diastereomers

10.20 The product of syn addition will have H and Cl both up or down (both on wedges or both on dashed wedges), while the product of anti addition will have one up and one down (one wedge, one dashed wedge).

10.21

a.
H+ would add here to form a 3° carbocation.   *or*   H+ would add here to form a 3° carbocation.

b.
H+ would add here to form a 3° carbocation.   *or*   H+ would add here to form a 3° carbocation.

c.
H+ would add here to form a 2° carbocation.   *or*   (+ cis isomer)

10.22

pent-1-ene   H₂O / H₂SO₄   **enantiomers**

10.23 Halogenation of an alkene adds two elements of X in an anti fashion.

a.   b.

10.24 To draw the products of halogenation of an alkene, remember that the halogen adds to both ends of the double bond but only anti addition occurs.

a. **enantiomers**   b. **achiral meso compound**

c.

diastereomers

10.25 Halohydrin formation adds the elements of X and OH across the double bond in an anti fashion. The reaction is regioselective, so X ends up on the carbon that had more H's to begin with.

a.

b.

Cl bonds to the carbon with more H's to begin with.

10.26 In hydroboration the boron atom is the electrophile and becomes bonded to the carbon atom that had more H's to begin with.

a.

C with more H's. B will add here.

c.

C with more H's. B will add here.

b.

C with more H's. B will add here.

10.27 The hydroboration–oxidation reaction occurs in two steps:
[1] Syn addition of BH_3, with the boron on the less substituted carbon atom
[2] OH replaces the BH_2 with retention of configuration.

a.

b.

c.

10.28 Remember that hydroboration results in addition of OH on the less substituted C.

a.

c.

b.

(*E* or *Z* isomer can be used.)

10.29

a.
$$\xrightarrow[\text{H}_2\text{SO}_4]{\text{H}_2\text{O}}$$

Hydration places the OH on the more substituted carbon.

$$\xrightarrow[\text{[2] H}_2\text{O}_2,\ \text{HO}^-]{\text{[1] BH}_3}$$

Hydroboration–oxidation places the OH on the less substituted carbon.

b.
$$\xrightarrow[\text{H}_2\text{SO}_4]{\text{H}_2\text{O}}$$

Hydration places the OH on the more substituted carbon.

$$\xrightarrow[\text{[2] H}_2\text{O}_2,\ \text{HO}^-]{\text{[1] BH}_3}$$

Hydroboration–oxidation places the OH on the less substituted carbon.

c.
$$\xrightarrow[\text{H}_2\text{SO}_4]{\text{H}_2\text{O}}$$

Hydration places the OH on the more substituted carbon.

$$\xrightarrow[\text{[2] H}_2\text{O}_2,\ \text{HO}^-]{\text{[1] BH}_3}$$

Hydroboration–oxidation places the OH on the less substituted carbon.

10.30 There are always two steps in this kind of question:
   [1] **Identify the functional group and decide what types of reactions it undergoes** (e.g., substitution, elimination, or addition).
   [2] **Look at the reagent and determine if it is an electrophile, nucleophile, acid, or base.**

a.

acid

HBr

alkene:
addition reactions

c.

acid:
catalyzes loss of H_2O

H_2SO_4

alcohol:
substitution and elimination

b.

nucleophile and base

NaOCH₃

2° alkyl halide:
substitution and elimination

(+ cis isomer)

10.31 To devise a synthesis:
[1] Look at the starting material and decide what reactions it can undergo.
[2] Look at the product and decide what reactions could make it.

a.
alkyl halide
Can undergo
substitution and elimination.

halohydrin:
Can form from an
alkene with Cl_2 and H_2O.

b.
alcohol
Can undergo
substitution and elimination.

alcohol
Can be formed by
substitution and addition.

OH is added to
the more substituted C.

(major
product)

OH is added to
the less substituted C.

10.32 Convert each ball-and-stick model to a skeletal structure and then name the molecule.

a.
6 C chain with a C=C → hexene
C=C at C2 → hex-2-ene
2 CH$_3$'s at C3 and C5
two higher priority groups on opposite sides → *E* isomer
Answer: **(*E*)-3,5-dimethylhex-2-ene**

b.
5 C ring with a C=C → cyclopentene
sec-butyl at C1
methyl at C2
Answer: **1-*sec*-butyl-2-methylcyclopentene**

10.33

a.
higher → priority

higher priority

A

The two higher priority groups are on the same side
of the C=C, making it a **Z** alkene.

b. Add H_2O in a Markovnikov fashion to form two products.

$$\xrightarrow[\text{H}_2\text{SO}_4]{\text{H}_2\text{O}}$$

OH

+

OH

10.34

8-methylnon-1-ene

a. Br_2 →

b. Br_2 / H_2O →

c. Br_2 / CH_3OH →

10.35 Use the directions from Answer 10.2 to calculate degrees of unsaturation.

a. C_6H_8
[1] maximum number of H's = $2n + 2 = 2(6) + 2 = 14$
[2] subtract actual from maximum = $14 - 8 = 6$
[3] divide by 2 = $6/2$ = **3 degrees of unsaturation**

b. $C_{40}H_{56}$
[1] maximum number of H's = $2n + 2 = 2(40) + 2 = 82$
[2] subtract actual from maximum = $82 - 56 = 26$
[3] divide by 2 = $26/2$ = **13 degrees of unsaturation**

c. $C_{10}H_{16}O_2$
Ignore both O's.
[1] maximum number of H's = $2n + 2 = 2(10) + 2 = 22$
[2] subtract actual from maximum = $22 - 16 = 6$
[3] divide by 2 = $6/2$ = **3 degrees of unsaturation**

d. C_8H_9Br
Because of Br, add one H (9 + 1 = 10 H's).
[1] maximum number of H's = $2n + 2 = 2(8) + 2 = 18$
[2] subtract actual from maximum = $18 - 10 = 8$
[3] divide by 2 = $8/2$ = **4 degrees of unsaturation**

e. C_8H_9ClO
Ignore the O; count Cl as one more H (9 + 1 = 10 H's).
[1] maximum number of H's = $2n + 2 = 2(8) + 2 = 18$
[2] subtract actual from maximum = $18 - 10 = 8$
[3] divide by 2 = $8/2$ = **4 degrees of unsaturation**

f. $C_7H_{11}N$
Because of N, subtract one H (11 − 1 = 10 H's).
[1] maximum number of H's = $2n + 2 = 2(7) + 2 = 16$
[2] subtract actual from maximum = $16 - 10 = 6$
[3] divide by 2 = $6/2$ = **3 degrees of unsaturation**

g. C_4H_8BrN
Because of Br, add one H, but subtract one for N
(8 + 1 − 1 = 8 H's).
[1] maximum number of H's = $2n + 2 = 2(4) + 2 = 10$
[2] subtract actual from maximum = $10 - 8 = 2$
[3] divide by 2 = $2/2$ = **1 degree of unsaturation**

h. $C_{10}H_{18}ClNO$
Add one H because of Cl and subtract 1 H for N
(18 + 1 − 1 = 18) .
[1] maximum number of H's = $2n + 2 = 2(10) + 2 = 22$
[2] subtract actual from maximum = $22 - 18 = 4$
[3] divide by 2 = $4/2$ = **2 degrees of unsaturation**

10.36 First determine the number of degrees of unsaturation in the compound. Then decide which combinations of rings and π bonds could exist.

$C_{10}H_{14}$
[1] maximum number of H's = $2n + 2 = 2(10) + 2 = 22$
[2] subtract actual from maximum = $22 - 14 = 8$
[3] divide by two = $8/2$ = **4 degrees of unsaturation**

possibilities:
4 π bonds
3 π bonds + 1 ring
2 π bonds + 2 rings
1 π bond + 3 rings
4 rings

10.37

a. enclomiphene
E

b. tamoxifen
Z

c. clavulanic acid
Z

10.38 Name the alkenes using the rules in Answers 10.4 and 10.6.

a. **(*E*)-6-isopropyl-3-methylnon-3-ene**

9 C chain with a double bond =
nonene

b. **2-isopropyl**

2-isopropyl-4-methylpent-1-ene

pent-1-ene

5 C chain with a double bond =
pentene

4-methyl

c. **5-*sec*-butyl-1,3,3-trimethylcyclohexene**

6 C ring with a double bond =
hexene

d. *E* **double bond** (higher priority groups on opposite sides with bold bonds)

4-isopropyl

(*E*)-4-isopropylhept-4-en-3-ol

3-ol

hept-4-ene

7 C chain with a double bond =
heptene

e. **cyclohex-2-ene** **5-*sec*-butylcyclohex-2-enol**

1-ol

5-*sec*-butyl

6 C ring with a double bond =
cyclohexene

f. **(*E*)-5-ethyl-3,4-dimethylnon-2-ene**

10.39 Use the directions from Answer 10.8.

a. (*E*)-4-ethyl**hept-3-ene**

↑
7 carbons

4-ethyl →

3

Higher priority groups on
opposite sides = *E*.

b. 3,3-dimethyl**cyclopentene**

↑
5 carbon ring

← **3,3-dimethyl**

1

c. 4-vinyl**cyclopentene**

↑
5 carbon ring

2 **1**

4

d. (*Z*)-3-isopropyl**hept-2-ene**

↑
7 carbons

2

3-isopropyl

Higher priority groups on
the same side = *Z*.

e. *cis*-3,4-dimethyl**cyclopentene**

↑
5 carbon ring

1

4 **3**

3,4-dimethyl

or

f. 1-isopropyl-4-propyl**cyclohexene**

↑
6 carbon ring

4-propyl

1-isopropyl

g. 3,4-dimethylcyclo**hex-2-enol**

OH

1 **2**

4 **3**

h. 3,5-diethyl**hex-5-en-3-ol**

HO

5 **3**

10.40

a.

(2*E*,4*S*)-4-methylnon-2-ene
A

(2*E*,4*R*)-4-methylnon-2-ene
B

(2*Z*,4*S*)-4-methylnon-2-ene
C

(2*Z*,4*R*)-4-methylnon-2-ene
D

b. **A** and **B** are enantiomers. **C** and **D** are enantiomers.
c. Pairs of diastereomers: **A** and **C**, **A** and **D**, **B** and **C**, **B** and **D**.

10.41

a.

H

(1*E*,4*R*)-1,4-dimethylcyclodecene

b.

H

(1*E*,4*S*)-1,4-dimethylcyclodecene
enantiomer

c.

(1*Z*,4*S*)-1,4-dimethylcyclodecene
diastereomer

(1*Z*,4*R*)-1,4-dimethylcyclodecene
diastereomer

10.42 Name the alkene from which the epoxide can be derived and add the word *oxide*.

a.

derived from →

1-ethyl

6 carbon ring
cyclohexene
1-ethylcyclohexene

1-ethylcyclohexene oxide

b.

derived from →

2-methyl

6 carbon chain
hexene
2-methylhex-2-ene

2-methylhex-2-ene oxide

c.

derived from →

7 carbon chain
heptene
(*E*)-hept-3-ene
or
***trans*-hept-3-ene**

(*E*)-hept-3-ene oxide
or
***trans*-hept-3-ene oxide**

d.

derived from →

5 carbon ring
cyclopentene
4-*tert*-butylcyclopentene

4-*tert*-butylcyclopentene oxide

10.43 a, b. *E,Z* and *R,S* designations are shown.

c. Nine double bonds that can be *E* or *Z*
Six tetrahedral stereogenic centers
Maximum possible number of stereoisomers = 2^{15}

10.44

stearic acid — **highest melting point**
no double bonds

elaidic acid — **intermediate melting point**
one *E* double bond

oleic acid — **lowest melting point**
one *Z* double bond

10.45

eleostearic acid

a. all trans double bonds
higher melting point

b. all cis double bonds
lower melting point

10.46

a. $\xrightarrow{\text{HBr}}$ (structure with Br)

b. $\xrightarrow[\text{H}_2\text{SO}_4]{\text{H}_2\text{O}}$ (structure with OH)

c. $\xrightarrow[\text{H}_2\text{SO}_4]{\text{CH}_3\text{CH}_2\text{OH}}$ (structure with O)

d. $\xrightarrow{\text{Cl}_2}$ (structure with Cl, Cl)

e. $\xrightarrow{\text{Br}_2, \text{H}_2\text{O}}$ (structure with Br, OH)

f. $\xrightarrow[\text{aqueous DMSO}]{\text{NBS}}$ (structure with Br, OH)

g. $\xrightarrow[\text{[2] H}_2\text{O}_2, \text{HO}^-]{\text{[1] BH}_3}$ (structure with OH)

10.47

a. (structure with Cl, Cl) ← (cyclohexene)

b. (structure with Br) ← (structure)

c. (structure with Cl) ← (structure) or (structure)

d. (structure with Br) ← (structure)

10.48 Hydroboration–oxidation results in addition of an OH group on the *less* substituted carbon, whereas acid-catalyzed addition of H_2O results in the addition of an OH group on the *more* substituted carbon.

a. (structure with OH) ← (structure)

hydroboration–oxidation

b. (structure with OH) ← (structure)

acid-catalyzed addition

or (structure)

c. (structure with OH) ← (structure)

hydroboration–oxidation

d. (structure with OH)

Both methods would give
product mixtures.

10.49

a.
HCl

H adds here
to less substituted C.

b.
H_2O

H_2SO_4

H adds here
to less substituted C.

c.
[1] BH_3

[2] H_2O_2, HO⁻

BH_3 adds here
to less substituted C.

d.
Br_2

H_2O

Br adds here
to less substituted C.

e.
[1] 9-BBN

[2] H_2O_2, HO⁻

OH adds here
to less substituted C.

f.
Br_2

10.50

or   or

(or *Z* isomer)   (or *E* isomer)

10.51

a.
Br_2

b.
Cl_2

H_2O

c.
NBS

DMSO, H_2O

10.52

a.
H_2O

H_2SO_4

b.
HI

c. only anti addition

d. only syn addition

e. only anti addition

f.

10.53 The alkene that forms the more stable carbocation when H$^+$ is added reacts faster. Both alkenes form 2° carbocations, but the carbocation from **B** is also resonance stabilized by the benzene ring. As a result, **B** reacts faster.

2° carbocation

A

B

+ 3 more resonance structures

10.54 Draw each reaction. (a) The cis isomer of oct-4-ene gives two enantiomers on addition of Br$_2$. (b) The trans isomer gives a meso compound.

a.

cis-oct-4-ene

(4R,5R)-4,5-dibromooctane   (4S,5S)-4,5-dibromooctane

enantiomers

b.

trans-oct-4-ene

(4R,5S)-4,5-dibromooctane

meso compound

10.55

By protonation of the alkene, the cis and trans isomers produce **identical** carbocation intermediates.

Both *cis*- and *trans*-hex-3-ene give the same racemic mixture of products, so the reaction is not stereospecific.

10.56

2° carbocation
+ Br⁻

3° carbocation

10.57

2° carbocation

3° carbocation

+ H_2SO_4

10.58

10.59

10.60 The isomerization reaction occurs by protonation and deprotonation.

2,3-dimethylbut-1-ene   2,3-dimethylbut-2-ene

10.61

Since two resonance structures can be drawn for the intermediate carbocation, two different products result from attack by Br⁻.

10.62

This carbocation is resonance stabilized by the O atom, and therefore preferentially forms and results in **B**.

This carbocation is formed preferentially and results in product **D**. It is not destabilized by an adjacent electron-withdrawing COOCH$_3$ group.

This carbocation is destabilized by the δ+ on the adjacent C, so it does not form.

10.63

10.64

a.

POCl$_3$, pyridine

OH adds to **more** substituted C.

b.

c.

d.

e.

f.

10.65

a.

b.

c.

(from b.)

d.

(from a.)

10.66 Having two rings joined together as in **A** and **B** creates a very rigid ring system and constrains bond angles. Evidently the bond angles around the C=C in **A** are close enough to the trigonal planar bond angle of 120°, so **A** is stable. With **B**, however, the C=C is located at a carbon shared by both rings and the bond angles around the C=C deviate greatly from the desired angle, so **B** is *not* a stable compound.

10.67 a. Br₂ adds in an anti fashion to form a meso dibromide. Rotate around the C–C bond to place H and Br anti periplanar in the second step. HBr can be eliminated in two ways, but both give the same product.

b. Br₂ addition forms two enantiomers. Anti periplanar elimination of H and Br gives the same alkene from both compounds.

c. The products in (a) and (b) are diastereomers.

10.68

10.69

10.70

10.71

nerol

10.72

Chapter 11 Alkynes

Chapter Review

General facts about alkynes

- Alkynes contain a carbon–carbon triple bond consisting of a strong σ bond and two weak π bonds. Each carbon is *sp* hybridized and linear (11.1).

$$H-C \equiv C-H$$
acetylene

180°

sp hybridized

- Alkynes are named using the suffix **-yne** (11.2).
- Alkynes have weak intermolecular forces, giving them low mp's and low bp's, and making them water insoluble (11.3).
- Since its weaker π bonds make an alkyne electron rich, alkynes undergo addition reactions with electrophiles (11.6).

Addition reactions of alkynes

[1] Hydrohalogenation—Addition of HX (X = Cl, Br, or I) (11.7)

$$R-\!\!\!\equiv \xrightarrow[\text{(2 equiv)}]{H-X}$$

geminal dihalide

- Markovnikov's rule is followed. H bonds to the *less* substituted C in order to form the more stable carbocation.

[2] Halogenation—Addition of X_2 (X = Cl or Br) (11.8)

$$R-\!\!\!\equiv \xrightarrow[\text{(2 equiv)}]{X-X}$$

tetrahalide

- Bridged halonium ions are formed as intermediates.
- Anti addition of X_2 occurs.

[3] Hydration—Addition of H_2O (11.9)

$$R-\!\!\!\equiv \xrightarrow[\substack{H_2SO_4 \\ HgSO_4}]{H_2O}$$

enol

ketone

- Markovnikov's rule is followed. H bonds to the *less* substituted C in order to form the more stable carbocation.
- The unstable enol that is first formed rearranges to a carbonyl group.

[4] Hydroboration–oxidation—Addition of H_2O (11.10)

$$R-\!\!\!\equiv \xrightarrow[\text{[2] } H_2O_2,\ HO^-]{\text{[1] } R_2BH} \left[\underset{\substack{\text{OH} \\ \text{enol}}}{R}\!\!\diagup\!\!\diagdown^{H} \rightleftharpoons \underset{\substack{O \\ \text{aldehyde}}}{R}\!\!\diagup\!\!\diagdown^{H} \right]$$

- The unstable enol, first formed after oxidation, rearranges to a carbonyl group.

Reactions involving acetylide anions

[1] Formation of acetylide anions from terminal alkynes (11.6B)

$$R-C\equiv C-H \ + \ :B \ \rightleftharpoons \ \boxed{R-C\equiv C:^-} \ + \ HB^+$$

- Typical bases used for the reaction are $NaNH_2$ and NaH.

[2] Reaction of acetylide anions with alkyl halides (11.11A)

$$H-C\equiv C:^- \ + \ R-X \ \longrightarrow \ H-C\equiv C-R \ + \ X^-$$

- The reaction follows an S_N2 mechanism.
- The reaction works best with CH_3X and RCH_2X.

[3] Reaction of acetylide anions with epoxides (11.11B)

$$H-C\equiv C:^- \ \xrightarrow[\text{[2] } H_2O]{\text{[1] } \triangle^O} \ \boxed{H-C\equiv C-CH_2CH_2OH}$$

- The reaction follows an S_N2 mechanism.
- Ring opening occurs from the back side at the *less* substituted end of the epoxide.

Practice Test on Chapter Review

1. Draw the structure for the compound with the following IUPAC name: 5-*tert*-butyl-6,6-dimethyl-non-3-yne.

2.a. Which of the following compounds is an enol tautomer of compound **A**?

4. **A** can be a tautomer of both compounds (1) and (2).
5. **A** can be a tautomer of compounds (1), (2), and (3).

b. Which of the following bases is strong enough to deprotonate $CH_3C \equiv CH$ (propyne, $pK_a = 25$)? The pK_a's of the conjugate acids of the bases are given in parentheses.

1. CH_3Li ($pK_a = 50$)
2. $NaOCH_3$ ($pK_a = 15.5$)
3. $NaOCOCH_3$ ($pK_a = 4.8$)
4. The bases in (1) and (2) are both strong enough.
5. The bases in (1), (2), and (3) are all strong enough.

3. Draw the organic products formed in the following reactions.

a.

[1] R_2BH

[2] H_2O_2, ^-OH

b.

[1] NaH

[2] (epoxide)

[3] H_2O

c.

HO H
D

[1] TsCl, pyridine

[2] $^-C \equiv CH$

d.

H_2O
H_2SO_4
$HgSO_4$

e.

2 HBr

4. Draw two different enol tautomers for the following compound.

5. What acetylide anion and alkyl halide are needed to make the following alkyne?

Answers to Practice Test

1.

2.a. 2
 b. 1

3.

a.

b.

c. HC≡C D

d.

e. Br Br

4.

(E + Z)

(E + Z)

5.

CH_3X +

Answers to Problems

11.1

nepheliosyne B

a. The most acidic H is part of the COOH group of the carboxylic acid.
b. The shortest C–C σ bond is formed from C atoms with a higher percent *s*-character. The labeled C–C bond is formed from an *sp* hybridized orbital and an *sp²* hybridized orbital.
c. Nepheliosyne B has four triple bonds and five double bonds = 13 degrees of unsaturation.
d. Six bonds (labeled with x) are formed from C*sp*–C*sp³*.
e. All triple bonds are internal except the C≡C at the end of the chain, which is boxed in.

11.2 To name an alkyne:
[1] Find the longest chain that contains both atoms of the triple bond, change the *-ane* ending of the parent name to *-yne*, and number the chain to give the first carbon of the triple bond the lower number.
[2] Name all substituents following the other rules of nomenclature.

a.

4,4-dipropylhept-1-yne

c.

(Number to give the lower number to the first site of unsaturation.)

4-ethyl-7,7-dimethyldec-1-en-5-yne

b.

5-chloro-5,6-dimethyloct-3-yne

d.

(The longest chain must contain both functional groups.)

3-isopropylocta-1,5-diyne

11.3 To work backwards from a name to a structure:
[1] Find the parent name and the functional group.
[2] Add the substituents to the appropriate carbon.

a. *trans*-2-ethynyl**cyclopentanol**
   5 C ring with OH at C1

OH ◄——— OH on C1

⋯C≡CH ◄——— ethynyl at C2   or

b. 4-*tert*-butyl**dec-5-yne**

    10 C chain with
      a triple bond

←*tert*-butyl at C4

triple bond at C5

c. 3,3,5-trimethyl**cyclononyne**

    9 C ring with a
   triple bond at C1

triple bond at C1

11.4 Two factors cause the boiling point increase. The linear *sp* hybridized C's of the alkyne allow for more van der Waals attraction between alkyne molecules. Also, a triple bond is more polarizable than a double bond, which increases the van der Waals forces between two alkyne molecules as well.

11.5 To convert an alkene to an alkyne:
    [1] Make a vicinal dihalide from the alkene by addition of X_2.
    [2] Add base to remove two equivalents of HX and form the alkyne.

a. (Br₂CH... with Na⁺ ⁻NH₂ → [Br alkene, not isolated] → Na⁺ ⁻NH₂ → alkyne)

b. (Cl-alkene with Na⁺ ⁻NH₂ → alkyne)

c. (alkene + Cl₂ → Cl,Cl dihalide → Na⁺ ⁻NH₂ (2 equiv) → alkyne)

11.6 Acetylene has a pK_a of 25, so **bases having a conjugate acid with a pK_a *above* 25** will be able to deprotonate it.

a. CH_3NH^- [pK_a (CH_3NH_2) = 40]
   $pK_a > 25 =$ **Can deprotonate acetylene.**

b. CO_3^{2-} [pK_a (HCO_3^-) = 10.2]
   $pK_a < 25 =$ **Cannot deprotonate acetylene.**

c. $CH_2=CH^-$ [pK_a $(CH_2=CH_2)$ = 44]
   $pK_a > 25 =$ **Can deprotonate acetylene.**

d. $(CH_3)_3CO^-$ {pK_a [$(CH_3)_3COH$] = 18}
   $pK_a < 25 =$ **Cannot deprotonate acetylene.**

11.7 To draw the products of reactions with HX:
    • Add two moles of HX to the triple bond, following Markovnikov's rule.
    • Both X's end up on the more substituted C.

a.
2 HBr

b.
2 HBr

c.
2 HBr

11.8

a.

c.

b.

11.9 Addition of one equivalent of X_2 to alkynes forms trans dihalides. Addition of two equivalents of X_2 to alkynes forms tetrahalides.

a.
2 Br$_2$

b.
Cl$_2$ → **trans dihalide**

11.10

Cl$_2$ →

The two Cl atoms are electron withdrawing, making the π bond less electron rich and therefore less reactive with an electrophile.

11.11 To draw the keto form of each enol:

[1] Change the C–OH to a C=O at one end of the double bond.
[2] Add a proton at the other end of the double bond.

a.
new C–H bond

c.
new C–H bond

b.
new C–H bond

11.12 The treatment of alkynes with H_2O, H_2SO_4, and $HgSO_4$ yields ketones.

$$\xrightarrow[\text{H}_2\text{SO}_4,\ \text{HgSO}_4]{\text{H}_2\text{O}}$$

Two enols form.
(*E* and *Z* isomers)

two ketones after
tautomerization

11.13

a.

enol tautomers

b.

constitutional isomers,
but not tautomers

11.14 Reaction with H_2O, H_2SO_4, and $HgSO_4$ adds the oxygen to the *more* substituted carbon. Reaction with [1] R_2BH, [2] H_2O_2, ⁻OH adds the oxygen to the *less* substituted carbon.

a.
$$\xrightarrow[\text{H}_2\text{SO}_4,\ \text{HgSO}_4]{\text{H}_2\text{O}}$$
Forms a **ketone.** H_2O is added with the O atom on the *more* substituted carbon.

$$\xrightarrow[\text{[2] H}_2\text{O}_2,\ \text{HO}^-]{\text{[1] R}_2\text{BH}}$$
Forms an **aldehyde.** H_2O is added with the O atom on the *less* substituted carbon.

b.
$$\xrightarrow[\text{H}_2\text{SO}_4,\ \text{HgSO}_4]{\text{H}_2\text{O}}$$
Forms a **ketone.** H_2O is added with the O atom on the *more* substituted carbon.

$$\xrightarrow[\text{[2] H}_2\text{O}_2,\ \text{HO}^-]{\text{[1] R}_2\text{BH}}$$
Forms an **aldehyde.** H_2O is added with the O atom on the *less* substituted carbon.

11.15

a. $H-C\equiv C-H$ $\xrightarrow{\text{[1] NaH}}$ $H-C\equiv C:^-$ + H_2 $\xrightarrow{\text{[2]}}$ + NaCl

b. $\xrightarrow{\text{[1] NaH}}$ $C\equiv C:^-$ + H_2 $\xrightarrow{\text{[2]}}$ + NaBr

1° alkyl halide
substitution product

$\xrightarrow{\text{[1] NaNH}_2}$ $C\equiv C:^-$ + NH_3 $\xrightarrow{\text{[2]}}$ + NaCl

+

3° alkyl halide
elimination product

11.16

a. terminal alkyne only one possibility

b. internal alkyne two possibilities

c. internal alkyne only one possibility

The 3° alkyl halide would undergo elimination.

11.17

11.18

2,2,5,5-tetramethylhex-3-yne

The 3° alkyl halide is too crowded for nucleophilic substitution. Instead, it would undergo elimination with the acetylide anion.

11.19

a.

[1] ⁻:C≡C–H
[2] H₂O

Epoxide is drawn up, so the **acetylide anion attacks from *below* at *less* substituted C.**

b.

[1] ⁻:C≡C–H
[2] H₂O

enantiomers

Backside attack of the nucleophile (⁻C≡CH) occurs at either C because both ends are equally substituted.

11.20

a.

b.

c.

d.

e.

f.

11.21 To use a retrosynthetic analysis:

[1] **Count the number of carbon atoms** in the starting material and product.

[2] **Look at the functional groups** in the starting material and product.

- Determine what types of reactions can form the product.
- Determine what types of reactions the starting material can undergo.

[3] **Work backwards** from the product to make the starting material.

[4] Write out the synthesis in the synthetic direction.

$$\Longrightarrow \qquad HC\equiv CH$$

6 C's   2 C's

$$\Longrightarrow \qquad + \qquad Br \qquad \Longrightarrow \qquad HC\equiv C^- \quad + \qquad Br$$

$$HC\equiv C-H \xrightarrow{Na^+H:^-} HC\equiv C:^- \xrightarrow{Br} \xrightarrow{\quad} \xrightarrow[H]{} \xrightarrow{Na^+H:^-} \xrightarrow{Br} \xrightarrow{\quad}$$

11.22

product:
4 carbons, aldehyde functional group
(can be made by hydroboration–oxidation of
a terminal alkyne)

$$\Longrightarrow \qquad HC\equiv CH$$

starting material:
2 carbons, C≡C functional group
(can form an acetylide
anion by reaction with NaH)

Retrosynthetic
analysis:

$$\Longrightarrow \qquad \Longrightarrow \qquad H-C\equiv C-H$$

Forward direction:

$$H-C\equiv C-H \xrightarrow{Na^+H:^-} :C\equiv C-H \xrightarrow{Br} \xrightarrow[\text{[2] } H_2O_2,\ HO^-]{\text{[1] } R_2BH}$$

11.23

a.

6 C alkyne → hexyne
C≡C at C1 → hex-1-yne
CH_3 and CH_2CH_3 at C3
3-ethyl-3-methylhex-1-yne

hex-1-yne

b.

12 C chain with 2 C≡C's → dodecadiyne
C≡C's at C3 and C5 → dodeca-3,5-diyne
CH_3 at C11
11-methyldodeca-3,5-diyne

11.24

a.

keto form   enol form

b.

enol form   keto form

11.25

a, b.

erlotinib

most acidic C–H proton

shortest C–C single bond
C_{sp}–C_{sp^2}

c.
d. * = *sp* hybridized C
e. 3 < 1 < 2

(1)   (2)

phomallenic acid C

(3)

most acidic proton

11.26 Use the rules from Answer 11.2 to name the alkynes.

a.
5-ethyl-2-methylhept-3-yne

d.
octa-2,5-diyne

b.
3-ethyl
non-4-yne
6-ethyl
7-methyl
3,6-diethyl-7-methylnon-4-yne

e.
5-ethyl
4-ethyl
(E)-4,5-diethyldec-2-en-6-yne

c.
hex-1-yne
3-ethyl
3-ethylhex-1-yne

f.
6-methyl
ethynyl
1-ethynyl-6-methylcyclohexene

11.27 Use the directions from Answer 11.3 to draw each structure.

a. 5,6-dimethyl**hept-2-yne**

hept-2-yne →

b. 5-*tert*-butyl-6,6-dimethyl**non-3-yne**

6,6-dimethyl

non-3-yne

5-*tert*-butyl

c. (*S*)-4-chloro**pent-2-yne**

Cl ← 4-chloro
       S configuration

pent-2-yne

d. *cis*-1-ethynyl-2-methyl**cyclopentane**

1-ethynyl

C≡CH   or   C≡CH

2-methyl

e. 3,4-dimethyl**octa-1,5-diyne**

3,4-dimethyl

1

diyne   5

diyne

f. (*Z*)-6-methyl**oct-6-en-1-yne**

6-methyl

Z

1

11.28 Keto–enol tautomers are constitutional isomers in equilibrium that differ in the location of a double bond and a hydrogen. The OH in an enol must be bonded to a C=C.

a.
OH

and

• C=C
• OH on C=C

• C=O
• one more CH bond

keto–enol tautomers

b.
O

and

OH

OH is not bonded
to the C=C.

NOT keto–enol tautomers

c.
OH

and

H
O

• C=C
• OH on C=C

• C=O
• one more CH bond

keto–enol tautomers

d.
O

and

OH ← OH is not bonded
to the C=C.

NOT keto–enol tautomers

11.29 **To draw the enol form of each keto form:** [1] Change the C=O to a C–OH. [2] Change one single C–C bond to a double bond, making sure the OH group is bonded to the C=C. Use the directions from Answer 11.11 to draw each keto form.

a.
O
H

→

OH
H

b.
O

→

OH

+

OH

c.
OH

→

O
H

d.
OH

→

O

11.30 Tautomers are constitutional isomers that are in equilibrium and differ in the location of a double bond and a hydrogen atom.

11.31

11.32

11.33

11.34

a.

$$H_2O$$
$$H_2SO_4, HgSO_4$$

b.

[1] R$_2$BH
[2] H$_2$O$_2$, HO$^-$

c.

HCl
(2 equiv)

d.

[1] NaH
[2] CH$_3$CH$_2$Br

11.35 Reaction rate (which is determined by E_a) and enthalpy ($\Delta H°$) are not related. More exothermic reactions are not necessarily faster. Because the addition of HX to an alkene forms a more stable carbocation in an endothermic, rate-determining step, this carbocation is formed faster by the Hammond postulate.

R–C≡C–R'   HX

sp hybridized carbocation
less stable
slower reaction

sp² hybridized carbocation
more stable
faster reaction

11.36

a.

b.

c.

d.

11.37

a.

b.

11.38

a.

b.

c.

d.

e. $HC\equiv C^- + D_2O \longrightarrow HC\equiv CD + DO^-$

f.

g.

h.

i.

j.

11.39

11.40

most acidic H

[1] NaNH₂
[2] CH₃I

NaNH₂ will remove the proton from the OH because it is more acidic.

11.41

a. stereogenic center at the site of reaction — inversion

b. stereogenic center NOT at the site of reaction — Configuration is retained.

c. identical

d. enantiomers

11.42

A — TsCl / pyridine — retention — HC≡C⁻ / S_N2 — inversion — B

A — PBr₃ / S_N2 — inversion — HC≡C⁻ / S_N2 — inversion — C

11.43

a.

b.

[1] **D** = HC≡C⁻
[2] H₂O

These 2 C's are added.

E = **TsCl, pyridine**

new C–C bond **G**

[1] NaH
[2] CH₃I

[1] NaH
[2] **F** = CH₃CH₂Br

11.44

but-1-yne

major product
more substituted alkyne

but-2-yne

(*E* and *Z* isomers possible)

CH₃CH=C=CH₂

buta-1,2-diene

Reaction by-products:

2 N̈H₃ + 2 Br⁻

11.45 A carbanion is more stable when its lone pair is in an orbital with a higher percentage of the smaller *s* orbital. A carbocation is more stable when its positive charge is due to a vacant orbital with a lower percentage of the smaller *s* orbital. In HC≡C⁺, the positively charged C uses two *p* orbitals to form two π bonds. If the σ bond is formed using an *sp* hybrid orbital, the second hybrid orbital would have to remain vacant, a highly unstable situation.

HC≡C⁻

sp hybridized
higher % *s*-character
more stable

CH₂=CH⁻

sp² hybridized
lower % *s*-character
less stable

HC≡C⁺

sp hybridized
Vacant orbital has 50% *s*-character.
less stable

CH₂=CH⁺

sp hybridized
Vacant orbital is a *p* orbital.
more stable

11.46

a.

b.

11.47

a.

b.

resonance structures

11.48

11.49

11.50 The alkyl halides must be methyl or 1°.

11.51

b. $HC\equiv C-H$ $\xrightarrow{Na^+H^-}$ $HC\equiv C^-$ → (with propyl chloride) → [boxed structure with alkyne -H] $\xrightarrow{NaH}$ (pent-1-ynyl anion) $^-$

→ (with isobutyl chloride, Cl) ↓ → (branched alkyne product)

c. (alkyne)-H (from b.) $\xrightarrow[\text{[2] H}_2\text{O}_2, \text{HO}^-]{\text{[1] R}_2\text{BH}}$ (aldehyde, H)

d. (alkyne)-H (from b.) $\xrightarrow[\substack{\text{H}_2\text{SO}_4 \\ \text{HgSO}_4}]{\text{H}_2\text{O}}$ (ketone)

e. (alkyne anion) $^-$ (from b.) → (with propyl chloride, Cl) → (internal alkyne) $\xrightarrow[\text{H}_2\text{SO}_4, \text{HgSO}_4]{\text{H}_2\text{O}}$ (ketone)

11.52

a. (butene) $\xrightarrow{Cl_2}$ (2,3-dichloro with Cl, Cl) $\xrightarrow{2\ ^-NH_2}$ (alkyne)

b. (alkyne) (from a.) $\xrightarrow[\text{(2 equiv)}]{HBr}$ (dibromide, Br Br)

c. (butene) $\xrightarrow{Br_2}$ (dibromide, Br, Br)

d. (alkyne) (from a.) $\xrightarrow{NaH}$ (alkyne anion $^-$) $\xrightarrow[\text{[2] H}_2\text{O}]{\text{[1] epoxide}}$ (alkynol, OH)

e. $^-$(alkyne) (from d.) $\xrightarrow[\text{[2] H}_2\text{O}]{\text{[1] cyclohexene oxide}}$ (cyclohexanol with OH and alkynyl substituent)

(+ enantiomer)

11.53

$HC\equiv CH$ ↓ NaH

(cyclohexene) $\xrightarrow[\text{H}_2\text{O}]{Br_2}$ (cyclohexane with OH and Br) $\xrightarrow{NaH}$ (cyclohexene oxide) :$C\equiv CH$ → (cyclohexane with O$^-$ and alkynyl) $\xrightarrow{H_2O}$ (cyclohexanol with OH and alkynyl)

+ enantiomer   + enantiomer

11.54

a. $HC\equiv CH$ —NaH→ $HC\equiv C^-$ → (octynyl chain) —NaH→ (alkynyl anion) —$\nearrow$Br→ (product)

b. (from a.) —[1] epoxide / [2] H_2O→ (alkynol, OH) —[1] NaH / [2] $\nearrow$Br→ (ether product)

11.55

$\nearrow$Br —$K^+\,{}^-OC(CH_3)_3$→ $CH_2=CH_2$ —Br_2→ BrCH₂CH₂Br —$NaNH_2$ (2 equiv)→ $HC\equiv CH$ —NaH→ $HC\equiv C^-$ —$\nearrow$Br→ (alkyne) —NaH→ (anion) —$\nearrow$Br→ (internal alkyne) —H_2O, H_2SO_4, $HgSO_4$→ (ketone)

11.56

a. (ethanol) OH —H_2SO_4→ (alkene) —Br_2→ (dibromide, Br Br) —$2\,{}^-NH_2$→ (alkyne) —NaH→ (anion); (propanol) OH —$SOCl_2$→ (propyl Cl) → (internal alkyne)

b. (from a.) (alkene) —Cl_2 / H_2O→ (chlorohydrin, Cl OH) —NaH→ (epoxide) —[1] (from a.) / [2] H_2O→ (alkynol, HO)

11.57

(ethanol) OH —H_2SO_4→ $CH_2=CH_2$ —Br_2→ BrCH₂CH₂Br —$NaNH_2$ (2 equiv)→ $HC\equiv CH$ —NaH→ $HC\equiv C^-$; (ethanol) OH —PBr_3→ $\nearrow$Br → $HC\equiv C$— —[1] NaH / [2] epoxide / [3] H_2O→ (HO alkynol)

(ethanol) OH —H_2SO_4→ $CH_2=CH_2$ —Br_2 / H_2O→ (Br OH) —NaH→ (epoxide)

11.58 Two resonance structures can be drawn for an enol.

negative charge on C that is part of the enol C=C

Because the second resonance structure places an electron pair (and therefore a negative charge) on an enol carbon, the C=C is more nucleophilic than the C=C of an alkene for which no additional resonance forms can be drawn. Thus, the OH group donates electron density to the C=C by a resonance effect.

11.59

but-2-yne

11.60

11.61

11.62

Only this carbocation forms because it is resonance stabilized. The positive charge is delocalized on oxygen.

not resonance stabilized

(not formed)

re-draw

NOT

X   Y

11.63 A more stable internal alkyne can be isomerized to a less stable terminal alkyne under these reaction conditions because when $CH_3CH_2C{\equiv}CH$ is first formed, it contains an *sp* hybridized C–H bond, which is more acidic than any proton in $CH_3–C{\equiv}C–CH_3$. Under the reaction conditions, this proton is removed with base. Formation of the resulting acetylide anion drives the equilibrium to favor its formation. Protonation of this acetylide anion gives the less stable terminal alkyne.

but-2-yne

+ :NH₂

+ :NH₃

acetylide anion

+ :NH₃

but-1-yne

2,5-dimethylhex-3-yne

A

B

+ :NH₃

+ :NH₂

2,5-dimethylhexa-2,3-diene

In this case the reaction stops with formation of 2,5-dimethylhexa-2,3-diene because a terminal alkyne (with an acidic *sp* hybridized C–H bond) is not formed. Removal of the H in the diene re-forms the anion shown in resonance structures **A** and **B**.

11.64

+ H₂O

+ HCOOH

resonance structures

enol

11.65   In the presence of acid, (*R*)-α-methylbutyrophenone enolizes to form an achiral enol.

(*R*)-α-methylbutyrophenone

(+ 1 resonance structure)

(*E* and *Z* isomers) achiral enol

The achiral enol can then be protonated from above or below the plane to form a racemic mixture that is optically inactive.

below

above

+ H₃O⁺

+ H₃O⁺

racemic mixture

Chapter 12 Oxidation and Reduction

Chapter Review

Summary: Terms that describe reaction selectivity

- A **regioselective reaction** forms predominately or exclusively one constitutional isomer (Section 8.5).

- A **stereoselective reaction** forms predominately or exclusively one stereoisomer (Section 8.5).

- An **enantioselective reaction** forms predominately or exclusively one enantiomer (Section 12.15).

One enantiomer is favored.

Definitions of oxidation and reduction

Oxidation reactions result in:
- an increase in the number of C–Z bonds, *or*
- a decrease in the number of C–H bonds.

Reduction reactions result in:
- a decrease in the number of C–Z bonds, *or*
- an increase in the number of C–H bonds.

[Z = an element more electronegative than C]

Reduction reactions

[1] Reduction of alkenes—Catalytic hydrogenation (12.3)

- **Syn addition** of H_2 occurs.
- Increasing alkyl substitution on the C=C decreases the rate of reaction.

[2] Reduction of alkynes

a. $R-\!\!\!\equiv\!\!\!-R \xrightarrow[\text{Pd-C}]{2\,H_2}$

alkane

- Two equivalents of H_2 are added and four new C–H bonds are formed (12.5A).

b. $R-\!\!\!\equiv\!\!\!-R \xrightarrow[\substack{\text{Lindlar}\\\text{catalyst}}]{H_2}$

cis alkene

- **Syn addition** of H_2 occurs, forming a **cis** alkene (12.5B).
- The Lindlar catalyst is deactivated so that reaction stops after one equivalent of H_2 has been added.

c. $R-\!\!\!\equiv\!\!\!-R \xrightarrow[\text{NH}_3]{\text{Na}}$

trans alkene

- **Anti addition** of H_2 occurs, forming a **trans** alkene (12.5C).

[3] Reduction of alkyl halides (12.6)

$R-X \xrightarrow[\text{[2] }H_2O]{\text{[1] LiAlH}_4}$ R–H

alkane

- The reaction follows an S_N2 mechanism.
- CH_3X and RCH_2X react faster than more substituted RX.

[4] Reduction of epoxides (12.6)

$\xrightarrow[\text{[2] }H_2O]{\text{[1] LiAlH}_4}$

alcohol

- The reaction follows an S_N2 mechanism.
- In unsymmetrical epoxides, H^- (from LiAlH$_4$) attacks at the less substituted carbon.

Oxidation reactions

[1] Oxidation of alkenes

a. Epoxidation (12.8)

$\diagdown\!\!\diagup C\!=\!C\diagup\!\!\diagdown$ + $RCO_3H \longrightarrow$ epoxide

- The mechanism has **one step.**
- **Syn addition** of an O atom occurs.
- The reaction is stereospecific.

b. Anti dihydroxylation (12.9A)

$\diagdown\!\!\diagup C\!=\!C\diagup\!\!\diagdown$ $\xrightarrow[\text{[2] }H_2O\ (\text{H}^+\text{ or }HO^-)]{\text{[1] }RCO_3H}$

1,2-diol

- Ring opening of an epoxide intermediate with $^-$OH or H_2O forms a 1,2-diol with two OH groups added in an **anti** fashion.

c. Syn dihydroxylation (12.9B)

$$\text{alkene} \xrightarrow[\substack{\text{or} \\ [1]\ OsO_4, NMO;\ [2]\ NaHSO_3, H_2O \\ \text{or} \\ KMnO_4, H_2O, HO^-}]{[1]\ OsO_4;\ [2]\ NaHSO_3, H_2O} \text{1,2-diol}$$

• Each reagent adds two new C–O bonds to the C=C in a **syn** fashion.

d. Oxidative cleavage (12.10)

$$\xrightarrow[\substack{[2]\ Zn, H_2O\ \text{or} \\ CH_3SCH_3}]{[1]\ O_3}$$

ketone + aldehyde

• Both the σ and π bonds of the alkene are cleaved to form two carbonyl groups.

[2] Oxidative cleavage of alkynes (12.11)

a.   R————R' $\xrightarrow[\text{[2] } H_2O]{\text{[1] } O_3}$

internal alkyne

carboxylic acids

• The σ bond and both π bonds of the alkyne are cleaved.

b.   R————H $\xrightarrow[\text{[2] } H_2O]{\text{[1] } O_3}$ + CO_2

terminal alkyne

[3] Oxidation of alcohols (12.12, 12.13)

a.   R—OH $\xrightarrow[\substack{\text{or} \\ HCrO_4- \\ \text{Amberlyst A-26} \\ \text{resin}}]{PCC}$ aldehyde

1° alcohol

• Oxidation of a 1° alcohol with PCC or $HCrO_4^-$ (Amberlyst A-26 resin) stops at the aldehyde stage. Only one C–H bond is replaced by a C–O bond.

b.   R—OH $\xrightarrow[H_2SO_4, H_2O]{CrO_3}$ carboxylic acid

1° alcohol

• Oxidation of a 1° alcohol under harsher reaction conditions—CrO_3 (or $Na_2Cr_2O_7$ or $K_2Cr_2O_7$) + H_2O + H_2SO_4—affords a RCO_2H. Two C–H bonds are replaced by two C–O bonds.

c.   R—R $\xrightarrow[\substack{\text{or} \\ HCrO_4- \\ \text{Amberlyst A-26} \\ \text{resin}}]{PCC\ \text{or}\ CrO_3}$ ketone

2° alcohol

• A 2° alcohol has only one C–H bond on the carbon bearing the OH group, so all Cr^{6+} reagents—PCC, CrO_3, $Na_2Cr_2O_7$, $K_2Cr_2O_7$, or $HCrO_4^-$ (Amberlyst A-26 resin)—oxidize a 2° alcohol to a ketone.

[4] Asymmetric epoxidation of allylic alcohols (12.15)

with (–)-DET   or   with (+)-DET

Practice Test on Chapter Review

1.a. Compound **X** has a molecular formula of C_9H_{12} and contains no triple bonds. **X** is hydrogenated to a compound of molecular formula C_9H_{14} with excess H_2 and a palladium catalyst. What can be said about **X**?

1. **X** has four rings.
2. **X** has three rings and one double bond.
3. **X** has two rings and two double bonds.

4. **X** has one ring and three double bonds.
5. **X** has four double bonds.

b. Syn addition to an alkene occurs exclusively with which reagents?

1. OsO_4
2. $KMnO_4$, H_2O, ^-OH
3. mCPBA, then H_2O, ^-OH
4. Both reagents (1) and (2) give syn addition exclusively.
5. Reagents (1), (2), and (3) give syn addition exclusively.

c. Which of the following reagents adds to an alkene exclusively in an anti fashion?

1. Br_2
2. H_2, Pd-C
3. BH_3, then H_2O_2, ^-OH

4. Reagents (1) and (2) both add in an anti fashion.
5. Reagents (1), (2), and (3) all add in an anti fashion.

2. Label each statement as True (T) or False (F).

a. PCC oxidizes 1° alcohols to aldehydes.
b. CrO_3 oxidizes 2° alcohols to ketones.
c. Treatment of hex-2-yne with Na in NH_3 forms *cis*-hex-2-ene.
d. Reduction of propene oxide with $LiAlH_4$ forms propan-1-ol.
e. Ozonolysis of 2-methyloct-2-ene forms one ketone and one aldehyde.
f. mCPBA is an oxidizing agent that converts alkenes to trans diols.
g. Oct-1-en-5-yne reacts with H_2 and Pd-C, but does not react with H_2 and Lindlar catalyst.
h. Treatment of cyclohexene with OsO_4 affords an optically inactive product mixture that contains two enantiomers.

3. Label each reagent as an oxidizing agent, reducing agent, or neither.

a. O_3
b. $LiAlH_4$
c. mCPBA
d. H_2O, H_2SO_4
e. PCC
f. Na, NH_3

4. Draw the organic products formed in each reaction and indicate stereochemistry when necessary.

a. $\xrightarrow[\text{NH}_3]{\text{Na}}$

b. $\xrightarrow[\text{Pd–C}]{\text{H}_2}$

c. $\xrightarrow{\text{PCC}}$

d. $\xrightarrow[\text{[2] LiAlH}_4]{\text{[1] SOCl}_2}$

5 a. Fill in the appropriate starting material (including any needed stereochemistry) in the following reaction.

$\xrightarrow[\text{[2] H}_2\text{O, NaHSO}_3]{\text{[1] OsO}_4}$ + enantiomer

b. Fill in the appropriate reagent in the following reaction.

c. What starting material is needed for the following reaction?

$\xrightarrow[\text{[2] (CH}_3)_2\text{S}]{\text{[1] O}_3}$

Answers to Practice Test

1.a. 2   2.a. T   3.a. oxidizing   4.   5.
  b. 4   b. T   b. reducing
  c. 1   c. F   c. oxidizing
            d. F   d. neither
            e. T   e. oxidizing
            f. F   f. reducing
            g. F
            h. F

a.

a.

b.

+

b. Sharpless reagent
   (+)-DET

c.

c.

d.

Chapter 12: Answers to Problems

12.1 *Oxidation* results in an *increase* in the number of C–Z bonds (usually C–O bonds) *or* a *decrease* in the number of C–H bonds.
Reduction results in a *decrease* in the number of C–Z bonds (usually C–O bonds) *or* an *increase* in the number of C–H bonds.

a.   **reduction**

c.   **oxidation**
                                (two new C–H bonds,
                                three new C–O bonds)

b.   **oxidation**

d.   **neither**
                                (one new C–H bond and
                                one new C–Cl bond)

12.2 **Hydrogenation** is the addition of hydrogen. When alkenes are hydrogenated, they are *reduced* by the addition of H_2 to the π bond. To draw the alkane product, add a H to each C of the double bond.

a.

c.

b.

12.3 Draw the alkenes that form each alkane when hydrogenated.

12.4 Cis alkenes are less stable than trans alkenes, so they have larger heats of hydrogenation. Increasing alkyl substitution increases the stability of a C=C, thus decreasing the heat of hydrogenation.

a.

cis alkane
less stable
larger **heat of hydrogenation**

trans alkane

b.

trisubstituted

disubstituted
less stable
larger **heat of hydrogenation**

12.5 Hydrogenation products must be identical to use hydrogenation data to evaluate the relative stability of the starting materials.

2-methylpent-2-ene

Different products are formed.
Hydrogenation data can't be used to determine the relative stability of the starting materials.

3-methylpent-1-ene

12.6

a.

new stereogenic center

H

Two enantiomers are formed in equal amounts:

=   +

b.

H   +   H

diastereomers

c.

+

diastereomers

12.7

Compound	Molecular formula before hydrogenation	Molecular formula after hydrogenation	Number of rings	Number of π bonds
A	$C_{10}H_{12}$	$C_{10}H_{16}$	3	2
B	C_4H_8	C_4H_{10}	0	1
C	C_6H_8	C_6H_{12}	1	2

12.8

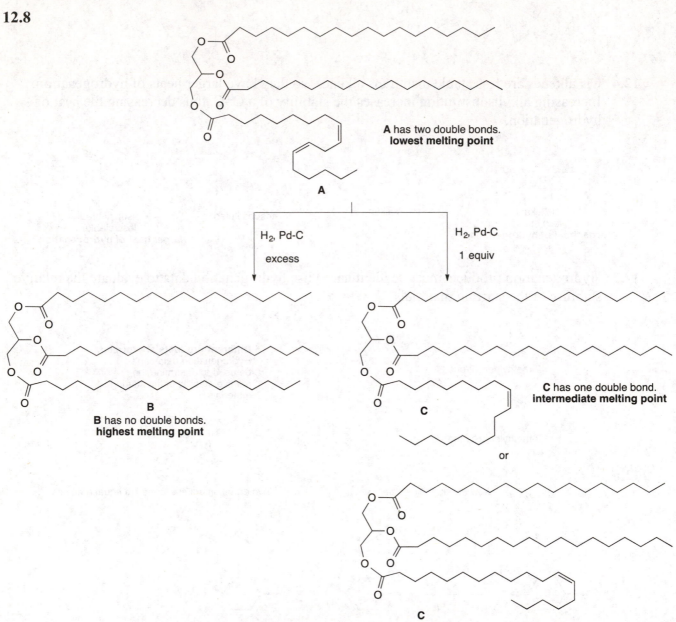

A has two double bonds.
lowest melting point

A

H$_2$, Pd-C

excess

H$_2$, Pd-C

1 equiv

B
B has no double bonds.
highest melting point

C has one double bond.
intermediate melting point

C

or

C

12.9 Hydrogenation of HC≡CCH$_2$CH$_2$CH$_3$ and CH$_3$C≡CCH$_2$CH$_3$ yields the same compound. The heat of hydrogenation is larger for HC≡CCH$_2$CH$_2$CH$_3$ than for CH$_3$C≡CCH$_2$CH$_3$ because internal alkynes are more stable (lower in energy) than terminal alkynes.

12.10 In the presence of the Lindlar catalyst, alkynes react with H_2 to form cis alkenes. Alkenes do not react with the Lindlar catalyst.

cis-jasmone

(perfume component isolated from jasmine flowers)

12.11 In the presence of Pd-C, H_2 adds to alkenes and alkynes to form alkanes. In the presence of the Lindlar catalyst, alkynes react with H_2 to form cis alkenes. Alkenes do not react with the Lindlar catalyst.

12.12 Use the directions from Answer 12.11.

12.13

12.14 LiAlH₄ reduces alkyl halides to alkanes and epoxides to alcohols.

a.   H replaces Cl.

b.

12.15 To draw the product, add an O atom across the π bond of the C=C.

a.

b.

c.

12.16 For epoxidation reactions:
- There are two possible products: O adds from above and below the double bond.
- Substituents on the C=C retain their original configuration in the products.

a. enantiomers

b. identical

c. enantiomers

12.17 Treatment of an alkene with a peroxyacid followed by H₂O, HO⁻ adds two hydroxy groups in an **anti** fashion. *cis*-But-2-ene and *trans*-but-2-ene yield different products of dihydroxylation. *cis*-But-2-ene gives a mixture of two enantiomers and *trans*-but-2-ene gives a meso compound. The reaction is stereospecific because two stereoisomeric starting materials give different products that are also stereoisomers of each other.

12.18 Treatment of an alkene with OsO_4 adds two hydroxy groups in a **syn** fashion. *cis*-But-2-ene and *trans*-but-2-ene yield different stereoisomers in this dihydroxylation, so the reaction is stereospecific.

12.19 To draw the oxidative cleavage products:
- **Locate all the π bonds** in the molecule.
- **Replace all C=C's with *two* C=O's**.

12.20 To find the alkene that yields the oxidative cleavage products:
- **Find the two carbonyl groups** in the products.
- **Join the two carbonyl carbons** together with a double bond. This is the double bond that was broken during ozonolysis.

12.21

a.

b.

c.

12.22 To draw the products of oxidative cleavage of alkynes:
- **Locate the triple bond.**
- For internal alkynes, **convert the *sp* hybridized C to COOH.**
- For terminal alkynes, the *sp* hybridized C–H becomes CO₂.

a.
internal alkyne

b.
internal alkyne
identical compounds

c.
terminal alkyne   internal alkyne

12.23

a.

b.

c.

d.

12.24 For the **oxidation of alcohols,** remember:
- **1° Alcohols** are oxidized to aldehydes with PCC.
- **1° Alcohols** are oxidized to carboxylic acids with oxidizing agents like CrO₃ or Na₂Cr₂O₇.
- **2° Alcohols** are oxidized to ketones with all Cr⁶⁺ reagents.

a. PCC

c. CrO₃, H₂SO₄, H₂O

b. PCC

d. CrO₃, H₂SO₄, H₂O

12.25 Upon treatment with $HCrO_4^-$–Amberlyst A-26 resin:
- **1° Alcohols** are oxidized to aldehydes.
- **2° Alcohols** are oxidized to ketones.

a. $HCrO_4^-$ / Amberlyst A-26 resin

c. $HCrO_4^-$ / Amberlyst A-26 resin

b. $HCrO_4^-$ / Amberlyst A-26 resin

12.26

a. NaOCl → product $+ NaCl + H_2O$

The by-products of the reaction with sodium hypochlorite are water and table salt (NaCl), as opposed to the by-products with $HCrO_4^-$– Amberlyst A-26 resin, which contain carcinogenic Cr^{3+} metal.

b. Oxidation with NaOCl has at least two advantages over oxidation with CrO_3, H_2SO_4 and H_2O. Because no Cr^{6+} is used as oxidant, there are no Cr by-products that must be disposed of. Also, CrO_3 oxidation is carried out in corrosive inorganic acids (H_2SO_4) and oxidation with NaOCl avoids this.

12.27 To draw the products of a **Sharpless epoxidation:**
- With the C=C horizontal, draw the allylic alcohol with the OH on the **top right** of the alkene.
- Add the new oxygen **above** the plane if (−)-DET is used and **below** the plane if (+)-DET is used.

a. $(CH_3)_3C-OOH$ / $Ti[OCH(CH_3)_2]_4$ / (+)-DET

(+)-DET adds O **below** the plane.

b. re-draw → $(CH_3)_3C-OOH$ / $Ti[OCH(CH_3)_2]_4$ / (−)-DET

(−)-DET adds O **above** the plane.

12.28 Sharpless epoxidation needs an *allylic alcohol* as the starting material. Alkenes with no allylic OH group will not undergo reaction with the Sharpless reagent.

This alkene is part of an **allylic alcohol** and will be epoxidized.

geraniol

This alkene is **not** part of an allylic alcohol and will not be epoxidized.

12.29

a.

$$A \xrightarrow[\text{Pd-C}]{H_2}$$

b.

$$A \xrightarrow{\text{mCPBA}}$$

c.

$$A \xrightarrow{\text{PCC}}$$ —CHO

d.

$$A \xrightarrow[\substack{H_2SO_4 \\ H_2O}]{CrO_3}$$ —COOH

e.

$$A \xrightarrow[\text{(+)-DET}]{\text{Sharpless reagent}}$$

12.30

$$\xrightarrow[\text{[2] CH}_3\text{SCH}_3]{\text{[1] O}_3}$$

12.31

$$HC \equiv CH \xrightarrow[\text{[2] CH}_3\text{CH}_2\text{Br}]{\text{[1] NaH}} \xrightarrow[\substack{\text{[2]} \\ \text{[3] H}_2\text{O}}]{\text{[1] NaH}} \xrightarrow[\text{NH}_3]{\text{Na}}$$

12.32 Use the rules from Answer 12.1.

a. (epoxycyclohexane) → (cyclohexanol, OH) **reduction**

b. HO—⟨benzene ring⟩—OH → O=⟨quinone⟩=O **oxidation**

c. (OCH₃, OCH₃ acetal) → (aldehyde with Br, H, O) **oxidation** (one new C–Br bond)

d. (OCH₃, OCH₃ cyclopentane) → (OCH₃ cyclopentene) **neither** (one fewer C–O bond and one fewer C–H bond)

12.33 Use the principles from Answer 12.2 and draw the products of syn addition of H₂ from above and below the C=C.

a. (alkene) $\xrightarrow{H_2 \text{, Pd-C}}$ (cis product) + (trans product)

b. (alkene) $\xrightarrow{H_2 \text{, Pd-C}}$ (cis product) + (trans product)

c. (alkene) $\xrightarrow{H_2 \text{, Pd-C}}$ (H product) + (H product)

12.34 Increasing alkyl substitution increases alkene stability, thus decreasing the heat of hydrogenation.

2-methylbut-2-ene
trisubstituted
smallest $\Delta H^o = -112$ kJ/mol

2-methylbut-1-ene
disubstituted
intermediate $\Delta H^o = -119$ kJ/mol

3-methylbut-1-ene
monosubstituted
largest $\Delta H^o = -127$ kJ/mol

12.35

A possible structure:

a. Compound **A**: molecular formula C_5H_8: hydrogenated to C_5H_{10}.
2 degrees of unsaturation, 1 is hydrogenated.
1 ring and 1 π bond ⎯⎯⎯⎯⎯⎯⎯⎯⎯⎯⎯→ (methylenecyclobutane)

b. Compound **B**: molecular formula $C_{10}H_{16}$: hydrogenated to $C_{10}H_{18}$.
3 degrees of unsaturation, 1 is hydrogenated.
2 rings and 1 π bond ⎯⎯⎯⎯⎯⎯⎯⎯⎯⎯⎯→ (bicyclic decalin-type alkene)

c. Compound **C**: molecular formula C_8H_8: hydrogenated to C_8H_{16}.
5 degrees of unsaturation, 4 are hydrogenated.
1 ring and 4 π bonds ⎯⎯⎯⎯⎯⎯⎯⎯→ (cyclooctatetraene)

12.36

a. **mono**substituted
   largest heat of hydrogenation
b. **fastest** reaction rate

c.
$$\xrightarrow[\text{[2] Zn, H}_2\text{O}]{\text{[1] O}_3}$$

A

a. **tetra**substituted
   smallest heat of hydrogenation
b. **slowest** reaction rate

c.
$$\xrightarrow[\text{[2] Zn, H}_2\text{O}]{\text{[1] O}_3}$$

identical

B

a. **tri**substituted
   intermediate heat of hydrogenation
b. **intermediate** reaction rate

c.
$$\xrightarrow[\text{[2] Zn, H}_2\text{O}]{\text{[1] O}_3}$$

C

12.37

a.

stearidonic acid $\xrightarrow[\text{Pd-C}]{\text{H}_2 \text{ (excess)}}$ stearic acid

b.
$\xrightarrow[\text{Pd-C}]{\text{H}_2 \text{ (1 equiv)}}$

c.

trans

one possibility

d.

stearidonic acid
4 cis C=C's

<

product in (b)
3 cis C=C's

<

product in (c)
2 cis and one trans C=C's

12.38

a. cyclopentene + H₂ / Pd-C → cyclopentane

b. cyclopentene + H₂ / Lindlar catalyst → no reaction

c. cyclopentene + Na / NH₃ → no reaction

d. cyclopentene + CH₃CO₃H → epoxide

e. cyclopentene + [1] CH₃CO₃H / [2] H₂O, HO⁻ → trans-1,2-cyclopentanediol (two enantiomers) **anti addition**

f. cyclopentene + [1] OsO₄ + NMO / [2] NaHSO₃, H₂O → cis-1,2-cyclopentanediol **syn addition**

g. cyclopentene + KMnO₄ / H₂O, HO⁻ → cis-1,2-cyclopentanediol **syn addition**

h. cyclopentene + [1] LiAlH₄ / [2] H₂O → no reaction

i. cyclopentene + [1] O₃ / [2] CH₃SCH₃ → pentanedial (OHC–CH₂CH₂CH₂–CHO)

j. cyclopentene + (CH₃)₃COOH / Ti[OCH(CH₃)₂]₄ / (−)-DET → no reaction

k. cyclopentene + mCPBA → epoxide

l. cyclopentene oxide + [1] LiAlH₄ / [2] H₂O → cyclopentanol

12.39

a. 3-methyl-2-penten-1-ol + H₂ / Pd-C → 3-methylpentan-1-ol (two enantiomers)

b. 3-methyl-2-penten-1-ol + mCPBA → epoxy alcohol (two diastereomers)

c. 3-methyl-2-penten-1-ol + PCC → 3-methyl-2-pentenal

d. 3-methyl-2-penten-1-ol + CrO₃ / H₂SO₄, H₂O → 3-methyl-2-pentenoic acid

e. 3-methyl-2-penten-1-ol + (CH₃)₃COOH / Ti[OCH(CH₃)₂]₄ / (+)-DET → epoxy alcohol

f. 3-methyl-2-penten-1-ol + (CH₃)₃COOH / Ti[OCH(CH₃)₂]₄ / (−)-DET → epoxy alcohol

g. 3-methyl-2-penten-1-ol + [1] PBr₃ → allylic bromide; then [2] LiAlH₄ / [3] H₂O → 3-methyl-2-pentene

h. 3-methyl-2-penten-1-ol + HCrO₄⁻ / Amberlyst A-26 → 3-methyl-2-pentenal

12.40

a.

b.

c.

d.

12.41

2° OH

(R)-glycerol phosphate

With NAD+ the 2° alcohol is oxidized to a ketone, and NAD+ is reduced to NADH.

12.42 LiAlH₄ attacks at the less substituted end of an unsymmetrical epoxide to form an alcohol with an OH on the more substituted carbon.

2-methylpentan-2-ol   or

Hydride attacks here.

12.43 The two sides of the C=C of **A** are different. D₂ adds only from above, so this side must be less sterically hindered. Other reagents will also add from the same side.

a.

    A   mCPBA

b.

    A   Br₂
            H₂O

H₂O

base

Since the OH of the bromohydrin is down, the resulting epoxide must also be down.

12.44 Alkenes treated with [1] OsO₄ followed by NaHSO₃ in H₂O will undergo **syn** addition, whereas alkenes treated with [2] CH₃CO₃H followed by ⁻OH in H₂O will undergo **anti** addition.

a. [1]

$$[1]\ OsO_4$$
$$[2]\ NaHSO_3,\ H_2O$$

[2]

$$[1]\ CH_3CO_3H$$
$$[2]\ ^-OH,\ H_2O$$
anti addition

rotate

b. [1]

$$[1]\ OsO_4$$
$$[2]\ NaHSO_3,\ H_2O$$
syn addition

rotate

+ enantiomer

[2]

$$[1]\ CH_3CO_3H$$
$$[2]\ ^-OH,\ H_2O$$

+ enantiomer

c. [1]

$$[1]\ OsO_4$$
$$[2]\ NaHSO_3,\ H_2O$$
syn addition

rotate

[2]

$$[1]\ CH_3CO_3H$$
$$[2]\ ^-OH,\ H_2O$$

12.45

+ AlH₃ + Li⁺

D⁻ (from LiAlD₄) opens the epoxide ring from the back side, so it is oriented on a wedge in the final product.

12.46

a.

b.

c. To form chiral **A**, Sharpless reagent with (+)-DET could be used.

12.47

12.48 Use the directions from Answer 12.19.

a.

b.

c.

d.

identical

12.49

a.

12.50 Use the directions from Answer 12.20.

12.51

12.52

a. **A** C_8H_{12} $\xrightarrow[\text{[2] } CH_3SCH_3]{\text{[1] } O_3}$

b. C_6H_{10} **B** $\xrightarrow[\text{Pd-C}]{H_2 \text{ (excess)}}$ $\xrightarrow[\text{[2] } CH_3I]{\text{[1] } NaNH_2}$ C_7H_{12} **C**

12.53

$C_{10}H_{16}$ $\xrightarrow[\text{Pd-C}]{H_2}$

3 degrees of unsaturation   2,6-dimethyloctane

The hydrogenation reaction tells you that both oximene and myrcene have 3 π bonds (and no rings). Use this carbon backbone and add in the double bonds based on the oxidative cleavage products.

Oximene: $(CH_3)_2C{=}O$ $CH_2{=}O$ $CH_2(CHO)_2$ $CH_3 \overset{O}{\underset{}{C}} CHO$ $\Longrightarrow$

Myrcene: $(CH_3)_2C{=}O$ $CH_2{=}O$ (2 equiv) $\Longrightarrow$

12.54

$C_{10}H_{18}O$ **A** $\xrightarrow{H_2SO_4}$ $C_{10}H_{16}$ **B**   + $C_{10}H_{16}$ **C** $\xrightarrow[\text{Pd-C}]{H_2}$ decalin

ozonolysis

D $C_{10}H_{16}O_2$ **E**

12.55   Hydrogenation of DHA forms $CH_3(CH_2)_{20}COOH$, so DHA is a 22-carbon fatty acid. The ozonolysis products show where the double bonds are located.

DHA

$\xrightarrow[\text{[2] Zn, } H_2O]{\text{[1] } O_3}$

+ +

(from portion **A**)   5 equiv (from portion **B**)   (from portion **C**)

12.56 The stereogenic center (labeled with *) in both structures can be *R* or *S*.

12.57

a.

re-draw

$(CH_3)_3COOH$
$Ti[OC(CH_3)_2]_4$
(–)-DET

b.

$(CH_3)_3COOH$
$Ti[OC(CH_3)_2]_4$
(+)-DET

12.58

re-draw

$(CH_3)_3COOH$
$Ti[OC(CH_3)_2]_4$
(–)-DET

major product
87%

+

minor product
13%

enantiomeric excess =
% one enantiomer – % second enantiomer

ee = 87% – 13% = **74%**

12.59

a.
Replace this O
to make an alkene.

(–)-DET

c.
Replace this O
to make an alkene.

(+)-DET

b.

(–)-DET

Replace this O
to make an alkene.

12.60

12.61

12.62

12.63

a.

b. $HC \equiv CH$ $\xrightarrow{\text{NaH}}$ $^-C \equiv CH$ $\xrightarrow{\text{Cl}\diagup}$ ⟶ $\xrightarrow{\text{NaH}}$ ⟶ $\xrightarrow{\text{Cl}\diagup}$ ⟶

$\xrightarrow[\text{Pd-C}]{\begin{array}{c}H_2\\(2\text{ equiv})\end{array}}$

c. $HC \equiv CH$ $\xrightarrow{\text{NaH}}$ $HC \equiv C^-$ $\xrightarrow{\text{CH}_3\text{Cl}}$ ⟶ $\xrightarrow{\text{NaH}}$ ⟶ $\xrightarrow{\text{CH}_3\text{Cl}}$ ⟶

$\xrightarrow[\text{Lindlar catalyst}]{H_2}$

$\xleftarrow{\text{mCPBA}}$

d. ⟶ $\xrightarrow{\text{Na, NH}_3}$ ⟶ $\xrightarrow[\text{H}_2\text{O, HO}^-]{\text{KMnO}_4}$ (+ enantiomer)

(from a.)

12.64

$\xrightarrow{\text{NaH}}$ $\xrightarrow[\text{Br}\diagdown\diagup\text{Br}]{}$

CH_3OH $\xrightarrow{\text{NaH}}$ CH_3O^-

$\xleftarrow{\text{PBr}_3}$ $\xleftarrow[\text{Pd-C}]{H_2}$

A

12.65

pent-1-ene $\xrightarrow{\text{Br}_2}$ $\xrightarrow{\text{2 NaNH}_2}$ $\xrightarrow{\text{NaH}}$

$\downarrow$ CH_3Cl

(E)-hex-2-ene $\xleftarrow[\text{NH}_3]{\text{Na}}$

12.66

a.

b.

(+ enantiomer)

c.

d.

(from c.)

12.67 Determine which 2 C's can come from HC≡CH and two alkyl halides that are unhindered (1° or CH₃X).

1-phenyl-5-methylhexane

In synthetic direction:

12.68

$$HC \equiv CH \xrightarrow{NaH} {}^-C \equiv CH \xrightarrow{\text{Br}} \text{(alkyne)} \xrightarrow{NaH} \text{(acetylide)} \xrightarrow{\text{Br}} \text{(internal alkyne)}$$

Na, NH₃ ↓

Cl₂
anti
addition

(3R,4S)-3,4-dichlorohexane

12.69

a.

$$\xrightarrow{OH} \xrightarrow{PBr_3} \xrightarrow{Br} \xrightarrow{K^+ \; ^-OC(CH_3)_3} CH_2=CH_2 \xrightarrow{Br_2} Br\text{—}Br \xrightarrow{2\,NaNH_2} HC \equiv CH$$

$$HC \equiv CH \xrightarrow{NaH} {}^-C \equiv CH \xrightarrow{\text{Br}} \text{(alkyne)} \xrightarrow{NaH} \text{(acetylide)} \xrightarrow{\text{Br}} \text{(internal alkyne)}$$

Na, NH₃ ↓

b.

(from a.) $\xrightarrow[\text{Lindlar catalyst}]{H_2}$ (cis alkene) $\xrightarrow{mCPBA}$ (epoxide) H ⟍ ◺ ⟋ H

c.

(from b.) $\xrightarrow[\text{[2] NaHSO}_3,\, H_2O]{\text{[1] OsO}_4}$ HO—OH $\xrightarrow{\text{rotate}}$ OH ... OH

12.70

12.71

The favored conformation
for both molecules places
the *tert*-butyl group equatorial.

A

A
This OH is axial and
will react **faster** because
the OH group is more
hindered.

B
This OH is equatorial
and will react **more slowly**
because the OH group
is less hindered.

B

12.72

R = alkyl group

mCPBA

12.73 The two OH's are added to opposite faces of the C=C, so anti addition occurs.

trans-but-2-ene

H_2O_2
HCO_2H

meso

cis-but-2-ene

H_2O_2
HCO_2H

two enantiomers

12.74

12.75

Chapter 13 Mass Spectrometry and Infrared Spectroscopy

Chapter Review

Mass spectrometry (MS)

- Mass spectrometry measures the molecular weight of a compound (13.1A).
- The mass of the molecular ion (**M**) = the molecular weight of a compound. Except for isotope peaks at M + 1 and M + 2, the molecular ion has the highest mass in a mass spectrum (13.1A).
- The base peak is the tallest peak in a mass spectrum (13.1A).
- A compound with an odd number of N atoms gives an odd molecular ion. A compound with an even number of N atoms (including zero) gives an even molecular ion (13.1B).
- Organic chlorides show two peaks for the molecular ion (M and M + 2) in a 3:1 ratio (13.2).
- Organic bromides show two peaks for the molecular ion (M and M + 2) in a 1:1 ratio (13.2).
- The fragmentation of radical cations formed in a mass spectrometer gives lower molecular weight fragments, often characteristic of a functional group (13.3).
- High-resolution mass spectrometry gives the molecular formula of a compound (13.4A).

Electromagnetic radiation

- The wavelength (λ) and frequency (ν) of electromagnetic radiation are *inversely* related by the following equations, where c is the speed of light: $\lambda = c/\nu$ or $\nu = c/\lambda$ (13.5).
- The energy (E) of a photon is proportional to its frequency; the higher the frequency, the higher the energy (h = Planck's constant): $\boldsymbol{E = h\nu}$ (13.5).

Infrared spectroscopy (IR, 13.6 and 13.7)

- Infrared spectroscopy identifies functional groups.
- IR absorptions are reported in wavenumbers:

$$\text{wavenumber} = \tilde{\nu} = 1/\lambda$$

- The functional group region from **4000–1500 cm^{-1}** is the most useful region of an IR spectrum.
- C–H, O–H, and N–H bonds absorb at high frequency, ≥ 2500 cm^{-1}.
- As bond strength increases, the wavenumber of an absorption increases; thus triple bonds absorb at higher wavenumber than double bonds.

C=C   C≡C

~1650 cm^{-1} ~2250 cm^{-1}

→

Increasing bond strength
Increasing $\tilde{\nu}$

- The higher the percent *s*-character, the stronger the bond, and the higher the wavenumber of an IR absorption.

C_{sp^3}–H	C_{sp^2}–H	C_{sp}–H
25% *s*-character	33% *s*-character	50% *s*-character
3000–2850 cm^{-1}	3150–3000 cm^{-1}	3300 cm^{-1}

Increasing percent *s*-character
Increasing $\tilde{\nu}$

Practice Test on Chapter Review

1.a. Which compound has a molecular ion at 112 and a peak at 1720 cm^{-1} in its IR spectrum?

  1. 2.   3.

4. Compounds (1) and (2) both fit these criteria.
5. Compounds (1), (2), and (3) all fit these criteria.

b. Which of the following compounds has peaks at 3300, 3000, and 2250 cm^{-1} in its IR spectrum?

1. 2. 3.

4. Compounds (1) and (2) have these peaks in their IR spectra.
5. Compounds (1), (2), and (3) all contain these peaks in their IR spectra.

c. What is the base peak in a mass spectrum?
1. the peak due to the radical cation formed when a molecule loses an electron
2. the tallest peak in the mass spectrum
3. the peak due to the fragment with the largest m/z ratio
4. Both (1) and (2) describe the base peak.
5. Statements (1), (2), and (3) all describe the base peak.

d. Which compounds are possible structures for a molecule that has a molecular ion at 150 in its mass spectrum?

1. 2. 3.

4. Both (1) and (2) are possible structures.
5. Compounds (1), (2), and (3) are all possible structures.

e. Which compounds exhibit prominent M + 2 peaks in their mass spectra?

1. 2. 3.

4. Both (1) and (2) show M + 2 peaks.
5. Compounds (1), (2), and (3) all show M + 2 peaks.

2. Answer each question with the number that corresponds to one of the following regions of an IR spectrum.

$$1.\ 4000–2500\ cm^{-1}$$
$$2.\ 2500–2000\ cm^{-1}$$
$$3.\ 2000–1500\ cm^{-1}$$
$$4.\ < 1500\ cm^{-1}$$

a. This region is called the fingerprint region of an IR spectrum.
b. The OH group of propan-1-ol absorbs in this region.
c. The C=N of $C_6H_5CH_2CH=NCH_3$ absorbs in this region.
d. An unsymmetrical C≡C absorbs in this region.
e. An sp hybridized C–H bond absorbs in this region.
f. Ethyl benzoate ($C_6H_5CO_2CH_2CH_3$) absorbs in all regions of the IR except this one.

3. Answer True (T) or False (F).
a. IR spectroscopy is useful for determining the molecular weight of a compound.
b. A C–H bond that absorbs at 3140 cm^{-1} is stronger than a C–H bond that absorbs at 2950 cm^{-1}.
c. A compound with a molecular ion at 109 contains a N atom.
d. A compound with a base peak at 57 must contain a N atom.
e. But-2-yne shows an IR absorption at 2250 cm^{-1}.
f. Propan-1-ol shows an IR absorption at 3200–3600 cm^{-1}.
g. An ether shows no IR absorptions at 3200–3600 or 1700 cm^{-1}.
h. In its mass spectrum, a compound that has a molecular ion with two peaks of approximately equal intensity at 124 and 126 contains chlorine.

Answers to Practice Test

1.a. 4
 b. 1
 c. 2
 d. 4
 e. 4

2.a. 4
 b. 1
 c. 3
 d. 2
 e. 1
 f. 2

3.a. F
 b. T
 c. T
 d. F

e. F
f. T
g. T
h. F

Answers to Problems

13.1   The molecular ion formed from each compound is equal to its molecular weight.

a. C_3H_6O	b. $C_{10}H_{20}$	c. $C_8H_8O_2$	d. $C_{10}H_{15}N$
molecular weight = **58**	molecular weight = **140**	molecular weight = **136**	molecular weight = **149**
molecular ion (m/z) = **58**	molecular ion (m/z) = **140**	molecular ion (m/z) = **136**	molecular ion (m/z) = **149**

13.2   Some possible formulas for each molecular ion:
   a.   Molecular ion at 72: C_5H_{12}, C_4H_8O, $C_3H_4O_2$
   b.   Molecular ion at 100: C_8H_4, C_7H_{16}, $C_6H_{12}O$, $C_5H_8O_2$
   c.   Molecular ion at 73: $C_4H_{11}N$, $C_2H_7N_3$

13.3   Use the molecular ion to propose molecular formulas that contain C, H, and O. Then determine which formula has five degrees of unsaturation.

M at 218: $C_{18}H_2$ (not enough H's)

$$C_{17}H_{14} \xrightarrow[+\,O]{-\,CH_4} C_{16}H_{10}O \text{ (not enough H's)}$$

$$C_{16}H_{26} \xrightarrow[+\,O]{-\,CH_4} C_{15}H_{22}O \text{ (five degrees of unsaturation)}$$

Answer

13.4   To calculate the molecular ions you would expect for compounds with Cl, calculate the molecular weight using each of the two most common isotopes of Cl (^{35}Cl and ^{37}Cl).

a. $C_4H_9{}^{35}Cl$ = **92**
    $C_4H_9{}^{37}Cl$ = **94**
    Two peaks in 3:1 ratio at m/z 92 and 94

c. $C_4H_{11}N$ = **73**
    One peak at m/z 73

b. C_3H_7F = **62**
    One peak at m/z 62

d. $C_4H_4N_2$ = **80**
    One peak at m/z 80

13.5   Convert the ball-and-stick model to a skeletal structure and determine the molecular formula. Calculate the molecular weight using each of the two common isotopes for Br (^{79}Br and ^{81}Br).

$C_6H_{11}{}^{79}Br$ = **162**
$C_6H_{11}{}^{81}Br$ = **164**
Two peaks in a 1:1 ratio at m/z 162 and 164

$C_6H_{11}Br$

13.6   After calculating the mass of the molecular ion, draw the structure and determine which C–C bond is broken to form fragments of the appropriate mass-to-charge ratio.

[1]

m/z = 100

Cleave bond [1].

e⁻

m/z = 43

+

Cleave bond [1].

e⁻

m/z = 57

+

13.7

Break this bond.

m/z = 114

e⁻

m/z = 57
This 3° carbocation is more
stable than others that can form,
and is therefore the most
abundant fragment.

+

13.8

a.

OH

[1]   [2]

+ •

Cleave bond [1].

CH₃•   +   HO⁺

m/z = 59

Cleave bond [2].

•CH₂CH₃   +

⁺OH

m/z = 45

b.

OH

+ •

−H₂O

( )⁺ •   +   ( )⁺ •

m/z = 56   *m/z* = 56

13.9

a.

O

[1]   [2]

O⁺

(from cleavage of bond [2])

+

O=

⁺

(from cleavage of bond [1])

b.

OH

⁺CH₂OH

c.

O

H

O

⁺

+   HC⁺=O

13.10 Use the exact mass values given in Table 13.1 to calculate the exact mass of each compound.

$C_7H_5NO_3$

mass: 151.0270

$C_8H_9NO_2$

mass: 151.0634
compound **X**

$C_{10}H_{17}N$

mass: 151.1362

13.11

benzene
C_6H_6 *m/z* = 78

toluene
C_7H_8 *m/z* = 92

p-xylene
C_8H_{10} *m/z* = 106

GC–MS analysis:
Three peaks in the gas chromatogram.
Order of peaks: benzene, toluene, *p*-xylene,
in order of increasing bp.
Molecular ions observed in the three mass spectra: 78, 92, 106.

13.12 **Wavelength and frequency are inversely proportional.** The higher frequency light will have a shorter wavelength.
   a. Light having $\lambda = 10^2$ nm has a higher ν than light with $\lambda = 10^4$ nm.
   b. Light having $\lambda = 100$ nm has a higher ν than light with $\lambda = 100$ μm.
   c. Blue light has a higher ν than red light.

13.13 The **energy of a photon** is *proportional* to its **frequency,** and inversely proportional to its wavelength.
   a. Light having $\nu = 10^8$ Hz is of higher energy than light having $\nu = 10^4$ Hz.
   b. Light having $\lambda = 10$ nm is of higher energy than light having $\lambda = 1000$ nm.
   c. Blue light is of higher energy than red light.

13.14 Higher wavenumbers are proportional to higher frequencies and higher energies.
   a. IR light with a wavenumber of 3000 cm^{-1} is higher in energy than IR light with a wavenumber of 1500 cm^{-1}.
   b. IR light having $\lambda = 10$ μm is higher in energy than IR light having $\lambda = 20$ μm.

13.15 Stronger bonds absorb at a higher wavenumber. Bonds to lighter atoms (H versus D) absorb at higher wavenumber.

a.   or   stronger bond
higher wavenumber

b.   or   lighter atom H
higher wavenumber

13.16 Cyclopentane and pent-1-ene are both composed of C–C and C–H bonds, but pent-1-ene also has a C=C bond. This difference will give the IR of pent-1-ene an additional peak at 1650 cm^{-1} (for the C=C). Pent-1-ene will also show C–H absorptions for sp^2 hybridized C–H bonds at 3150–3000 cm^{-1}.

13.17 Look at the functional groups in each compound below to explain how each IR is different.

A

C=O peak at ~1700 cm^{-1}

B

C=C peak at 1650 cm^{-1}
C_{sp^2}–H at 3150–3000 cm^{-1}

C

O–H peak at 3200–3600 cm^{-1}

13.18 a. Compound **A** has peaks at ~3150 (sp^2 hybridized C–H), 3000–2850 (sp^3 hybridized C–H), and 1650 (C=C) cm^{-1}.
   b. Compound **B** has a peak at 3000–2850 (sp^3 hybridized C–H) cm^{-1}.

13.19 All compounds show an absorption at 3000–2850 cm^{-1} due to the sp^3 hybridized C–H bonds. Additional peaks in the functional group region for each compound are shown.

a.

no additional peaks

d.

C=O bond at ~1700 cm^{-1}

b.

O–H bond at 3600–3200 cm^{-1}

e.

O–H at 3600–3200 cm^{-1}
N–H at 3500–3200 cm^{-1}
C_{sp^2}–H at 3150–3000 cm^{-1}
C=O at ~1700 cm^{-1}
C=C at 1650 cm^{-1}
aromatic ring at 1600, 1500 cm^{-1}

c.

C_{sp^2}–H at 3150–3000 cm^{-1}
C=C bond at 1650 cm^{-1}

13.20

C_{sp^2}–H at 3150–3000 cm^{-1}
C_{sp^3}–H at 3000–2850 cm^{-1}
C=O at ~1700 cm^{-1}

C=C at 1650 cm^{-1}

O–H above 3000 cm^{-1}

[The OH of a COOH is much broader than the OH of an alcohol and occurs at 3500–2500 cm^{-1} (see Chapter 19).]

13.21 Possible structures are (a) $CH_3COOCH_2CH_3$ and (c) $CH_3CH_2COOCH_3$. Compounds (b) and (d) also have an OH group that would give a strong absorption at ~3600–3200 cm^{-1}, which is absent in the IR spectrum of **X,** thus excluding them as possibilities.

13.22

a. Hydrocarbon with a molecular ion at $m/z = 68$
   IR absorptions at 3310 cm^{-1} = Csp–H bond
           3000–2850 cm^{-1} = Csp^3–H bonds
           2120 cm^{-1} = C≡C bond
   Molecular formula: C_5H_8

b. Compound with C, H, and O with a molecular ion at $m/z = 60$
   IR absorptions at 3600–3200 cm^{-1} = O–H bond
           3000–2850 cm^{-1} = Csp^3–H bonds
   Molecular formula: C_3H_8O

13.23

a. Csp^2–H at 3000–3150 cm^{-1}
   Csp^3–H at 2850–3000 cm^{-1}
   C=C at 1650 cm^{-1}

b. O–H at 3200–3600 cm^{-1}
   Csp^2–H at 3000–3150 cm^{-1}
   Csp^3–H at 2850–3000 cm^{-1}
   C=C at 1650 cm^{-1}

13.24

Cleave bond [1].   ·CH$_3$ + $m/z = 127$

Cleave bond [2].   + $m/z = 85$

Cleave bond [3].   ·CH$_2$CH$_3$ + $m/z = 113$

$m/z = 142$

13.25

a. molecular formula: C_6H_6
   molecular ion (m/z): **78**

c. molecular formula: $C_5H_{10}O$
   molecular ion (m/z): **86**

e. molecular formula: $C_8H_{17}Br$
   molecular ions (m/z): **192, 194**

b. molecular formula: $C_{10}H_{16}$
   molecular ion (m/z): **136**

d. molecular formula: $C_5H_{11}Cl$
   molecular ions (m/z): **106, 108**

13.26

C_9H_{12}	$C_9H_{10}O$	$C_8H_{10}O$
molecular weight = 120	molecular weight = 134	molecular weight = 122

13.27 Examples are given for each molecular ion.
  a. molecular ion 102: C_8H_6, $C_6H_{14}O$, $C_5H_{10}O_2$, $C_5H_{14}N_2$
  b. molecular ion 98: C_8H_2, C_7H_{14}, $C_6H_{10}O$, $C_5H_6O_2$
  c. molecular ion 119: C_8H_9N, $C_6H_5N_3$
  d. molecular ion 74: C_6H_2, $C_4H_{10}O$, $C_3H_6O_2$

13.28 Likely molecular formula, C_8H_{16} (one degree of unsaturation—one ring or one π bond).

Four structures with m/z = 112

13.29 Use the molecular ion to propose molecular formulas for a compound with only C and H. Then determine which formula has four degrees of unsaturation.

M at 204: C_{17} (no H's)

$C_{16}H_{12}$ (not enough H's)

$C_{15}H_{24}$ (four degress of unsaturation)
Answer

13.30 Use the directions from Answer 13.3 and propose a formula with two degrees of unsaturation.

M at 154: $C_{12}H_{10}$ (not enough H's)

$C_{11}H_{22} \xrightarrow[+O]{-CH_4} C_{10}H_{18}O$ (two degrees of unsaturation)
Answer

13.31

B

$C_4H_7O_2Cl$

molecular weight: **122, 124**
should show 2 peaks for the
molecular ion with a **3:1 ratio**

Mass spectrum [1]

C

$C_8H_{10}O$

molecular weight: **122**

Mass spectrum [2]

A

C_3H_7Br

molecular weight: **122, 124**
should show 2 peaks for the
molecular ion with a **1:1 ratio**

Mass spectrum [3]

13.32

Possible structures
C_7H_{12}
(exact mass 96.0940)

13.33

a.

(from cleavage of bond [1])   (from cleavage of bond [2])

b.

(from cleavage of bond [1])   (from cleavage of bond [2])

c.

(from cleavage of bond [1])   (from cleavage of bond [2])

13.34 2,2-Dimethylbutane shows a peak at $m/z = 57$ in it s mass spectrum due to a 3° carbocation. 2,3-Dimethylbutane shows a peak at $m/z = 43$ due to a 2° carbocation.

2,2-dimethylbutane $m/z = 57$

2,3-dimethylbutane $m/z = 43$

13.35

a.

$m/z = 104$ $m/z = 122$ $m/z = 91$ (resonance-stabilized carbocation)

b.

$m/z = 68$ $-H_2O$ [1] [3] [2] $m/z = 86$

Cleave bond [1]. $m/z = 71$

Cleave bond [2]. $m/z = 41$

Cleave bond [3]. $^+CH_2-OH$ $m/z = 31$

13.36

Ketone **A**

$m/z = 128$

α cleavage

·CH₂CH₃   + $m/z = 99$

This is ketone **A** because α cleavage gives a fragment with m/z of 99.

Ketone **B**

$m/z = 128$

α cleavage

·CH₃   + $m/z = 113$

This is ketone **B** because α cleavage gives a fragment with m/z of 113.

13.37 One possible structure is drawn for each set of data:

a. a compound that contains a benzene ring and has a molecular ion at $m/z = 107$

C_7H_9N

b. a hydrocarbon that contains only sp^3 hybridized carbons and a molecular ion at $m/z = 84$

C_6H_{12}

c. a compound that contains a carbonyl group and gives a molecular ion at $m/z = 114$

$C_7H_{14}O$

d. a compound that contains C, H, N, and O and has an exact mass for the molecular ion at 101.0841

$C_5H_{11}NO$

13.38 Use the values given in Table 13.1 to calculate the exact mass of each compound. $C_8H_{11}NO_2$ (exact mass 153.0790) is the correct molecular formula.

13.39 Alpha cleavage of a 1° alcohol (RCH₂OH) forms an alkyl radical (R•) and a resonance-stabilized carbocation with $m/z = 31$.

$+CH_2OH \longleftrightarrow CH_2=\overset{+}{O}H$ $m/z = 31$

resonance-stabilized carbocation

13.40 An ether fragments by α cleavage because the resulting carbocation is resonance stabilized.

13.41

a. stronger bond, higher ν̃ absorption

b. stronger bond, higher ν̃ absorption

c. stronger bond, higher ν̃ absorption

13.42 Locate the functional groups in each compound. Use Table 13.2 to determine what IR absorptions each would have.

a.
C_{sp}–H at 3300 cm^{-1}
C_{sp^3}–H at 2850–3000 cm^{-1}
C–C triple bond at 2250 cm^{-1}

b.
O–H at 3200–3600 cm^{-1}
C_{sp^3}–H at 2850–3000 cm^{-1}

c.
C_{sp^3}–H at 2850–3000 cm^{-1}
C=O at 1700 cm^{-1}

d.
O–H at > 3000 cm^{-1}
C_{sp^2}–H at 3000–3150 cm^{-1}
C=O at ~1700 cm^{-1}
phenyl group at 1600, 1500 cm^{-1}

The OH of the RCOOH is even broader than the OH of an alcohol (3500–2500 cm^{-1}), as we will learn in Chapter 19.

13.43

a.
and

C=C bond
1650 cm^{-1}
C_{sp^2}–H at 3150–3000 cm^{-1}

C≡C bond
2250 cm^{-1}
C_{sp}–H at 3300 cm^{-1}

c.
and

no C=O bond

C=O bond
~1700 cm^{-1}

b.
and

O–H bond
> 3000 cm^{-1}
[See note on OH in Answer 13.42.]

no O–H bond

d.
and

C_{sp}–H bond
3300 cm^{-1}

13.44 The IR absorptions above 1500 cm⁻¹ are different for each of the narcotics.

morphine
• O–H bond at ~3200–3600 cm⁻¹
• no C=O bond

heroin
• C=O bond at ~1700 cm⁻¹
• no O–H bond

oxycodone
• C=O bond at ~1700 cm⁻¹
• O–H bond at ~3200–3600 cm⁻¹

13.45 The three compounds show differences in their IR spectra.

cyclohexanone
C=O at ~1700 cm⁻¹

cyclohex-2-enol
C=C at ~1650 cm⁻¹
O–H at ~3200–3600 cm⁻¹
C$_{sp^2}$–H at ~3150–3000 cm⁻¹

cyclohexanol
O–H at ~3200–3600 cm⁻¹

13.46 Look for a **change in functional groups** from starting material to product to see how IR could be used to determine when the reaction is complete.

a. $\xrightarrow{H_2 \\ Pd}$ Loss of the C=C will be visible in the IR by disappearance of the peak at 1650 cm⁻¹.

b. $\xrightarrow{PCC}$ Loss of the O–H group will be visible in the IR by disappearance of the peak at 3200–3600 cm⁻¹ and appearance of the C=O at ~1700 cm⁻¹.

c. $\xrightarrow{[1] O_3 \\ [2] CH_3SCH_3}$ Loss of the C=C will be visible in the IR by disappearance of the peak at 1650 cm⁻¹ and appearance of the C=O at ~1700 cm⁻¹.

d. $\xrightarrow{[1] NaH \\ [2] CH_3Br}$ Loss of the O–H will be visible in the IR by disappearance of the peak at 3200–3600 cm⁻¹.

13.47 In addition to Csp^3–H at ~3000–2850 cm^{-1}:

Spectrum [1]:

(B)
C=C peak at 1650 cm^{-1}
Csp^2–H at ~3150 cm^{-1}

Spectrum [2]:

HO

(F)
OH at 3600–3200 cm^{-1}

Spectrum [3]:

(D)
No other peaks above 1500 cm^{-1}

Spectrum [4]:

(C)
Csp^2–H at ~3150 cm^{-1}
Phenyl peaks at 1600 and 1500 cm^{-1}

Spectrum [5]:

OH **(A)**
OH at ~3500–2500 cm^{-1}
C=O at ~1700 cm^{-1}

Spectrum [6]:

(E)
C=O at ~1700 cm^{-1}

13.48

a. Compound with a molecular ion at $m/z = 72$
   IR absorption at 1725 cm^{-1} = C=O bond
   Molecular formula: C_4H_8O

b. Compound with a molecular ion at $m/z = 55$
   The odd molecular ion means an odd number
   of N's present. Molecular formula: C_3H_5N
   IR absorption at 2250 cm^{-1} = C≡N bond

   ≡N

c. Compound with a molecular ion at $m/z = 74$
   IR absorption at 3600–3200 cm^{-1} = O–H bond
   Molecular formula: $C_4H_{10}O$

   OH

   or

   OH

   OH or OH

13.49

Chiral hydrocarbon with a molecular ion at $m/z = 82$
Molecular formula: C_6H_{10}
   IR absorptions at 3300 cm^{-1} = Csp–H bond
                    3000–2850 cm^{-1} = Csp^3–H bonds
                    2250 cm^{-1} = C≡C bond

stereogenic center

Two possible enantiomers:

H   or   H

13.50 The chiral compound **Y** has a strong absorption at 2970–2840 cm^{-1} in its IR spectrum due to sp^3 hybridized C–H bonds. The two peaks of equal intensity at 136 and 138 indicate the presence of a Br atom. The molecular formula is C_4H_9Br. Only one constitutional isomer of this molecular formula has a stereogenic center:

13.51 The molecular ion of 192 suggests $C_{12}H_{16}O_2$ as a possible molecular formula. IR absorption at 1721 cm^{-1} is due to a C=O, and the absorptions around 3000 cm^{-1} are due to C_{sp^2}–H and C_{sp^3}–H. The compound is an ester, formed in the following manner.

13.52

m/z = 92; molecular formula C_7H_8
IR absorptions at:
3150–2950 cm^{-1} = C_{sp^3}–H and C_{sp^2}–H bonds
1605 cm^{-1} and 1496 cm^{-1} due to phenyl group

13.53

13.54 The molecular ion of 144 suggests $C_8H_{16}O_2$ as a possible molecular formula for **X**. The IR absorption at 1739 cm^{-1} is due to a C=O, and the absorptions at less than 3000 cm^{-1} are due to C_{sp^3}–H.

2-methylpropan-1-ol

$C_8H_{16}O_2$
X
possible structure

13.55

J

$C_6H_{12}O$

$m/z = 100$

IR absorption at 2962 cm^{-1} = Csp^3–H bonds

1718 cm^{-1} = C=O bond

fragments:

α cleavage product
$m/z = 43$

α cleavage product
$m/z = 85$

The fragment at $m/z = 57$ could be due to $(C_4H_9)^+$ or $(C_3H_5O)^+$.

13.56

K

C_7H_9N

$m/z = 107$

IR absorptions at 3373 and 3290 cm^{-1} = N–H

3062 cm^{-1} = Csp^2–H bonds
2920 cm^{-1} = Csp^3–H bonds
1600 cm^{-1} = benzene ring

The odd molecular ion indicates the
presence of a N atom.

L

C_7H_6O

$m/z = 106$

$m/z = 105$ $m/z = 77$

IR absorption at 3068 cm^{-1} = Csp^2–H bonds on ring

2820 cm^{-1} and 2736 cm^{-1} = C–H of RCHO (Appendix E)

1703 cm^{-1} = C=O bond
1600 cm^{-1} = aromatic ring

13.57

Possible structures of **P**:

C_7H_7ClO

$m/z = 142, 144$

IR absorption at 3096–2837 cm^{-1} = Csp^3–H bonds and Csp^2–H bonds

1582 cm^{-1} and 1494 cm^{-1} = benzene ring

The peak at M + 2 shows the presence of Cl or Br. Since Cl$_2$ is a
reactant, the compound presumably contains Cl.

13.58   The mass spectrum has a molecular ion at 71. The odd mass suggests the presence of an odd
number of N atoms; likely formula, C$_4$H$_9$N. The IR absorption at ~3300 cm^{-1} is due to N–H and
the 3000–2850 cm^{-1} is due to sp^3 hybridized C–H bonds.

W

13.59 Because the carbonyl absorption of an amide is at lower wavenumber than the carbonyl absorption of an ester, the C=O of the amide must be weaker and have more single bond character. This can be explained by resonance. Although both an ester and amide are resonance stabilized, the N atom of the amide is more basic, making it more willing to donate its electron pair.

Because the amide carbonyl has more single bond character, the bond is weaker and it absorbs at lower wavenumber.

13.60 The α,β-unsaturated carbonyl compound has three resonance structures, two of which place a single bond between the C and O atoms. This means that the C–O bond has partial single bond character, making it weaker than a regular C=O bond, and moving the absorption to lower wavenumber.

three resonance structures for cyclohex-2-enone

13.61 If a ketone carbonyl absorbs at lower wavenumber than an aldehyde carbonyl, the ketone carbonyl is weaker and has more single bond character. This can be explained by the fact that R groups are electron donating and stabilize an adjacent (+) charge.

ketone

The two R groups stabilize the (+) charge, so this form contributes more to the hybrid than the charge-separated resonance form of an aldehyde.

One R group stabilizes the (+) charge less.

As a result, the charge-separated resonance form of a ketone, which contains a C–O single bond, contributes more to the hybrid of a ketone, making the C=O weaker and shifting the absorption to lower wavenumber.

13.62

a, b.

A
molecular ion at 154
$C_{10}H_{18}O$
IR at 1730 cm^{-1} (C=O)

citronellol

isopulegone

+ Cr^{4+} + H–B$^+$

+ H–B$^+$

+ :B

a, c.

isopulegone

+ $H_2\ddot{O}$:

B

+ $^-\!\ddot{O}H$

Chapter 14 Nuclear Magnetic Resonance Spectroscopy

Chapter Review

¹H NMR spectroscopy

[1] The **number of signals** equals the number of different types of protons (14.2).

[2] The **position of a signal** (its chemical shift) is determined by shielding and deshielding effects.
- Shielding shifts an absorption upfield; deshielding shifts an absorption downfield.
- Electronegative atoms withdraw electron density, deshield a nucleus, and shift an absorption downfield (14.3).

- Loosely held π electrons can either shield or deshield a nucleus. Protons on benzene rings and double bonds are deshielded and absorb downfield, whereas protons on triple bonds are shielded and absorb upfield (14.4).

[3] The **area under an NMR signal** is proportional to the number of absorbing protons (14.5).

[4] **Spin–spin splitting** tells about nearby nonequivalent protons (14.6–14.8).
- Equivalent protons do not split each other's signals.
- A set of n nonequivalent protons on the same carbon or adjacent carbons split an NMR signal into $n + 1$ peaks.
- OH and NH protons do not cause splitting (14.9).
- When an absorbing proton has two sets of nearby nonequivalent protons that are equivalent to each other, use the $n + 1$ rule to determine splitting.
- When an absorbing proton has two sets of nearby nonequivalent protons that are not equivalent to each other, the number of peaks in the NMR signal = $(n + 1)(m + 1)$. In flexible alkyl chains, peak overlap often occurs, resulting in $n + m + 1$ peaks in an NMR signal.

¹³C NMR spectroscopy (14.11)

[1] The number of signals equals the number of different types of carbon atoms. All signals are single lines.

[2] The relative position of ¹³C signals is determined by shielding and deshielding effects.
- Carbons that are sp^3 hybridized are shielded and absorb upfield.
- Electronegative elements (N, O, and X) shift absorptions downfield.
- The carbons of alkenes and benzene rings absorb downfield.
- Carbonyl carbons are highly deshielded, and absorb farther downfield than other carbon types.

Practice Test on Chapter Review

1.a. Which of the following statements is true about ¹H NMR absorptions?
1. A signal that occurs at 1800 Hz on a 300 MHz NMR spectrometer occurs at 3000 Hz on a 500 MHz NMR spectrometer.
2. A signal that occurs at 3.3 ppm on a 60 MHz NMR absorbs at 198 Hz upfield from TMS.
3. A signal that occurs at 600 Hz is downfield from a signal that occurs at 800 Hz.
4. Statements (1) and (2) are both true.
5. Statements (1), (2), and (3) are all true.

b. Which of the following statements is true about ¹H NMR spectroscopy?
1. Electronegative elements shield a nucleus so an absorption shifts downfield.
2. A triplet is due to a proton that has four adjacent nonequivalent protons.
3. Circulating π electrons create a magnetic field that reinforces the applied field in the vicinity of the protons in benzene.
4. Statements (1) and (2) are both true.
5. Statements (1), (2), and (3) are all true.

2. How many different types of protons does each of the following molecules contain?

a.   c.   e.   g.

b.   d.   f.   h.

3. Into how many peaks will each of the circled protons be split in a proton NMR spectrum?

a.   c.   e.   g.

b.   d.   f.

4. How many lines are presents in the ^{13}C NMR spectrum of each compound?

a. CH$_3$O

c.   Cl

e.

b.

d.

f.

5. With reference to the ^1H NMR absorptions in the following compound, (a) which proton absorbs farthest upfield; (b) which proton absorbs farthest downfield?

H$_a$ —H$_b$ H$_d$
    O
O   H$_c$

6. With reference to the ^{13}C NMR absorptions in the following compound, (a) which carbon absorbs farthest downfield; (b) which carbon absorbs farthest upfield?

C$_b$ O
C$_a$ →   F   C$_c$ C$_d$

Answers to Practice Test

1. a. 1	2. a. 5	3. a. 8	4. a. 6	5. a. H$_b$	6. a. C$_c$
b. 3	b. 5	b. 3	b. 4	b. H$_c$	b. C$_a$
	c. 4	c. 7	c. 6		
	d. 5	d. 4	d. 4		
	e. 4	e. 4	e. 4		
	f. 5	f. 3	f. 5		
	g. 3	g. 8			
	h. 9				

Answers to Problems

14.1 Use the formula δ = [observed chemical shift (Hz)/ν of the NMR (MHz)] to calculate the chemical shifts.

a. **CH$_3$ protons:**
   δ = [1715 Hz] / [500 MHz]
   = **3.43 ppm**

OH proton:
   δ = [1830 Hz] / [500 MHz]
   = **3.66 ppm**

b. The positive direction of the δ scale is downfield from TMS. The CH$_3$ protons absorb upfield from the OH proton.

14.2 Calculate the chemical shifts as in Answer 14.1.

a. one signal:
   δ = [1017 Hz] / [300 MHz]
   = **3.39 ppm**

second signal:
   δ = [1065 Hz] / [300 MHz]
   = **3.55 ppm**

b. one signal:
   3.39 = [x Hz] / [500 MHz]
   x = **1695 Hz**

second signal:
   3.55 = [x Hz] / [500 MHz]
   x = **1775 Hz**

14.3 To determine if two H's are equivalent replace each by an atom X. If this yields the same compound or mirror images, the two H's are equivalent. Each kind of H will give one NMR signal.

a.
2 kinds of H's
2 NMR signals

c.
2 kinds of H's
2 NMR signals

e.
4 kinds of H's
4 NMR signals

g.
8 kinds of H's
8 NMR signals

b.
2 kinds of H's
2 NMR signals

d.
2 kinds of H's
2 NMR signals

f.
4 kinds of H's
4 NMR signals

h.
4 kinds of H's
4 NMR signals

14.4 Draw in all of the H's and compare them. If two H's are cis and trans to the same group, they are equivalent.

a. 4 identical H's

2 NMR signals

b.

4 NMR signals

c.

3 NMR signals

14.5

a.
3 NMR signals

b.
4 NMR signals

c.
6 NMR signals

d.
3 NMR signals

14.6 If replacement of H by X yields the same compound, the protons are **homotopic**.
If replacement of H with X yields enantiomers, the protons are **enantiotopic**.
If replacement of H with X yields diastereomers, the protons are **diastereotopic**. In general, if the compound has **one stereogenic center**, the protons in a CH$_2$ group are **diastereotopic**.

a.

replacement of H with X

enantiomers =
enantiotopic H's

b.

replacement of either H with X

no stereogenic center
homotopic H's

c.

Pick one configuration at the existing stereogenic center.

replacement of H with X

diastereomers =
diastereotopic H's

14.7 The two protons of a CH₂ group are different from each other if the compound has one stereogenic center. Replace one proton with X and compare the products.

a. The stereogenic center makes the H's in the CH₂ group diastereotopic and therefore different from each other.

5 NMR signals

b.

5 NMR signals

c.

7 NMR signals

14.8 Decreased electron density deshields a nucleus and the absorption goes downfield. Absorption also shifts downfield with increasing alkyl substitution.

a. FC**H₂**CH₂C**H₂**Cl
F is more electronegative than Cl. The CH₂ group adjacent to the F is more deshielded and the H's will absorb farther downfield.

b. CH₃C**H₂**CH₂C**H₂**OCH₃
The CH₂ group adjacent to the O will absorb farther downfield because it is closer to the electronegative O atom.

c. C**H₃**OC(CH₃)₃
The CH₃ group bonded to the O atom will absorb farther downfield.

14.9

a.

3 types of protons:
$H_b < H_c < H_a$

b.

3 types of protons:
$H_c < H_a < H_b$

c.

3 types of protons:
$H_c < H_a < H_b$

14.10 a. False. When a nucleus is strongly shielded, the effective field is smaller than the applied field and the absorption shifts upfield.

b. True.

c. False. A nucleus that is strongly deshielded requires a higher field strength for resonance. Alternatively, a nucleus that is strongly *shielded* requires a lower field strength for resonance.

d. False. A nucleus that is strongly shielded absorbs at a smaller δ value. Alternatively, a nucleus that is strongly *deshielded* absorbs at a larger δ value.

14.11

a.

H_c protons are shielded because they are bonded to an sp^3 C.
H_a is shielded because it is bonded to an sp C.
H_b protons are deshielded because they are bonded to an sp^2 C.

$$H_c < H_a < H_b$$

b.

H_c protons are shielded because they are bonded to an sp^3 C.
H_a protons are deshielded slightly because the CH_3 group is bonded to a C=O.
H_b protons are deshielded because the CH_2 group is bonded to an O atom.

$$H_c < H_a < H_b$$

14.12 An integration ratio of 2:3 means that there are two types of hydrogens in the compound, and that the ratio of one type to another type is 2:3.

a.

2 types of H's
3:2 - YES

b.

2 types of H's
6:2 or 3:1 - no

c.

2 types of H's
6:4 or 3:2 - YES

d.

2 types of H's
6:4 or 3:2 - YES

14.13

downfield absorption
closer to O

$CH_3O_2CCH_2CH_2CO_2CH_3$

A
ratio of absorbing signals 2:3

Signal [1] = **4 H = 2.64**
Signal [2] = **6 H = 3.69** ← 6 H's with downfield absorption

downfield absorption
closer to O

$CH_3CO_2CH_2CH_2O_2CCH_3$

B
ratio of absorbing signals 3:2

Signal [1] = **6 H = 2.09**
Signal [2] = **4 H = 4.27** ← 4 H's with downfield absorption

14.14 To determine the **splitting pattern** for a molecule:
- Determine the number of different kinds of protons.
- Nonequivalent protons on the same C or adjacent C's split each other.
- Apply the *n* + 1 rule.

a.

H_a: 3 peaks - triplet
H_b: 4 peaks - quartet

c.

H_a: 1 peak - singlet
H_b: 3 peaks - triplet
H_c: 3 peaks - triplet

e.

H_a: 2 peaks - doublet
H_b: 2 peaks - doublet

b.

H_a: 2 peaks - doublet
H_b: 4 peaks - quartet

d.

H_a: 2 peaks - doublet
H_b: 2 peaks - doublet

f.

H_a: 2 peaks - doublet
H_b: 3 peaks - triplet

14.15 Use the directions from Answer 14.14.

a.
H_a: quartet
H_b: triplet
2 NMR signals

b.
H_a and H_d are both singlets.
H_b: triplet
H_c: triplet
4 NMR signals

c.
H_a: doublet
H_b: quartet
2 NMR signals

d.
H_a: triplet
H_b: doublet
H_c: singlet
3 NMR signals

14.16 CH_3CH_2Cl

2 units

3 units

3   1

chemical shift (ppm)

There are two kinds of protons, and they can split each other. The CH_3 signal will be split by the CH_2 protons into $2 + 1 = 3$ peaks. It will be upfield from the CH_2 protons because it is farther from the Cl. The CH_2 signal will be split by the CH_3 protons into $3 + 1 = 4$ peaks. It will be downfield from the CH_3 protons because the CH_2 protons are closer to the Cl. The ratio of integration units will be 3:2.

14.17

a.
split by 6 equivalent H's
$6 + 1 = $ **7 peaks**

b.
H_a: split by 2 H's
3 peaks
H_c: split by 4 equivalent H's
5 peaks
H_b: split by 2 sets of H's
$(3 + 1)(2 + 1) = $ **12 peaks (maximum)**
Since this is a flexible alkyl chain, the signal due to H_b will have peak overlap, and
$3 + 2 + 1 = $ **6 peaks** will likely be visible.

c.
H_a: split by 1 H
2 peaks
H_b: split by 2 sets of H's
$(1 + 1)(2 + 1) = $ **6 peaks**

d.
H_a: split by 2 different H's
$(1+1)(1+1) = $ **4 peaks**
H_b: split by 2 different H's
$(1+1)(1+1) = $ **4 peaks**
H_c: split by 2 different H's
$(1+1)(1+1) = $ **4 peaks**

14.18

a.
H_a: singlet at ~3 ppm
H_b: quartet at ~3.5 ppm
H_c: triplet at ~1 ppm

b.
H_a: triplet at ~1 ppm
H_b: quartet at ~2 ppm
H_c: septet at ~3.5 ppm
H_d: doublet at ~1 ppm

c.
H_a: singlet at ~3 ppm
H_b: triplet at ~3.5 ppm
H_c: quintet at ~1.5 ppm

d.
H_a: triplet at ~1 ppm
H_b: multiplet (8 peaks) at ~2.5 ppm
H_c: triplet at ~5 ppm

14.19

trans-1,3-dichloropropene

2 H$_c$ protons

J_{ab} = 13.1 Hz

J_{bc} = 7.2 Hz

Splitting diagram for H$_b$

1 **trans** H$_a$ proton splits H$_b$ into
1 + 1 = 2 peaks
a doublet

2 H$_c$ protons split H$_b$ into
2 + 1 = 3 peaks
Now it's a doublet of triplets.

14.20

C$_3$H$_4$Cl$_2$

A
H$_a$: 1.75 ppm, doublet, 3 H, *J* = 6.9 Hz
H$_b$: 5.89 ppm, quartet, 1 H, *J* = 6.9 Hz

doublet

singlet

doublet

B
signal at 4.16 ppm, singlet, 2 H
signal at 5.42 ppm, doublet, 1 H, *J* = 1.9 Hz
signal at 5.59 ppm, doublet, 1 H, *J* = 1.9 Hz

14.21 Remember that OH (or NH) protons do not split other signals, and are not split by adjacent protons.

a.
singlet
singlet
singlet
3 NMR signals

b.
triplet   triplet
singlet
12 peaks (maximum)
6 peaks (more likely, resulting
from peak overlap)
4 NMR signals

c.
doublet
singlet
7 peaks
3 NMR signals

14.22

H$_d$
5 H's on
benzene ring

A

H$_a$: doublet at ~1.4 due to the CH$_3$ group, split into two peaks by one adjacent nonequivalent H (H$_c$).
H$_b$: singlet at ~2.7 due to the OH group. OH protons are not split by nor do they split adjacent protons.
H$_c$: quartet at ~4.7 due to the CH group, split into four peaks by the adjacent CH$_3$ group.
H$_d$: multiplets at ~7.2–7.4 due to five protons on the benzene ring.

14.23

palau'amine

H$_a$: one adjacent nonequivalent H, so two peaks
H$_b$: one adjacent nonequivalent H, so two peaks
H$_c$: H$_c$ is located on a N atom, so there is no splitting and it appears as one peak.
H$_d$: H$_d$ has one nonequivalent H on the same carbon and one on an adjacent carbon, so it is split into (1 + 1)(1 + 1) = 4 peaks (a doublet of doublets).

14.24 Use these steps to propose a structure consistent with the molecular formula, IR, and NMR data.
- Calculate the **degrees of unsaturation.**
- Use the IR data to determine what types of **functional groups** are present.
- Determine the number of different **types of protons.**
- Calculate the **number of H's** giving rise to each signal.
- Analyze the **splitting pattern** and put the molecule together.
- Use the **chemical shift** information to check the structure.

- Molecular formula $C_7H_{14}O_2$

  $2n + 2 = 2(7) + 2 = 16$
  $16 - 14 = 2/2 = 1$ **degree of unsaturation**
  1 π bond or 1 ring

- IR peak at 1740 cm^{-1}

  C=O absorption is around 1700 cm^{-1} (causes the degree of unsaturation).
  No signal at 3200–3600 cm^{-1} means there is no O–H bond.

- NMR data:

absorption	ppm	relative area	
singlet	1.2	9	--------→ **9 H's**
triplet	1.3	3	--------→ **3 H's** (probably a CH_3 group)
quartet	4.1	2	--------→ **2 H's** (probably a CH_2 group)

- 3 kinds of H's
- number of H's per signal

  Because the sum of the relative areas equals the number of absorbing H's (9 + 3 + 2 = 14), the relative area shows the actual number of absorbing H's: 9 H's, 3 H's and 2 H's.
- splitting pattern

  The singlet (9 H) is likely from a *tert*-butyl group:

  $$\begin{array}{c} CH_3 \\ | \\ -C-CH_3 \\ | \\ CH_3 \end{array}$$

  The CH_3 and CH_2 groups split each other: CH_3-CH_2-

- Join the pieces together.

or

Pick this structure due to the chemical shift data.
The CH_2 group is shifted downfield (4 ppm), so it is close to the electron-withdrawing O.

14.25

- Molecular formula: C_3H_8O ➢ Calculate degrees of unsaturation

$$2n + 2 = 2(3) + 2 = 8$$
$$8 - 8 = \textbf{0 degrees of unsaturation}$$

- IR peak at 3200–3600 cm^{-1} ➢ Peak at 3200–3600 cm^{-1} is due to an **O–H bond.**
- NMR data:
  - doublet at ~1.2 (6 H)
  - singlet at ~2.2 (1 H)
  - septet at ~4 (1 H)

3 types of H's
septet from 1 H ◄——— split by 6 H's
──► **singlet** from 1 H
doublet from 6 H's ◄——— split by 1 H
from the O–H proton

➢ Put information together:

14.26 Identify each compound from the ^{1}H NMR data.

a. $CH_2{=}CHCOCH_3$ $\xrightarrow{\text{HCl}}$ [structure] triplet at 3.6, CH$_3$ singlet, A triplet at 3.05

b. $(CH_3)_2C{=}O$ $\xrightarrow[\text{H}_2\text{O}]{\text{base}}$ [structure B] singlet at 2.5, singlet at 1.3, singlet at 3.8, singlet at 2.2

14.27 Each different kind of carbon atom will give a different ^{13}C NMR signal.

a. [structure] C_b C_a / C_a C_b
2 kinds of C's
2 ^{13}C NMR signals

b. [structure]
Each C is different.
4 kinds of C's
4 ^{13}C NMR signals

c. [structure] C_b C_b / C_a C_c C_c C_a
same groups on both sides of O
3 kinds of C's
3 ^{13}C NMR signals

d. [structure]
Each C is different.
4 kinds of C's
4 ^{13}C NMR signals

14.28

a. [structure] Cl, Cl, H_c, H_d
2 different H's H_a and H_b
4 ^{1}H NMR signals

[structure] H_b, Cl, Cl, H_a, H_a
2 different H's
2 ^{1}H NMR signals

[structure] H_b Cl, Cl, H_c, H_a
3 ^{1}H NMR signals

[structure] Cl, Cl
all H's identical
1 ^{1}H NMR signal

b. [structure] Cl, Cl
Each C is different.
3 kinds of C's
3 ^{13}C NMR signals

[structure] C_b, Cl, Cl, C_a, C_a
2 kinds of C's
2 ^{13}C NMR signals

[structure] Cl, Cl
Each C is different.
3 kinds of C's
3 ^{13}C NMR signals

[structure] Cl, Cl, C_a, C_b, C_a
2 kinds of C's
2 ^{13}C NMR signals

c. Although the number of ^{13}C signals cannot be used to distinguish these isomers, each isomer exhibits a different number of signals in its ^1H NMR spectrum. As a result, the isomers are distinguishable by ^1H NMR spectroscopy.

14.29

 These 2 C's are different because they are cis and trans to different groups.

Every carbon is different so there are 10 lines for the 10 C atoms.

14.30 Electronegative elements shift absorptions downfield. The carbons of alkenes, benzene rings, and carbonyl groups are also shifted downfield.

a.

The C closer to the electronegative O will be farther downfield.

b.

The C of the CBr_2 group has two bonds to electronegative Br atoms and will be farther downfield.

c.

The carbonyl carbon is highly deshielded and will be farther downfield.

d.

The C atom that is part of the double bond will be farther downfield.

14.31

a. In order of lowest to highest chemical shift:

$$C_d < C_a < C_c < C_b$$

b. In order of lowest to highest chemical shift:

$$C_a < C_b < C_c$$

14.32

- molecular formula $C_4H_8O_2$
  $2n + 2 = 2(4) + 2 = 10$
  $10 - 8 = 2/2 = \textbf{1 degree of unsaturation}$

- no IR peaks at 3200–3600 or 1700 cm^{-1}
  no O–H or C=O

- ^1H NMR spectrum at 3.69 ppm
  only one kind of proton

- ^{13}C NMR spectrum at 67 ppm
  only one kind of carbon

This structure satisifies all the data. One ring is one degree of unsaturation. All carbons and protons are identical.

14.33

- molecular formula C_4H_8O
  $2n + 2 = 2(4) + 2 = 10$
  $10 - 8 = 2/2 = \textbf{1 degree of unsaturation}$

- ^{13}C NMR signal at > 160 ppm due to
  C=O

- molecular formula C_4H_8O
  $2n + 2 = 2(4) + 2 = 10$
  $10 - 8 = 2/2 = \textbf{1 degree of unsaturation}$

- all ^{13}C NMR signals at < 160 ppm
  NO C=O

14.34

A

a. 4 ^{1}H NMR signals
b. 5 ^{13}C NMR signals (including the 4° C)

B

a. 6 ^{1}H NMR signals
b. 7 ^{13}C NMR signals (including the carbonyl C)

14.35

C

a. 4 ^{1}H NMR signals
b. H_a: 1 adjacent H, so 2 peaks
   H_b: 2 adjacent H's, so 3 peaks
   H_c: 3 adjacent H's, so 4 peaks
   H_d: 2 adjacent H's, so 3 peaks

D

a. 5 ^{1}H NMR signals
b. H_a: singlet
   H_b: 2 adjacent H's, so 3 peaks
   H_c: 2 adjacent H's, so 3 peaks
   H_d: 1 nonequivalent H on the same C, so 2 peaks
   H_e: 1 nonequivalent H on the same C, so 2 peaks

14.36 Use the directions from Answer 14.3.

a.
2 kinds of H's

b.
7 kinds of H's

c.
5 kinds of H's

d.
3 kinds of H's

e.
3 kinds of H's

f.
H_a →
H_b →
H_c
→ H_d
Br
4 kinds of H's

g.
OH ← H_c
← H_f
H_a H_b two different H's
H_d and H_e
6 kinds of H's

h.
H_a and H_b →
O
→ H_d
← H_c
4 kinds of H's

i.
H_a
H_b →
H_c →
H_b
3 kinds of H's

j.
two different H's →
on each C
H_c and H_d
← H_a
← H_b
← H_b
← H_a
4 kinds of H's

14.37

caffeine
4 NMR signals

vanillin
6 NMR signals

thymol
7 NMR signals

capsaicin
15 NMR signals

14.38

δ **(in ppm) = [observed chemical shift (Hz)] / ν of the NMR (MHz)]**

a. $2.5 = x$ Hz/300 MHz
$x =$ **750 Hz**
b. ppm = 1200 Hz/300 MHz
= **4 ppm**
c. $2.0 = x$ Hz/300 MHz
$x =$ **600 Hz**

14.39 a. The chemical shift in δ is independent of the operating frequency, so there is no change in δ when the ν is increased.

b. When the operating ν increases, the ν of an absorption increases as well, because the two quantities are proportional.

c. Coupling constants are independent of the operating ν, so the *J* value in Hz remains the same.

14.40 Use the directions from Answer 14.8.

a.

Br H_a H_b F

More electronegative F
deshields the H's.
farther downfield

b.

H_a H_b

Increasing alkyl substitution
farther downfield

14.41

H_a H_b H_b H_a

$H_a : H_b = 3:2$

H_a H_b H_a

$H_a : H_b = 3:1$

different ratio of peak areas

14.42

a.

H_a H_b

H_a protons split by 1 H = **doublet**
H_b proton split by 3 H's = **quartet**

b.

H_a H_b

both CH_2 groups split
each other = **triplets**

c.

H_b H_a

H_a protons split by 1 H = **doublet**
H_b proton split by 2 H's = **triplet**

d.

H_a protons split by 1 H = **doublet**
H_b proton split by 6 H's = **septet**
H_c protons split by 3 H's = **quartet**
H_d protons split by 2 H's = **triplet**

e.

H_a protons split by 2 CH_2 groups = **quintet**
H_b protons split by 2 H's = **triplet**

f.

H_a protons split by 2 H's = **triplet**
H_b protons split by CH_3 + CH_2 protons = **12 peaks** (maximum)
H_c protons split by 2 different CH_2 groups = **9 peaks** (maximum)
H_d protons split by 2 H's = **triplet**
Since H_b and H_c are located in a flexible alkyl chain, it is likely that peak overlap occurs, so that the following is observed: H_b (3 + 2 + 1 = 6 peaks), H_c (2 + 2 + 1 = 5 peaks).

g.

H_a protons split by 2 H's = **triplet**
H_c protons split by 2 H's = **triplet**
H_b protons split by CH_3 + CH_2 protons = **12 peaks** (maximum)
Since H_b is located in a flexible alkyl chain, it is likely that peak overlap occurs, so that only 3 + 2 + 1 = 6 peaks will be observed.

h.

H_a: split by CH_3 group + H_b = **8 peaks** (maximum)
H_b: split by 2 H's = **triplet**

i.

H_a: split by 1 H = **doublet**
H_b: split by 1 H = **doublet**

j.

H_a: split by H_b + H_c = **doublet of doublets** (4 peaks)
H_b: split by H_a + H_c = **doublet of doublets** (4 peaks)
H_c: split by CH_3, H_a + H_b = **16 peaks**

14.43

H_a: split by 1 H = **doublet**
H_b: split by 1 H = **doublet**

H_a and H_b are geminal.

H_a: split by 1 H = **doublet**
H_b: split by 1 H = **doublet**

H_a and H_b are trans.

Both compounds exhibit two doublets for the H's on the C=C, but the coupling constants ($J_{geminal}$ and J_{trans}) are different. $J_{geminal}$ is much smaller than J_{trans} (0–3 Hz versus 11–18 Hz).

14.44

a.

H_a: 1 adjacent H_b = **doublet**
H_b: 1 adjacent H_a = **doublet**
H_c: no adjacent H's = **singlet**

b.

H_a: no adjacent H's = **singlet**
H_b: 1 adjacent H_c = **doublet**
H_c: 1 adjacent H_b and 2 adjacent H_d's (1+1)(2 + 1) = **6 peaks**
H_d: 1 adjacent H_c = **doublet**

14.45

Ha: doublet of doublets at 5.7 ppm. Two large J values are seen for the H's cis ($J_{ab} = 11.8$ Hz) and trans ($J_{ac} = 18$ Hz) to Ha.

Hb: doublet of doublets at ~6.2 ppm. One large J value is seen for the cis H ($J_{ab} = 11.8$ Hz). The geminal coupling ($J_{bc} = 0.9$ Hz) is hard to see.

Hc: doublet of doublets at ~6.6 ppm. One large J value is seen for the trans H ($J_{ac} = 18$ Hz). The geminal coupling ($J_{bc} = 0.9$ Hz) is hard to see.

H_a

Splitting diagram for H_a

1 **trans** H_c proton splits H_a into
1 + 1 = 2 peaks
a doublet

J_{ac} = the coupling constant between H_a and H_c

1 **cis** H_b proton splits H_a into
1 + 1 = 2 peaks
Now it's a doublet of doublets.

J_{ab} = the coupling constant between H_a and H_b

14.46

Four constitutional isomers of C_4H_9Br:

Br

Br

Br

Br

4 different C's   4 different C's   2 different C's   3 different C's

14.47 The O atom of an ester donates electron density, so the carbonyl carbon has less δ+, making it less deshielded than the carbonyl carbon of an aldehyde or ketone. Therefore, the carbonyl carbon of an aldehyde or ketone is more deshielded and absorbs farther downfield.

14.48

a.
5 signals

b.
3 signals

c.
7 signals

d.
3 signals

e.
7 signals

f.
5 signals

g.
4 signals

h.
3 signals

14.49

a.

C_a C_b C_c

$C_a < C_b < C_c$

b.

C_a C_b C_c

$C_b < C_c < C_a$

14.50

a. $CH_3CH_2CH_2CH_2OH$

19 ppm   62 ppm

14 ppm   35 ppm

b. $(CH_3)_2CHCHO$

16 ppm   205 ppm

41 ppm

c. $CH_2=CHCH(OH)CH_3$

143 ppm   23 ppm

113 ppm   69 ppm

14.51

a.

C_a C_c C_b C_b C_c C_a

3 different C's
3 signals

b.

C_a C_a C_b C_b C_c C_c

3 signals

C_a C_b C_c C_d C_b C_a

4 signals

C_a C_b C_c C_d C_e C_a

5 signals

14.52

1-hydroxybutan-2-one

A

4-hydroxybutan-2-one

B

The answers for parts (a)–(d) are the same for both compounds.
a. molecular ion for $C_4H_8O_2$ = 88
b. IR absorptions at 3200–3600 (OH), ~3000 (CH), and ~1700 (C=O) cm^{-1}.
c. Four lines in ^{13}C NMR spectrum
d. Four signals in ^{1}H NMR spectrum

e.

quartet at ~ 2.1 ppm

singlet at ~3.5 ppm
singlet anywhere in
the 1–5 ppm region

triplet at ~1.0 ppm

A

triplet at ~3.5 ppm

singlet anywhere in
the 1–5 ppm region

singlet at ~2.0 ppm

triplet at ~2.1 ppm

B

14.53 Use the directions from Answer 14.24.

a. $C_4H_8Br_2$: 0 degrees of unsaturation
IR peak at 3000–2850 cm^{-1}: Csp^3–H bonds
NMR: singlet at 1.87 ppm (6 H) (2 CH$_3$ groups)
singlet at 3.86 ppm (2 H) (CH$_2$ group)

b. $C_3H_6Br_2$: 0 degrees of unsaturation
IR peak at 3000–2850 cm^{-1}: Csp^3–H bonds
NMR: quintet at 2.4 ppm (split by 2 CH$_2$ groups)
triplet at 3.5 ppm (split by 2 H's)

c. $C_5H_{10}O_2$: 1 degree of unsaturation
IR peak at 1740 cm^{-1}: C=O
NMR: triplet at 1.15 ppm (3 H) (CH$_3$ split by 2 H's)
triplet at 1.25 ppm (3 H) (CH$_3$ split by 2 H's)
quartet at 2.30 ppm (2 H) (CH$_2$ split by 3 H's)
quartet at 4.72 ppm (2 H) (CH$_2$ split by 3 H's)

d. $C_6H_{14}O$: 0 degrees of unsaturation
IR peak at 3600–3200 cm^{-1}: O–H
NMR: triplet at 0.8 ppm (6 H) (2 CH$_3$ groups
split by CH$_2$ groups)
singlet at 1.0 ppm (3 H) (CH$_3$)
quartet at 1.5 ppm (4 H) (2 CH$_2$ groups split
by CH$_3$ groups)
singlet at 1.6 ppm (1 H) (O–H proton)

e. $C_6H_{14}O$: 0 degrees of unsaturation
IR peak at 3000–2850 cm^{-1}: Csp^3–H bonds
NMR: doublet at 1.10 ppm (relative area = 6)
(from 12 H's)
septet at 3.60 ppm (relative area = 1)
(from 2 H's)

f. C_3H_6O: 1 degree of unsaturation
IR peak at 1730 cm^{-1}: C=O
NMR: triplet at 1.11 ppm
multiplet at 2.46 ppm
triplet at 9.79 ppm

14.54

Two isomers of $C_9H_{10}O$: 5 degrees of unsaturation (benzene ring likely)

Compound A:
IR absorption at 1742 cm^{-1}: C=O
NMR data:
Absorptions:
singlet at 2.15 (3 H) (CH$_3$ group)
singlet at 3.70 (2 H) (CH$_2$ group)
broad singlet at 7.20 (5 H)
(likely a monosubstituted benzene ring)

Compound B:
IR absorption at 1688 cm^{-1}: C=O
NMR data:
Absorptions:
triplet at 1.22 (3 H) (CH$_3$ group split by 2 H's)
quartet at 2.98 (2 H) (CH$_2$ group split by 3 H's)
multiplet at 7.28–7.95 (5 H)
(likely a monosubstituted benzene ring)

14.55 IR absorptions:
3088–2897 cm^{-1}: sp^2 and sp^3 hybridized C–H
1740 cm^{-1}: C=O
1606 cm^{-1}: benzene ring

triplet at 2.91

5 H
multiplet
7.20–7.35

singlet at 2.02

triplet at 4.25

$C_{10}H_{12}O_2$
W

14.56 IR absorption at 1713 cm^{-1} is due to C=O.

doublet
at 1.09

triplet at 2.43

triplet at 0.91

septet at 2.60

multiplet at 1.6

$C_7H_{14}O$
V

14.57

Compound C:
  molecular ion 146 (molecular formula $C_6H_{10}O_4$)
  IR absorption at 1762 cm^{-1}: **C=O**
  ^{1}H NMR data:
    Absorptions:
    H_a: doublet at 1.47 (3 H) (CH$_3$ group adjacent to CH)
    H_b: singlet at 2.07 (6 H) (2 CH$_3$ groups)
    H_c: quartet at 6.84 (1 H adjacent to CH$_3$)

14.58

$$CH_3CC CH_3 \xrightarrow[\text{[2] } H_2O]{\text{[1] LiC}\equiv\text{CH}}$$

$CH_3-C-C\equiv CH$

D

Compound D:
  molecular ion 84 (molecular formula C_5H_8O)
  IR absorptions at 3600–3200 cm^{-1}: OH
               3303 cm^{-1}: Csp–H
               2938 cm^{-1}: Csp^3–H
               2120 cm^{-1}: C≡C
  ^{1}H NMR data:
    Absorptions:
    H_a: singlet at 1.53 (6 H) (2 CH$_3$ groups)
    H_b: singlet at 2.37 (1 H)
    H_c: singlet at 2.43 (1 H)   alkynyl CH and OH

14.59

Compound E:
$C_4H_8O_2$:
  1 degree of unsaturation
IR absorption at 1743 cm^{-1}: **C=O**
NMR data:

    H_a: quartet at 4.1 (**2 H**)
    H_b: singlet at 2.0 (**3 H**)
    H_c: triplet at 1.4 (**3 H**)

Compound F:
$C_4H_8O_2$:
  1 degree of unsaturation
IR absorption at 1730 cm^{-1}: **C=O**
NMR data:

    H_a: singlet at 4.1 (**2 H**)
    H_b: singlet at 3.4 (**3 H**)
    H_c: singlet at 2.1 (**3 H**)

14.60

Compound H:
$C_8H_{11}N$:
  4 degrees of unsaturation
IR absorptions at 3365 cm^{-1}: N–H
                3284 cm^{-1}: N–H
                3026 cm^{-1}: Csp^2–H
                2932 cm^{-1}: Csp^3–H
                1603 cm^{-1}: due to benzene
                1497 cm^{-1}: due to benzene

NMR data:

  multiplet at 7.2–7.4 ppm, **5 H** on a benzene ring
  H_a: triplet at 2.9 ppm, **2 H**, split by 2 H's
  H_b: triplet at 2.8 ppm, **2 H**, split by 2 H's
  H_c: singlet at 1.1 ppm, **2 H**, no splitting (on NH$_2$)

Compound I:
$C_8H_{11}N$:
  4 degrees of unsaturation
IR absorptions at 3367 cm^{-1}: N–H
                3286 cm^{-1}: N–H
                3027 cm^{-1}: Csp^2–H
                2962 cm^{-1}: Csp^3–H
                1604 cm^{-1}: due to benzene
                1492 cm^{-1}: due to benzene

NMR data:

  multiplet at 7.2–7.4 ppm, **5 H** on a benzene ring
  H_a: quartet at 4.1 ppm, **1 H**, split by 3H's
  H_b: singlet at 1.45 ppm, **2 H**, no splitting (NH$_2$)
  H_c: doublet at 1.4 ppm, **3 H**, split by 1 H

14.61

a. $C_9H_{10}O_2$:
    5 degrees of unsaturation
    IR absorption at 1718 cm^{-1}: **C=O**
    NMR data:

      multiplet at 7.4–8.1 ppm, **5 H** on a benzene ring
      quartet at 4.4 ppm, **2 H**, split by 3 H's
      triplet at 1.3 ppm, **3 H**, split by 2 H's

downfield due to the O atom

b. C_9H_{12}:
    4 degrees of unsaturation
    IR absorption at 2850–3150 cm^{-1}:
    C–H bonds
    NMR data:

      singlet at 7.1–7.4 ppm, **5 H**, benzene
      septet at 2.8 ppm, **1 H**, split by 6 H's
      doublet at 1.3 ppm, **6 H**, split by 1 H

14.62 IR absorption at 1717 cm^{-1} is due to a C=O.

singlet at 1.02 ⟶
CH$_3$ CH$_3$ O
CH$_3$ CH$_3$ ⟵ singlet at 2.13
     H H
⟶ singlet at 2.33

C$_7$H$_{14}$O
R

14.63 IR absorption at 1730 cm^{-1} is due to a C=O. Eight lines in the ^{13}C NMR spectrum means there are eight different types of C.

triplet at 4.20

H$_b$ O
H$_a$
H H
(CH$_3$)$_2$N   O   H ⟵ singlet at 9.97
      H H
      H$_a$ H$_b$

singlet at 2.32

triplet at 3.05

H$_a$ and H$_b$ appear as two doublets.

14.64

a. Compound **J** has a molecular ion at 72: molecular formula **C$_4$H$_8$O**
   1 degree of unsaturation
   IR spectrum at 1710 cm^{-1}: **C=O**
   ^{1}H NMR data (ppm):
   1.0 (triplet, 3 H), split by 2 H's
   2.1 (singlet, 3 H)
   2.4 (quartet, 2 H), split by 3 H's

b. Compound **K** has a molecular ion at 88: molecular formula **C$_5$H$_{12}$O**
   0 degrees of unsaturation
   IR spectrum at 3600–3200 cm^{-1}: **O–H bond**
   ^{1}H NMR data (ppm):
   0.9 (triplet, 3 H), split by 2 H's
   1.2 (singlet, 6 H), due to 2 CH$_3$ groups
   1.5 (quartet, 2 H), split by 3 H's
   1.6 (singlet, 1 H), due to the OH proton

14.65

Compound **L** has a molecular ion at 90: molecular formula **C$_4$H$_{10}$O$_2$**
   0 degrees of unsaturation
   IR absorptions at 2992 and 2941 cm^{-1}: Csp^3–H
   ^{1}H NMR data (ppm):
   H$_a$: 1.2 (doublet, 3 H), split by 1 H
   H$_b$: 3.3 (singlet, 6 H), due to 2 CH$_3$ groups
   H$_c$: 4.8 (quartet, 1 H), split by 3 adjacent H's

O
H + CH$_3$—OH $\xrightarrow{H^+}$

H$_c$
O ⟵ H$_b$
H$_a$ O ⟵ H$_b$
L

14.66

14.67

Compound **O** has a molecular formula **C₁₀H₁₂O.**

5 degrees of unsaturation

IR absorption at 1687 cm⁻¹

¹H NMR data (ppm):

H_a: 1.0 (triplet, 3 H), due to CH_3 group, split by 2 adjacent H's

H_b: 1.7 (sextet, 2 H), split by CH_3 and CH_2 groups

H_c: 2.9 (triplet, 2 H), split by 2 H's

7.4–8.0 (multiplet, 5 H), benzene ring

14.68

Compound **P** has a molecular formula **C₅H₉ClO₂.**

1 degree of unsaturation

¹³C NMR shows 5 different C's, including a C=O.

¹H NMR data (ppm):

H_a: 1.3 (triplet, 3 H), split by 2 H's

H_b: 2.8 (triplet, 2 H), split by 2 H's

H_c: 3.7 (triplet, 2 H), split by 2 H's

H_d: 4.2 (quartet, 2 H), split by CH_3 group

14.69

Compound **Q**: Molecular ion at 86.
Molecular formula: $C_5H_{10}O$:
    1 degree of unsaturation
IR absorption at ~1700 cm^{-1}: **C=O**
NMR data:
    H_a: doublet at 1.1 ppm, 2 CH$_3$ groups split by 1 H
    H_b: singlet at 2.1 ppm, CH$_3$ group
    H_c: septet at 2.6 ppm, 1 H split by 6 H's

14.70

C_6H_{12}:
    1 degree of unsaturation

^{1}H NMR of **T** (ppm):
    H_a: 1.01 (singlet, 9 H)
    H_b: 4.82 (doublet of doublets, 1 H, J = 10, 1.7 Hz)
    H_c: 4.93 (doublet of doublets, 1 H, J = 18, 1.7 Hz)
    H_d: 5.83 (doublet of doublets, 1 H, J = 18, 10 Hz)

^{1}H NMR of **U**: 1.60 (singlet) ppm.

All H's are identical, so there is only one singlet in the NMR.

14.71

a.

$C_6H_{12}O_2$:
    1 degree of unsaturation
IR peak at 1740 cm^{-1}: **C=O**
^{1}**H NMR** 2 signals: 2 types of H's
^{13}**C NMR:** 4 signals: 4 kinds of C's, including one at ~170 ppm due a C=O

b.

C_6H_{10}:
    2 degrees of unsaturation
IR peak at 3000 cm^{-1}: **Csp^3–H bonds**
    peak at 3300 cm^{-1}: **C$_{sp}$–H bond**
    peak at ~2150 cm^{-1}: **C≡C bond**
^{13}**C NMR:** 4 signals: 4 kinds of C's

14.72

a. Because **A** has no absorptions at 1700 cm^{-1} or 3600–3200 cm^{-1}, it has no C=O or OH. An oxygen-containing compound without these functional groups is an ether (or an epoxide). Because **B** is formed from a reaction with HCl, **A** must contain an epoxide, because ethers are unreactive with HCl.

doublet at 3.5 ⟶   CH$_3$ ⟵ doublet at 1.4

H ⟵ quartet of doublets at 3.0

singlet at 3.8 ⟶ CH$_3$O

2 doublets at 6.9 and 7.2

A

b. Epoxide **A** is equally substituted by R groups on both C's. With HCl, the epoxide is protonated first and then backside attack by Cl⁻ forms the chlorohydrin. Attack at the C adjacent to the benzene ring is preferred because the δ+ in the transition state at this carbon can be delocalized on the benzene ring.

The benzene ring and CH₃ group must be trans in the epoxide to give the correct configuration at the two stereogenic centers in the product.

14.73 A second resonance structure for *N,N*-dimethylformamide places the two CH₃ groups in different environments. One CH₃ group is cis to the O atom, and one is cis to the H atom. This gives rise to two different absorptions for the CH₃ groups.

14.74

18-Annulene has 18 π electrons that create an **induced magnetic field** similar to the 6 π electrons of benzene. 18-Annulene has 12 protons that are oriented on the outside of the ring (labeled H₀), and 6 protons that are oriented inside the ring (labeled Hᵢ). The induced magnetic field reinforces the external field in the vicinity of the protons on the outside of the ring. These H₀ protons are deshielded, so they absorb downfield (8.9 ppm). In contrast, the induced magnetic field is opposite in direction to the applied magnetic field in the vicinity of the protons on the inside of the ring. This shields the Hᵢ protons, so they absorb very far upfield, at –1.8 ppm, which is even higher than TMS.

14.75

3-methylbutan-2-ol

Replace a CH$_3$ group with X.

Replace C$_a$.   or   Replace C$_b$.

The CH$_3$ groups are not equivalent to each other, because replacement of each by X forms two diastereomers.

Thus, every C in this compound is different and there are five ^{13}C signals.

14.76

$$CH_3-\overset{\overset{O}{\|}}{\underset{\underset{H_a}{\underset{|}{OCH_3}}}{P}}-OCH_3$$

H$_b$ H$_a$

One P atom splits each nearby CH$_3$ into a doublet by the $n + 1$ rule, making two doublets.

All 6 H$_a$ protons are equivalent.

14.77 a. Splitting pattern:

$J_{ab} = 11$ Hz

$J_{bc} = 4$ Hz

3 Hz

b. Three resonance structures can be drawn for cyclohex-2-enone.

A   **B**   **C**

Resonance structure **C** places a (+) charge on one C of the C=C, deshielding the H attached to it and shifting the absorption downfield.

Chapter 15 Radical Reactions

Chapter Review

General features of radicals
- A radical is a reactive intermediate with an unpaired electron (15.1).
- A carbon radical is sp^2 hybridized and trigonal planar (15.1).
- The stability of a radical increases as the number of C's bonded to the radical carbon increases (15.1).

- Allylic radicals are stabilized by resonance, making them more stable than 3° radicals (15.10).

Radical reactions
[1] Halogenation of alkanes (15.4)

$$R-H \xrightarrow[\substack{h\nu \text{ or } \Delta \\ X = Cl \text{ or } Br}]{X_2} R-X \text{ (alkyl halide)}$$

- The reaction follows a radical chain mechanism.
- The weaker the C–H bond, the more readily the H is replaced by X.
- Chlorination is faster and less selective than bromination (15.6).
- Radical substitution results in racemization at a stereogenic center (15.8).

[2] Allylic halogenation (15.10)

$$\xrightarrow[h\nu \text{ or ROOR}]{NBS} \text{(allylic halide)}$$

- The reaction follows a radical chain mechanism.

[3] Radical addition of HBr to an alkene (15.13)

$$R\diagup\!\!\diagdown \xrightarrow[\substack{h\nu, \Delta, \text{ or} \\ ROOR}]{HBr} R\text{—}Br \text{ (alkyl bromide)}$$

- A radical addition mechanism is followed.
- Br bonds to the less substituted carbon atom to form the more substituted, more stable radical.

[4] Radical polymerization of alkenes (15.14)

polymer

- A radical addition mechanism is followed.

Practice Test on Chapter Review

1.a. Which alkyl halide(s) can be made in good yield by radical halogenation of an alkane?

4. Both (1) and (2) can be made in good yield.
5. Compounds (1), (2), and (3) can all be made in good yield.

b. In which of the following reactions will rearrangement *not* occur?
1. halogenation of an alkane with Cl_2 and heat
2. addition of Cl_2 to an alkene
3. addition of HCl to an alkene
4. Rearrangements do not occur in reactions (1) and (2).
5. Rearrangements do not occur in reactions (1), (2), and (3).

c. Which labeled H is most easily abstracted in a radical halogenation reaction?

1. H_a
2. H_b
3. H_c
4. H_d
5. H_e

d. Which of the labeled C–H bonds in the following compound has the smallest bond dissociation energy?

1. $C–H_a$
2. $C–H_b$
3. $C–H_c$
4. $C–H_d$
5. $C–H_e$

2. (a) Which radical is the most stable? (b) Which radical is the least stable?

A   B   C   D

3. Draw all of the organic products formed in each reaction. Indicate stereochemistry in part (c).

a.
$$\xrightarrow[\text{ROOR}]{\text{HBr}}$$

b.
$$\xrightarrow[\text{h}\nu]{\text{NBS}}$$

c.
$$\xrightarrow[\text{h}\nu]{\text{Br}_2}$$

4. What monomer is needed to make the following polymer?

5. In each box, fill in the appropriate reagents needed to carry out the given reaction. This question involves reactions from Chapter 15, as well as previous chapters.

Answers to Practice Test

1.a. 2
 b. 4
 c. 2
 d. 3

2.a. **A**
 b. **B**

3.a.

3.b. (+ cis isomer)

3.c.

4.

5.

Answers to Problems

15.1 1° Radicals are on C's bonded to one other C; 2° radicals are on C's bonded to two other C's; 3° radicals are on C's bonded to three other C's.

a. **2° radical**

b. **3° radical**

c. **2° radical**

d. **1° radical**

15.2 The stability of a radical increases as the number of alkyl groups bonded to the radical carbon increases. Draw the most stable radical.

a.   b.   c.   d.

15.3 Reaction of a radical with:
- an alkane abstracts a hydrogen atom and creates a new carbon radical.
- an alkene generates a new bond to one carbon and a new carbon radical.
- another radical forms a bond.

a.

b. $CH_2{=}CH_2$ → $\dot{C}H_2{-}CH_2$

c. :$\ddot{C}l$· → :$\ddot{C}l{-}\ddot{C}l$:

d. :$\ddot{C}l$· + ·$\ddot{O}{-}\ddot{O}$· → :$\ddot{C}l{-}\ddot{O}{-}\ddot{O}$·

15.4 **Monochlorination** is a radical substitution reaction in which a Cl replaces a H, thus generating an alkyl halide.

a. Cl_2, Δ

b. Cl_2, Δ

c. Cl_2, Δ

15.5

A

15.6

Initiation:

Propagation:

Termination:

or

or

15.7 The rate-determining step for halogenation reactions is formation of CH₃• + HX.

$$CH_3{-}H \ + \ \cdot\ddot{I}: \longrightarrow \ \cdot CH_3 \ + \ H{-}\ddot{I}:$$

1 bond broken
+435 kJ/mol

1 bond formed
–297 kJ/mol

$\Delta H° = +138$ kJ/mol

This reaction is more endothermic and has a higher E_a than a similar reaction with Cl₂ or Br₂.

15.8 The **weakest C–H bond** in each alkane is the **most readily cleaved** during radical halogenation.

a.

H
3°
most reactive

b.

H
3°
most reactive

c.

H
2°
most reactive

15.9 To draw the product of bromination:

- Draw out the starting material and find the most reactive C–H bond (on the most substituted C).
- The major product is formed by **cleavage of the *weakest* C–H bond.**

15.10 If 1° C–H and 3° C–H bonds were equally reactive there would be nine times as much (CH₃)₂CHCH₂Cl as (CH₃)₃CCl because the ratio of 1° to 3° H's is 9:1. The fact that the ratio is only 63:37 shows that the 1° C–H bond is less reactive than the 3° C–H bond. (CH₃)₂CHCH₂Cl is still the major product, though, because there are nine 1° C–H bonds and only one 3° C–H bond.

15.11

15.12

15.13 The reaction does not occur at the stereogenic center, so leave it as is.

15.14

15.15

Chain propagation:

$$:\ddot{O}=\ddot{N}\cdot \ + \ O_3 \longrightarrow :\ddot{O}=\ddot{N}-\ddot{\underset{..}{O}}\cdot \ + \ O_2$$

$$:\ddot{O}=\ddot{N}-\ddot{\underset{..}{O}}\cdot \ + \ \cdot\ddot{\underset{..}{O}}\cdot \longrightarrow :\ddot{O}=\ddot{N}\cdot \ + \ O_2$$

The radical is re-formed.

15.16 Draw the resonance structure by moving the π bond and the unpaired electron. The hybrid is drawn with dashed lines for bonds that are in one resonance structure but not another. The symbol δ· is used on any atom that has an unpaired electron in any resonance structure.

a.

c.

b.

d.

15.17 Reaction of an alkene with NBS or Br₂ + hν yields allylic substitution products.

a.

b.

c.

15.18

a.

c.

b.

15.19

15.20 Reaction of an alkene with NBS + $h\nu$ yields allylic substitution products.

a. one possible product: **high yield**

b.   c.

Cannot be made in high yield by allylic halogenation.
Any alkene starting material would yield a mixture of allylic halides.

15.21 The weakest C–H bond is most readily cleaved. To draw the hydroperoxide products, add OOH to each carbon that bears a radical in one of the resonance structures.

linoleic acid

This allylic C–H bond is most readily cleaved.

hydroperoxide products:

(E/Z isomers are possible.)

15.22

rosmarinic acid

15.23

a.

b.

c.

15.24 In addition of HBr under radical conditions:

- Br· adds first to form the more stable radical.
- Then H· is added to the carbon radical.

2 radical possibilities:

1° radical
less stable

or

3° radical
***more* stable**
This radical forms.

15.25

a. HBr / ROOR

c. Br₂

b. HBr

15.26

a.

polystyrene

b.

poly(vinyl acetate)

15.27

Initiation:

Propagation:

Repeat Step [3] over and over.

Termination:

[one possibility]

15.28 With Cl₂, each H of the starting material can be replaced by Cl. With Br₂, cleavage of the weakest C–H bond is preferred.

15.29

Abstraction of the phenol H
produces a resonance-stabilized radical.

BHA

15.30

a. increasing bond strength: 2 < 3 < 1

b. and c.

| 1° radical | 2° radical | 3° radical |
| **least stable** | **intermediate stability** | **most stable** |

d. increasing ease of H abstraction: 1 < 3 < 2

15.31 Use the directions from Answer 15.2 to rank the radicals.

2° radical
least stable

3° radical
intermediate stability

allylic radical
most stable

15.32 Draw the radical formed by cleavage of the benzylic C–H bond. Then draw all of the resonance structures. Having more resonance structures (five in this case) makes the radical more stable, and the benzylic C–H bond weaker.

benzylic C–H bond
bond dissociation energy = 356 kJ/mol

15.33

H_d
H_c
H_b H_a

H_a = bonded to an sp^3 3° carbon
H_b = bonded to an allylic carbon
H_c = bonded to an sp^3 1° carbon
H_d = bonded to an sp^3 2° carbon

Increasing ease of abstraction:
$H_c < H_d < H_a < H_b$

15.34

a.

b.

c.

15.35 To draw the product of bromination:
- Draw out the starting material and find the most reactive C–H bond (on the most substituted C).
- The major product is formed by **cleavage of the *weakest* C–H bond.**

a.

b.

c.

15.36 Draw all of the alkane isomers of C_6H_{14} and their products on chlorination. Then determine which letter corresponds to which alkane.

[* = stereogenic center]

15.37 Halogenation replaces a C–H bond with a C–X bond. To find the alkane needed to make each of the alkyl halides, replace the X with a H.

a.

b.

c.

15.38 For an alkane to yield one major product on monohalogenation with Cl$_2$, all of the hydrogens must be identical in the starting material. For an alkane to yield one major product on bromination, it must have a more substituted carbon in the starting material.

a.

This compound can be formed
in high yield from an alkane.

b.

three different
C–H bonds

c.

Br on 2° carbon
The product with Br on 3° carbon
will form predominantly.

These two compounds cannot be
formed in high yield from an alkane.

15.39 A single constitutional isomer is formed in both halogenations in (b) and (c). Bromination often forms a single product by cleavage of the weakest C–H bond. For chlorination to form a single product, the starting material must have only one kind of H that reacts. In (b), a single chlorination product is formed because there is only one type of sp^3 hybridized C–H bond.

a.

$$\xrightarrow[hv]{Cl_2}$$

+

c.

$$\xrightarrow[hv]{Cl_2}$$

$$\xrightarrow[\Delta]{Br_2}$$

$$\xrightarrow[\Delta]{Br_2}$$

b.

$$\xrightarrow[hv]{Cl_2}$$

$$\xrightarrow[\Delta]{Br_2}$$

15.40 In bromination, the predominant (or exclusive) product is formed by cleavage of the weaker C–H bond to form the more stable radical intermediate.

$$\xrightarrow[\Delta]{Br_2}$$

weaker bond
ΔH° = 356 kJ/mol

stronger bond
ΔH° = 460 kJ/mol

C

As usual, more product is formed by
homolysis of the weaker bond.

D

NOT formed

15.41 Chlorination is not selective, so a mixture of products results. Bromination is selective, and the major product is formed by cleavage of the weakest C–H bond.

a.

$$\xrightarrow[\Delta]{Cl_2}$$

+   +   +   +

Y

b.

c.

15.42 Draw the resonance structures by moving the π bonds and the radical.

a.

b.

c.

15.43 Reaction of an alkene with NBS + hν yields allylic substitution products.

a.

b.

(+ Z isomer)

c.

15.44

15.45

a.

b.

c.

d.

e.

f.

(+ Z isomer)

15.46

a.

b.

c.

15.47

A   **B**   **C**   **D**   cyclohexanone   acetone

15.48

a.

b.

c.

d.

15.49

a.

(R)-2-chloropentane

b. There would be seven fractions, because each molecule drawn has different physical properties.

c. Fractions **A, B, D, E,** and **G** would show optical activity.

15.50

15.51

15.52

a.

b.

15.53

a.

CH$_3$—C(CH$_3$)(H)—CH$_3$ + Br$_2$ $\xrightarrow{h\nu}$ CH$_3$—C(CH$_3$)(Br)—CH$_3$ + HBr

C–H bond broken	Br–Br bond broken		C–Br bond formed	H–Br bond formed
+381 kJ/mol	+192 kJ/mol		–272 kJ/mol	–368 kJ/mol

total bonds broken = +573 kJ/mol   total bonds formed = –640 kJ/mol $\Delta H° = -67$ kJ/mol

b. **Initiation:**

:Br—Br: $\xrightarrow[\text{or } \Delta]{h\nu}$:Br· + ·Br:

Propagation:

(CH$_3$)$_3$C—H + ·Br: ⟶ (CH$_3$)$_3$C· + H—Br:

(CH$_3$)$_3$C· + :Br—Br: ⟶ (CH$_3$)$_3$C—Br + ·Br:

c. $\Delta H° =$ (bonds broken) – (bonds formed)
   = (+381 kJ/mol) + (–368 kJ/mol)
   = +13 kJ/mol
   $\Delta H° =$ (bonds broken) – (bonds formed)
   = (+192 kJ/mol) + (–272 kJ/mol)
   = –80 kJ/mol

Termination:
(one possibility)

:Br· + ·Br: ⟶ :Br—Br:

d, e.

Transition state 1:

$$\left[\begin{array}{c} CH_3 \\ CH_3-\overset{\delta\cdot}{C}-CH_3 \\ | \\ H{-}{-}{-}\ddot{B}r{:} \\ \delta\cdot \end{array} \right]^{\ddagger}$$

Transition state 2:

$$\left[\begin{array}{c} CH_3 \\ CH_3-\overset{\delta\cdot}{C}-CH_3 \\ | \\ {:}\ddot{B}r{-}{-}{-}\ddot{B}r{:} \\ \delta\cdot \end{array} \right]^{\ddagger}$$

15.54

Initiation:

Propagation:

Termination:
(one possibility)

$$:\overset{..}{\underset{..}{Br}}\cdot \;+\; \cdot\overset{..}{\underset{..}{Br}}: \;\longrightarrow\; :\overset{..}{\underset{..}{Br}}-\overset{..}{\underset{..}{Br}}:$$

15.55 Calculate the $\Delta H°$ for the propagation steps of the reaction of CH$_4$ with I$_2$ to show why it does not occur at an appreciable rate.

$$CH_3-H \;+\; \cdot\overset{..}{\underset{..}{I}}: \;\longrightarrow\; \overset{\cdot}{C}H_3 \;+\; H-\overset{..}{\underset{..}{I}}: \qquad \Delta H° = +138 \text{ kJ/mol}$$

+435 kJ/mol   −297 kJ/mol

> This step is highly endothermic, making it difficult for chain propagation to occur over and over again.

$$\overset{\cdot}{C}H_3 \;+\; :\overset{..}{\underset{..}{I}}-\overset{..}{\underset{..}{I}}: \;\longrightarrow\; CH_3-I \;+\; \cdot\overset{..}{\underset{..}{I}}: \qquad \Delta H° = -83 \text{ kJ/mol}$$

+151 kJ/mol   −234 kJ/mol

15.56

3,3-dimethylbut-1-ene   2° carbocation   1,2-CH$_3$ shift   3° carbocation   2-bromo-2,3-dimethylbutane

+ Br$^-$

3,3-dimethylbut-1-ene   HBr + peroxide   The 2° radical does NOT rearrange.   1-bromo-3,3-dimethylbutane   + Br·

Addition of HBr without added peroxide occurs by an ionic mechanism and forms a 2° carbocation, which rearranges to a more stable 3° carbocation. The addition of H$^+$ occurs first, followed by Br$^-$. Addition of HBr with added peroxide occurs by a radical mechanism and forms a 2° radical that does not rearrange. In the radical mechanism, Br· adds first, followed by H·.

15.57

a.

b.
(from a.)

d.
(from c.)

c.
(from a.)

e.
(from b.)

15.58

major
product

1-methylcyclohexene
oxide

15.59

$$HC \equiv CH \xrightarrow{NaH} HC \equiv C^- \xrightarrow{CH_3CH_2Br} \text{...} \xrightarrow[\text{Lindlar catalyst}]{H_2} \text{...} \xrightarrow[\text{ROOR}]{HBr} \text{...Br}$$

15.60

a. $CH_3-CH_3 \xrightarrow[h\nu]{Br_2} CH_3CH_2Br \xrightarrow{K^+\ ^-OC(CH_3)_3} CH_2=CH_2 \xrightarrow{Br_2} BrCH_2-CH_2Br \xrightarrow{2\ NaNH_2} HC \equiv CH$

b. $CH_2=CH_2 \xrightarrow{mCPBA}$ $\xrightarrow[\text{[2] } H_2O]{\text{[1] } HC \equiv C^-}$ ... —OH
(from a.)

$HC \equiv CH$ (from a.)
| NaH

c. $HC \equiv CH \xrightarrow{NaH} HC \equiv C^- \xrightarrow[\text{(from a.)}]{CH_3CH_2Br}$ ... $\xrightarrow{NaH}$ ...
(from a.)

$\xrightarrow[\text{(from a.)}]{CH_3CH_2Br}$

$\xrightarrow[NH_3]{Na}$ ...

d. $\xrightarrow[\substack{H_2SO_4 \\ HgSO_4}]{H_2O}$
(from c.)

15.61

$\xrightarrow[h\nu]{Cl_2}$ ...Cl $\xrightarrow{K^+\ ^-OC(CH_3)_3}$ ... $\xrightarrow[\text{[2] } (CH_3)_2S]{\text{[1] } O_3}$ OHC ... CHO

15.62

hexane-2,3-diol

15.63

O$_2$ abstracts a H here.

arachidonic acid

+ HÖÖ·

This process is repeated.

R—H

5-HPETE

another molecule of
arachidonic acid

15.64

+ HÖÖ·

Then, repeat Steps
[2] and [3].

15.65

a.

(+ cis isomer)

b.

O₂ abstracts a H here.

Repeat Steps [2] and [3] again and again.

15.66 For resonance structures **A–F,** an additional resonance form can be drawn that moves the position of the three π bonds in the ring bonded to two OH groups.

a.

A B C

F E D

b. Homolysis of the indicated OH bond is preferred because it allows the resulting radical to delocalize over both benzene rings. Cleavage of one of the other OH bonds gives a radical that delocalizes over only one of the benzene rings.

15.67 Abstraction of the labeled H forms a highly resonance-stabilized radical. Four of the possible resonance structures are drawn.

vitamin C

X

15.68 The monomers used in radical polymerization always contain double bonds.

a.

polyisobutylene

b.

poly(ethyl acrylate)
(used in Latex paints)

15.69

a.

methyl methacrylate

PMMA

b.

hydroxyethyl methacrylate

poly-HEMA

15.70 Polystyrene has H atoms bonded to benzylic carbons—that is, carbons bonded directly to a benzene ring. These C–H bonds are unusually weak because the radical that results from homolysis is resonance stabilized.

No such resonance stabilization is possible for the radical that results from C–H bond cleavage in polyethylene.

15.71

Overall reaction:

Initiation:

carbon radical

Propagation:

Repeat Step [3] over and over. | new C–C bond |

Termination:

[one possibility]

15.72

a. CH_3O-

A

b. The OCH_3 group stabilizes an intermediate carbocation by resonance. This makes **A** react faster than styrene in cationic polymerization.

three of the possible resonance structures

15.73

alternating copolymer

15.74

singlet at 2.23   singlet at 4.04

doublet at 1.69

doublet at 5.85

multiplet at 4.34

C   A   B

15.75

triplet

quintet
minor product

Molecular formula $C_3H_6Cl_2$
Integration: relative area 2:1
Because the compound has 6 H's and the sum of the relative areas is 3, each absorption is due to twice as many H's.
    One signal is due to 4 H's.
    The second signal is due to 2 H's (2 x 1).
^{1}H NMR data:
    quintet at 2.2 (2 H's) split by 4 H's
    triplet at 3.7 (4 H's) split by 2 H's

15.76

+ H–ÖR

+ :Ö: (Repeat Steps [2] and [3].)

15.77

a. The triphenylmethyl radical is highly resonance stabilized, because the radical can be delocalized on each of the benzene rings. As an example using one ring:

In addition, the radical is very sterically hindered, making it difficult to undergo reactions.

b. First, draw the resonance form of the radical that places the unpaired electron on the C that forms the new C–C bond.

c. Hexaphenylethane formation would require two very crowded 3° radicals to combine. The formation of **A** results from a radical on one of the six-membered rings, which is much more accessible for reaction.

d. The ^{1}H NMR spectrum of hexaphenylethane should show signals only in the aromatic region (7–8 ppm), whereas the ^{1}H NMR spectrum of **A** will also have signals for the sp^2 hybridized C–H bonds (4.5–6.0 ppm) of the alkenes, as well as the single H on the sp^3 hybridized carbon. The ^{13}C NMR spectrum of hexaphenylethane should consist of lines due to the 4° C's and the aromatic C's. For **A**, the ^{13}C NMR spectrum will also have lines for the sp^3 and sp^2 hybridized C's that are not contained in the aromatic rings.

15.78

Initiation: R_3SnH + $\cdot Z$ ⟶ $R_3Sn\cdot$ + HZ

Propagation:

15.79

Chapter 16: Conjugation, Resonance, and Dienes

Chapter Review

Conjugation and delocalization of electron density

- The overlap of *p* orbitals on three or more adjacent atoms allows electron density to delocalize, thus adding stability (16.1).
- An allyl carbocation ($CH_2=CHCH_2^+$) is more stable than a 1° carbocation because of *p* orbital overlap (16.2).
- In a system X=Y–Z:, Z is generally sp^2 hybridized to allow the lone pair to occupy a *p* orbital, making the system conjugated (16.5).

Four common examples of resonance (16.3)

[1] The three atom "allyl" system: X=Y–Z ⟷ X–Y=Z   * = +, –, •, or ••

[2] Conjugated double bonds:

[3] Cations having a positive charge adjacent to a lone pair:

[4] Double bonds having one atom more electronegative than the other: X=Y ⟷ $\overset{+}{X}$–$\overset{-}{Y}$: [Electronegativity of Y > X]

Rules on evaluating the relative "stability" of resonance structures (16.4)

[1] Structures with more bonds and fewer charges are better.

all neutral atoms
one more bond
bettter resonance structure

[2] Structures in which every atom has an octet are better.

All second row elements have an octet.

[3] Structures that place a negative charge on a more electronegative element are better.

The (–) charge is on the more electronegative O atom.

better resonance structure

The unusual properties of conjugated dienes

[1] The C–C σ bond joining the two double bonds is unusually short (16.8).

[2] Conjugated dienes are more stable than similar isolated dienes. $\Delta H°$ of hydrogenation is smaller for a conjugated diene than for an isolated diene converted to the same product (16.9).

[3] The reactions are unusual:

- Electrophilic addition affords products of 1,2-addition and 1,4-addition (16.10, 16.11).
- Conjugated dienes undergo the Diels–Alder reaction, a reaction that does not occur with isolated dienes (16.12–16.14).

[4] Conjugated dienes absorb UV light in the 200–400 nm region. As the number of conjugated π bonds increases, the absorption shifts to longer wavelength (16.15).

Reactions of conjugated dienes

[1] Electrophilic addition of HX (X = halogen) (16.10, 16.11)

1,2-product
kinetic product

1,4-product
thermodynamic product

- The mechanism has two steps.
- Markovnikov's rule is followed. Addition of H⁺ forms the *more* stable allylic carbocation.
- The 1,2-product is the kinetic product. When H⁺ adds to the double bond, X⁻ adds to the end of the allylic carbocation to which it is closer (C2 not C4). The kinetic product is formed faster at low temperature.
- The thermodynamic product has the more substituted, more stable double bond. The thermodynamic product predominates at equilibrium. With buta-1,3-diene, the thermodynamic product is the 1,4-product.

[2] Diels–Alder reaction (16.12–16.14)

1,3-diene   dienophile

The three new bonds are labeled in **bold**.

- The reaction forms two σ bonds and one π bond in a six-membered ring.
- The reaction is initiated by heat.
- The mechanism is concerted: all bonds are broken and formed in a single step.
- The diene must react in the *s*-cis conformation (16.13A).
- Electron-withdrawing groups in the dienophile increase the reaction rate (16.13B).
- The stereochemistry of the dienophile is retained in the product (16.13C).
- Endo products are preferred (16.13D).

Practice Test on Chapter Review

1.a. Which of the following statements is true about the Diels–Alder reaction?
 1. The reaction is faster with electron-donating groups in the dienophile.
 2. The reaction is endothermic.
 3. The diene must adopt the *s*-cis conformation.
 4. Statements (1) and (2) are true.
 5. Statements (1), (2), and (3) are all true.

 b. Which of the following statements is true about the absorption of ultraviolet light by unsaturated systems?
 1. Penta-1,4-diene requires light having a wavelength < 200 nm for electron promotion.
 2. Cyclohexa-1,3-diene absorbs ultraviolet light with a wavelength > 200 nm.
 3. As the number of conjugated π bonds increases, the energy difference between the excited state and ground state decreases.
 4. Statements (1) and (2) are true.
 5. Statements (1), (2), and (3) are all true.

 c. Which of the following compounds contains a labeled carbon atom that is sp^2 hybridized?

 1. **A** only
 2. **B** only
 3. **C** only
 4. **A** and **B**
 5. **A**, **B**, and **C**

 d. Which of the following represent valid resonance structures for **A**?

 4. Both (1) and (2) are valid resonance structures.
 5. Structures (1), (2), and (3) are all valid.

2. Name the following compounds and indicate the conformation around the σ bond that joins the two double bonds.

a. b.

3.a. Consider the four hydrocarbons (**A–D**) drawn below. [1] Which compound absorbs the *shortest* wavelength of ultraviolet light? [2] Which compound absorbs the *longest* wavelength of ultraviolet light?

A   **B**   **C**   **D**

b. Consider the four dienes (**A–D**) drawn below. [1] Which diene is *most* reactive in the Diels–Alder reaction? [2] Which diene is the *least* reactive in the Diels–Alder reaction?

A   **B**   **C**   **D**

4. Draw the organic products formed in each reaction. In part (b), label the kinetic and thermodynamic products.

a. O=⬡=O + (diene with OCH₃) → Δ, indicate stereochemistry

b. (bicyclohexenyl) → HBr (1 equiv)

c. (bicyclic diene) → [1] Δ   [2] CH₃O₂C⟍═⟋CO₂CH₃   indicate stereochemistry

5. What diene and dienophile are needed to synthesize the following Diels–Alder adducts?

a.

b.

Answers to Practice Test

1.a. 3
   b. 5
   c. 4
   d. 1

2.a. (4*Z*,6*E*)-6,7-diethyl-2-methyldeca-4,6-diene, *s*-trans

   b. (2*Z*,4*E*)-3-ethyl-6,6-dimethylnona-2,4-diene, *s*-trans

3.a. [1] **A**; [2] **C**
   b. [1] **D**; [2] **A**

4.a.

(both H's up or both H's down)

b.

kinetic

+

thermodynamic

c.

5.a.

+

b.

Answers to Problems

16.1   **Isolated dienes** have two double bonds separated by two or more σ bonds.
Conjugated dienes have two double bonds separated by only one σ bond.

a.

One σ bond separates
two double bonds =
conjugated diene.

b.

Two σ bonds separate
two double bonds =
isolated diene.

c.

One σ bond separates
two double bonds =
conjugated diene.

d.

Four σ bonds separate
two double bonds =
isolated diene.

16.2   **Conjugation** occurs when there are overlapping *p* orbitals on three or more adjacent atoms.
Double bonds separated by two σ bonds are not conjugated.

a.

Six of the carbon atoms are *sp²*
hybridized. Each π bond is
separated by only one σ bond.
conjugated

c.

The two π bonds are
separated by only one σ bond.
conjugated

e.

This carbon is not *sp²* hybridized.
NOT conjugated

b.

The three π bonds are
separated by two or three σ bonds.
NOT conjugated

d.

Three adjacent carbon atoms are *sp²* hybridized
and have an unhybridized *p* orbital.
conjugated

16.3   Two resonance structures differ only in the placement of electrons. All σ bonds stay in the
same place. Nonbonded electrons and π bonds can be moved. To draw the hybrid:
 • Use a dashed line between atoms that have a π bond in one resonance structure and not the
    other.
 • Use a δ symbol for atoms with a charge or radical in one structure but not the other.

a.

resonance hybrid:

$\delta+$

$\delta+$

The + charge is delocalized
on two carbons.

b.

resonance hybrid:

$\delta+$

$\delta+$

The + charge is delocalized
on two carbons.

c.

resonance hybrid:

$\delta+$ $\delta+$

The + charge is delocalized
on two carbons.

16.4 S_N1 reactions proceed via a carbocation intermediate. Draw the carbocation formed on loss of Cl and compare. The more stable the carbocation, the faster the S_N1 reaction.

$CH_2{=}CHCH_2Cl$
3-chloroprop-1-ene
more reactive

$CH_3CH_2CH_2Cl$ is a 1° halide, which does not react by an S_N1 reaction because cleavage of the C–Cl bond forms a highly unstable 1° carbocation.

$\overset{+}{C}H_2{-}CH{=}CH_2 \longleftrightarrow CH_2{=}CH{-}\overset{+}{C}H_2$

resonance-stabilized carbocation

Two resonance structures delocalize the positive charge on 2 C's making 3-chloroprop-1-ene more reactive.

$CH_3CH_2CH_2Cl$
1-chloropropane
less reactive

$CH_3CH_2\overset{+}{C}H_2$
only one Lewis structure
very unstable

16.5

isopentenyl diphosphate   +   **X**   →   farnesyl diphosphate

Five C's of farnesyl diphosphate come from isopentenyl diphosphate, so the remaining 10 C's come from **X**. If the reaction is analogous to the formation of geranyl diphosphate, **X** must contain 10 C's and have an allylic diphosphate.

16.6

a.
Move the charge
and the double bond.

b.
Move the charge
and the double bond.

c.
Move the lone pair.

d.
Move the charge
and the double bond.

16.7   To compare the resonance structures remember:
- Resonance structures with **more bonds** are better.
- Resonance structures in which **every atom has an octet** are better.
- Resonance structures with **neutral atoms** are better than those with charge separation.
- Resonance structures that place a **negative charge on a more electronegative atom** are better.

16.8

16.9   Remember that in any allyl system, there must be *p* orbitals to delocalize the lone pair.

a.

sp² hybridized
trigonal planar geometry

b.

sp² hybridized
trigonal planar geometry

c.

sp² hybridized
trigonal planar geometry

16.10   The **s-cis** conformation has two double bonds on the **same side** of the single bond.
The **s-trans** conformation has two double bonds on **opposite sides** of the single bond.

a. (2*E*,4*E*)-**octa-2,4-diene** in the *s*-trans conformation

double bonds on opposite sides
s-trans

b. (3*E*,5*Z*)- **nona-3,5-diene** in the *s*-cis conformation

Z
E

double bonds on the same side
s-cis

c. (3*Z*,5*Z*)-4,5-dimethyl**deca-3,5-diene** in both the *s*-cis and *s*-trans conformations

Z

Z

s-cis →

s-trans →

Z

Z

16.11

s-cis

conjugated

conjugated →

s-trans

conjugated →

E E

OH

Z

isolated

Z

isolated

O

isolated

OH

HO

Z

Z

isolated

isolated

16.12 Bond length depends on hybridization and percent *s*-character. Bonds with a higher percent *s*-character have smaller orbitals and are shorter.

HC≡C–C≡CH

sp hybridized carbons
50% *s*-character
shortest bond

CH₂=CH–CH=CH₂

sp² hybridized carbons
33% *s*-character
intermediate length

CH₃–CH₃

sp³ hybridized carbons
25% *s*-character
longest bond

16.13 Two equivalent resonance structures delocalize the π bond and the negative charge.

[:O: ⟷ :Ö:⁻]
  :O:⁻   :O:

hybrid: :Ö δ– ← These bond lengths are equal
          :O δ– ← because they are identical.

16.14 The **less stable** (higher energy) **diene** has the **larger heat of hydrogenation.** Isolated dienes are higher in energy than conjugated dienes, so they will have a larger heat of hydrogenation.

a.

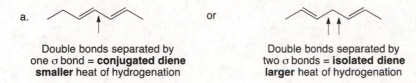

or

Double bonds separated by
one σ bond = **conjugated diene**
smaller heat of hydrogenation

Double bonds separated by
two σ bonds = **isolated diene**
larger heat of hydrogenation

b.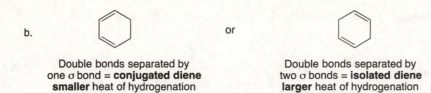

| Double bonds separated by one σ bond = **conjugated diene** **smaller** heat of hydrogenation | or | Double bonds separated by two σ bonds = **isolated diene** **larger** heat of hydrogenation |

16.15 Isolated dienes are higher in energy than conjugated dienes. Compare the location of the double bonds in the compounds below.

0 conjugated double bonds
least stable

2 conjugated double bonds
intermediate stability

3 conjugated double bonds
most stable

16.16 Conjugated dienes react with HX to form 1,2- and 1,4-products.

a.

HCl

Cl
(+ *Z* isomer)
1,2-product

+

Cl
(+ *Z* isomer)
1,4-product

b.

HCl

Cl

isolated diene

c.

HCl

Cl

d.

HCl

Cl

A

+

Cl

B

+

Cl

C

This double bond is more reactive, so **C** is probably a minor product because it results from HCl addition to the less reactive double bond.

16.17 The mechanism for addition of DCl has two steps:
[1] **Addition of D⁺** forms a resonance-stabilized carbocation.
[2] **Nucleophilic attack of Cl⁻** forms 1,2- and 1,4-products.

[1]

D—Cl

:Cl:⁻

D

D

+ :Cl:⁻

[2]

[2]

Cl

D

+

Cl

D

16.18 Label the products as 1,2- or 1,4-products. The 1,2-product is the kinetic product, and the 1,4-product, which has the more substituted double bond, is the thermodynamic product.

This is C1.

The H is added here.

The H is added here.

HCl

1,2-product
kinetic product

+

Cl
1,4-product
thermodynamic product

This is C4.

16.19 To draw the products of a Diels–Alder reaction:
[1] Find the 1,3-diene and the dienophile.
[2] Arrange them so the diene is on the left and the dienophile is on the right.
[3] Cleave three bonds and use arrows to show where the new bonds will be formed.

a.

diene   dienophile   re-draw   Δ

b.

diene
Rotate to
make it s-cis.   dienophile   re-draw   Δ

c.

dienophile   diene
Rotate to
make it s-cis.   re-draw   Δ

16.20 For a diene to be reactive in a Diels–Alder reaction, **a diene must be able to adopt an s-cis conformation.**

s-trans
**cannot rotate
unreactive**

rotate

s-cis
most reactive

The diene is always in the
s-cis conformation.

s-cis
reactive

16.21 Zingiberene reacts much faster than β-sesquiphellandrene as a Diels–Alder diene because its diene is constrained in the s-cis conformation. The diene in β-sesquiphellandrene is constrained in the s-trans conformation, so it is unreactive in the Diels–Alder.

16.22 Electron-withdrawing substituents in the dienophile increase the reaction rate.

no electron-withdrawing groups
least reactive

one electron-withdrawing group
intermediate reactivity

two electron-withdrawing groups
most reactive

16.23 A cis dienophile forms a cis-substituted cyclohexene.
A trans dienophile forms a trans-substituted cyclohexene.

a.

cis dienophile

cis-substituted products

b.

trans dienophile

trans-substituted products

c.

cis dienophile

identical

cis-substituted product

16.24 The **endo product** (with the substituents under the plane of the new six-membered ring) is the preferred product.

a.

endo substituent

b.

both groups **endo**

16.25 To find the diene and dienophile needed to make each of the products:
[1] Find the six-membered ring with a C–C double bond.
[2] Draw three arrows to work backwards.
[3] Follow the arrows to show the diene and dienophile.

16.26

16.27 Conjugated molecules absorb light at a longer wavelength than molecules that are not conjugated.

a.

conjugated
longer wavelength

not conjugated

b.

all double bonds conjugated
longer wavelength

or

one set of
conjugated dienes

16.28 Sunscreens contain conjugated systems to absorb UV radiation from sunlight. Look for conjugated systems in the compounds below.

a.

conjugated system
could be a sunscreen

b.

not a conjugated system

c.

conjugated system
could be a sunscreen

16.29

a. ← s-cis conformation

(2Z,4E)-3,4-dimethylhepta-2,4-diene

b. s-trans conformation

(2E,4Z)-3,4-dimethylocta-2,4-diene

16.30

a. (structure with OCH₃ and O) ⟹ (cyclopentadiene) + (diester with CO₂CH₃ and CH₃O₂C)

b. (bicyclic structure with H, O) ⟹ (diene) + (quinone with O)

16.31 Use the definition from Answer 16.1.

a. CN

2 multiple bonds with only
1 σ bond between
conjugated

b. This C is *sp²*.

1 π bond with
an adjacent
sp² hybridized atom
The lone pair occupies a *p* orbital,
so there are *p* orbitals on
three adjacent atoms.
conjugated

c. 1 π bond with
no adjacent
sp² hybridized atoms
NOT conjugated

d. 1 π bond with
no adjacent
sp² hybridized atoms
NOT conjugated

16.32

a. (resonance structures) ⟷ ⟷ ⟷

b. (resonance structures) ⟷ ⟷

c. (resonance structures) ⟷ ⟷

d. (resonance structures) ⟷ ⟷

e.

f.

16.33 No additional resonance structures can be drawn for compounds (a) and (d).

b.

c.

16.34

16.35

resonance hybrid:

Five resonance structures delocalize the negative charge on five C's, making them all equivalent.

All of the carbons are identical in the anion.

16.36

Draw the products of cleavage of the bond.

ethane $CH_3 — CH_3$

↓

$\cdot CH_3$ + $\cdot CH_3$

Two unstable radicals form.

but-1-ene

$\cdot CH_3$ + $\cdot$ ⟷ $\cdot$

One resonance-stabilized radical forms.
This makes the bond dissociation
energy lower because a more stable radical is formed.

16.37 Use the directions from Answer 16.10.

a. (*Z*)-penta-1,3-diene in the *s-trans* conformation

double bonds on opposite sides
s-trans

b. (2*E*,4*Z*)-1-bromo-3-methylhexa-2,4-diene

Br

c. (2*E*,4*E*,6*E*)-octa-2,4,6-triene

d. (2*E*,4*E*)-3-methylhexa-2,4-diene in the *s-cis* conformation

s-cis

16.38

2*E*,4*E*   2*E*,4*Z*   2*Z*,4*E*   2*Z*,4*Z*

16.39

a. and

(*E*)-hexa-1,3,5-triene
both *s-cis*

(*E*)-hexa-1,3,5-triene
both *s-trans*

different conformations

b. and

(*Z*)-hexa-1,3,5-triene
both *s-cis*

(*Z*)-hexa-1,3,5-triene
both *s-trans*

different conformations

c. and

Z

E

(*E*)-hexa-1,3,5-triene

(*Z*)-hexa-1,3,5-triene

different stereoisomers

16.40 Use the directions from Answer 16.14 and recall that more substituted double bonds are more stable.

Increasing heat of hydrogenation

conjugated diene
one tetra-, one disubstituted
double bond
smallest
heat of hydrogenation

conjugated diene
one di-, one trisubstituted
double bond
smaller intermediate
heat of hydrogenation

isolated diene
one di-, one trisubstituted
double bond
larger intermediate
heat of hydrogenation

isolated diene
both disubstituted
double bonds
largest
heat of hydrogenation

16.41 Conjugated dienes react with HX to form 1,2- and 1,4-products.

a.

HBr

(1 equiv)

Br

isolated diene

major product, formed by addition of HBr to the more substituted C=C

b.

1,2-product   **1,4-product**   (*E* and *Z* isomers can form.)

c.

1,2-product   **1,4-product**   **1,2-product**   (*E* and *Z* isomers)
                                                                 1,4-product

16.42

16.43

This cation forms because it is benzylic and resonance stabilized.

16.44   To draw the mechanism for the reaction of a diene with HBr and ROOR, recall from Chapter 15 that when an alkene is treated with HBr under these radical conditions, the Br ends up on the carbon with more H's to begin with.

Use each resonance structure to react with HBr.

16.45

a, b.

X

H adds here at C1.

Y

Cl added at C2.
1,2-product
kinetic product

C4

Z

Cl added at C4.
1,4-product
thermodynamic product

Y is the kinetic product because of the proximity effect. H and Cl add across two adjacent atoms.
Z is the thermodynamic product because it has a more stable trisubstituted double bond.

c.

Addition occurs at the labeled double bond due to the stability of the carbocation intermediate.

The two resonance structures for this allylic cation are 3° and 2° carbocations.

| more stable intermediate |
| **Addition occurs here.** |

If addition occurred at the other C=C, the following allylic carbocation would form:

The two resonance structures for this allylic cation are 1° and 2° carbocations.

less stable

16.46 Addition of HCl at the terminal double bond forms a carbocation that is highly resonance stabilized because it is both allylic and benzylic. Such stabilization does not occur when HCl is added to the other double bond. This gives rise to two products of electrophilic addition.

1,2-product

1,4-product

(+ three more resonance structures that delocalize the positive charge onto the benzene ring)

16.47 There are two possible products:

The 1,2-product is always the kinetic product because of the proximity effect. In this case, it is also the thermodynamic (more stable) product because it contains a more highly substituted C=C (trisubstituted) than the 1,4-product (disubstituted). Thus, the 1,2-product is the major product at high and low temperature.

16.48 The electron pairs on O can be donated to the double bond through resonance. This increases the electron density of the double bond, making it less electrophilic and therefore less reactive in a Diels–Alder reaction.

16.49 Use the directions from Answer 16.19.

16.50 Use the directions from Answer 16.25.

b.

c.

d.

16.51

This pathway is **preferred** because the dienophile has electron-withdrawing C=O groups that make it more reactive.

CH_2
$||$
CH_2
+

no electron-withdrawing groups
less reactive

16.52

a.

diene   **dienophile**

b.

diene   **dienophile**

16.53

For the 1,2-product, carbons with unlike charges would react. This is favored because the electron-rich and the electron-poor C's can bond.

For the 1,3-product, there are no partial charges of opposite sign on reacting carbons. This arrangement is less attractive.

1,2-disubstituted product
major

1,3-disubstituted product
minor

The major product is formed when the circled carbons with a δ+ and δ− react.

resonance hybrids:

16.54

X

Y

aldrin

This double bond is more electron rich, so it is epoxidized more readily.

mCPBA
1 equiv

Z
dieldrin

16.55

In each problem, the synthesis must begin with the preparation of cyclopentadiene from dicyclopentadiene.

dicyclopentadiene

2 cyclopentadiene

a.

COOCH₃

[1] OsO₄
[2] NaHSO₃, H₂O

CO₂CH₃

CO₂CH₃

b.

mCPBA

c.

16.56

a.

diene   dienophile

re-draw

Δ

b.

CO₂CH₃

diene   dienophile

re-draw

Δ

CO₂CH₃

CO₂CH₃

16.57 A transannular Diels–Alder reaction forms a tricyclic product from a monocyclic starting material.

Δ

16.58

A   B   C   D

a. **D** has the largest heat of hydrogenation because it contains no conjugated double bonds.
b. **C** has the smallest heat of hydrogenation because all three double bonds are conjugated.
c. **C** absorbs the longest wavelength of UV light because it has three C=C's in conjugation.
d. **C** reacts fastest in the Diels-Alder reaction because it contains a 1,3-diene constrained in the *s*-cis conformation. The 1,3-dienes in **A** and **B** are both *s*-trans.

16.59

two resonance
structures

16.60

a.

conjugated diene   HI   **1,2-product**   +   **1,4-product**

b.

diene   **dienophile**   Δ

c.

diene   +   **trans dienophile**   Δ

d.

conjugated diene   HBr   **1,2-product**   **1,4-product**

16.61

+ PP$_i$

re-draw

+ HB$^+$ B:

16.62 The more stable the carbocation, the faster the S$_N$1 reaction. The carbocation from **A** is more stable because the lone pairs on the O atom of the OCH$_3$ afford additional resonance stabilization.

more stable carbocation

additional resonance structure with
all atoms having an octet

less stable carbocation

This resonance structure is destabilized
because the (+) charge is adjacent to the
δ+ on the C=O atom.

16.63 The mechanism is E1, with formation of a resonance-stabilized carbocation.

16.64

a.

loss of H (from β_1 C) + Cl
conjugated
more stable
major product

loss of H (from β_2 C) + Cl
more substituted

b. Dehydrohalogenation generally forms the more stable product. In this reaction, loss of
H from the β_1 carbon forms a more stable conjugated diene, so this product is preferred
even though it does not contain the more substituted C=C.

16.65

singlet at 1.42 ppm

isoprene

mCPBA

Each H is a doublet of doublets in the 5.2–5.4 ppm region.

doublet of doublets at 5.5–5.7 ppm

two doublets at 2.7–2.9 ppm

16.66

1 H: doublet of doublets at 6.0 ppm

The IR shows an OH absorption at 3200–3600 cm^{-1}.

B

OH peak at 1.5 ppm

Each H is a doublet of doublets in the 4.9–5.2 ppm region.

6 H: singlet at 1.3 ppm

16.67

A   B

All double bonds in **A** are conjugated, whereas the double bonds in the circled ring in **B** are not conjugated. As a result, **A** absorbs at a longer wavelength of UV light.

16.68 **C** and **D** absorb UV light because they contain conjugated π systems.

C
conjugated

D
conjugated

16.69

The phenol makes ferulic acid an antioxidant. Loss of H forms a highly stabilized phenoxy radical that inhibits radical formation during oxidation.

ferulic acid

The highly conjugated π system makes it a sunscreen.

16.70 There are two possible modes of addition of HBr to allene. When H$^+$ adds to the terminal carbon, a 2° vinyl carbocation is formed, which affords 2-bromoprop-1-ene after nucleophilic attack.

[1] $CH_2=C=CH_2$

H—Br

+   :Br:$^-$

2° vinyl carbocation

Br

2-bromoprop-1-ene

When H$^+$ adds to the middle carbon, an intermediate carbocation with a (+) charge adjacent to the C=C is formed. This carbocation is not resonance stabilized (at least initially), because the two C=C's of allene are oriented 90° to each other, a geometry that does not allow for overlap of the C=C with the empty p orbital of the carbocation.

As a result, path [1] forms a more stable carbocation and is preferred.

16.71

cyclohexanamine

aniline

+ other resonance structures

N is surrounded by 3 atoms and 1 lone pair, so it is sp^3 hybridized.

N has a lone pair on an atom adjacent to a C=C. N must be sp^2 hybridized to delocalize the lone pair by resonance.

Basicity is a measure of how willing an atom is to donate an electron pair. The lone pair on the N in cyclohexanamine is localized on N, whereas the lone pair on the N in aniline is delocalized onto the benzene ring. As a result, the lone pair in cyclohexanamine is much more available for donation to a proton than the lone pair in aniline. This makes cyclohexanamine much more basic than aniline.

16.72

O$_3$ cleaves the C=C.

Endo product formed.

The Diels–Alder reaction establishes the stereochemistry of the four carbons on the six-membered ring. All four carbon atoms bonded to the six-membered ring are on the same side.

16.73

Diels–Alder reaction

A

loss of CO$_2$

B

16.74

16.75 A retro Diels–Alder reaction forms a conjugated diene. An intramolecular Diels–Alder reaction then forms **N**.

Chapter 17 Benzene and Aromatic Compounds

Chapter Review

Comparing aromatic, antiaromatic, and nonaromatic compounds (17.7)

- **Aromatic compound**
  - A cyclic, planar, completely conjugated compound that contains $4n + 2$ π electrons ($n = 0, 1, 2, 3$, and so forth).
  - An aromatic compound is more stable than a similar acyclic compound having the same number of π electrons.

- **Antiaromatic compound**
  - A cyclic, planar, completely conjugated compound that contains $4n$ π electrons ($n = 0, 1, 2, 3$, and so forth).
  - An antiaromatic compound is less stable than a similar acyclic compound having the same number of π electrons.

- **Nonaromatic compound**
  - A compound that lacks one (or more) of the requirements to be aromatic or antiaromatic.

Properties of aromatic compounds

- Every carbon in an aromatic ring has a p orbital to delocalize electron density (17.2).
- Aromatic compounds are unusually stable. $\Delta H°$ for hydrogenation is much less than expected, given the number of degrees of unsaturation (17.6).
- Aromatic compounds do not undergo the usual addition reactions of alkenes (17.6).
- ^{1}H NMR spectra show highly deshielded protons because of ring currents (17.4).

Examples of aromatic compounds with 6 π electrons (17.8)

benzene   pyridine   pyrrole   cyclopentadienyl anion   tropylium cation

Examples of compounds that are not aromatic (17.8)

not cyclic   not planar   not completely conjugated

Practice Test on Chapter Review

1. Give the IUPAC name for each of the following compounds.

a.

b.

c.

2. Label each compound as aromatic, nonaromatic, or antiaromatic. Choose only **one** possibility.
Assume all completely conjugated rings are planar.

a.   c.   e.   g.   i.

b.   d.   f.   h.   j.

3. Answer the following questions about compounds **A–E** drawn below.

A   B   C   D   E

a. How is nitrogen N_a in compound **A** hybridized?
b. In what type of orbital does the lone pair on N_a reside?
c. How is nitrogen N_b in compound **B** hybridized?
d. In what type of orbital does the lone pair on N_b reside?
e. Which of the labeled bonds in compound **C** is the shortest?
f. Which of the labeled bonds in compound **C** is the longest?
g. When considering both compounds **D** and **E,** which of the labeled hydrogen atoms (H_a, H_b, H_c, or H_d) is the most acidic? Give only **one** answer.
h. When considering both compounds **D** and **E,** which of the labeled hydrogen atoms (H_a, H_b, H_c, or H_d) is the least acidic? Give only **one** answer.

Answers to Practice Test

1.a. 2-*sec*-butyl-5-nitrophenol
  b. *o*-isobutyltoluene
  c. 2-*tert*-butyl-4-nitrophenol

2.a. nonaromatic
  b. aromatic
  c. nonaromatic
  d. antiaromatic
  e. aromatic
  f. nonaromatic
  g. aromatic
  h. aromatic
  i. antiaromatic
  j. aromatic

3.a. sp^2
  b. p
  c. sp^2
  d. sp^2
  e. 4
  f. 3
  g. H_a
  h. H_c

Answers to Problems

17.1 Move the electrons in the π bonds to draw all major resonance structures.

diphenhydramine

17.2 Look at the hybridization of the atoms involved in each bond. Carbons in a benzene ring are surrounded by three groups and are sp^2 hybridized.

C_{sp^2}–C_{sp^3}

C_{sp^2}–H_{1s}

C_{sp^2}–C_{sp^2}
C_p–C_p
shortest of all
the indicated bonds

C_{sp^2}–C_{sp^2}

C_{sp^2}–C_{sp^2}
C_p–C_p

17.3

- To name a benzene ring with **one substituent,** name the substituent and add the word *benzene*.
- To name a **disubstituted ring,** select the correct prefix (ortho = 1,2; meta = 1,3; para = 1,4) and alphabetize the substituents. Use a common name if it is a derivative of that monosubstituted benzene.
- To name a **polysubstituted ring,** number the ring to give the lowest possible numbers and then follow other rules of nomenclature.

17.4 Work backwards to draw the structures from the names.

a. isobutylbenzene

c. *cis*-1,2-diphenylcyclohexane

e. 4-chloro-1,2-diethylbenzene

b. *o*-dichlorobenzene

d. *m*-bromoaniline

f. 3-*tert*-butyl-2-ethyltoluene

17.5

propofol

17.6

Molecular formula $C_{10}H_{14}O_2$: 4 degrees of unsaturation
IR absorption at 3150–2850 cm^{-1}: sp^2 and sp^3 hybridized C–H bonds
NMR absorptions (ppm):
    1.4 (triplet, 6 H)
    4.0 (quartet, 4 H)
    6.8 (singlet, 4 H)

17.7 Count the different types of carbons to determine the number of ^{13}C NMR signals.

a.
4 types of C's in the benzene ring
6 signals

b.
All C's are different.
7 signals

c.
4 signals

17.8 The less stable compound has a larger heat of hydrogenation.

A
benzene ring, more stable
smaller $\Delta H°$

B
no benzene ring, less stable
larger $\Delta H°$

17.9 The protons on sp^2 hybridized carbons in aromatic hydrocarbons are highly deshielded and absorb at 6.5–8 ppm, whereas hydrocarbons that are not aromatic show an absorption at 4.5–6 ppm, typical of protons bonded to the C=C of an alkene.

a.
aromatic ring
H's ~6.5–8 ppm

b.
not aromatic
alkene H's ~4.5–6 ppm

c.
protons on isolated C=C
~4.5–6 ppm
protons on benzene ring
~6.5–8 ppm

17.10 To be aromatic, a ring must have $4n + 2$ π electrons.

16 π e$^-$
$4n$
$4(4) = 16$
antiaromatic

20 π e$^-$
$4n$
$4(5) = 20$
antiaromatic

22 π e$^-$
$4n + 2$
$4(5) + 2 = 22$
aromatic

17.11

17.12 In determining if a heterocycle is aromatic, count a nonbonded electron pair if it makes the ring aromatic in calculating $4n + 2$. Lone pairs on atoms already part of a multiple bond cannot be delocalized in a ring, and so they are never counted in determining aromaticity.

a.

Count one lone pair from O.
$4n + 2 = 4(1) + 2 = 6$
aromatic

b.

no lone pair from O
$4n + 2 = 4(1) + 2 = 6$
aromatic

c.

With one lone pair from each O there would be 8 π electrons. If O's are sp^3 hybridized, the ring is not completely conjugated.
not aromatic

d.

Both N atoms are part of a double bond, so the lone pairs cannot be counted: there are 6 π electrons.
$4n + 2 = 4(1) + 2 = 6$
aromatic

17.13

quinine
(antimalarial drug)

N is sp^3 hybridized and the lone pair is in an sp^3 hybrid orbital.

N is sp^2 hybridized and the lone pair is *not* part of the aromatic ring. This means it occupies an sp^2 hybrid orbital.

17.14

a. The five-membered ring is aromatic because it has 6 π electrons, two from each π bond and two from the N atom that is not part of a double bond.

b, c.

sitagliptin

sp^2 hybridized N lone pair in p orbital

sp^2 hybridized N lone pair in p orbital

sp^2 hybridized N lone pair in sp^2 orbital

sp^2 hybridized N lone pair in sp^2 orbital

sp^3 hybridized N lone pair in sp^3 orbital

17.15 Compare the conjugate base of cyclohepta-1,3,5-triene with the conjugate base of cyclopentadiene. Remember that the compound with the more stable conjugate base will have a lower pK_a.

cyclohepta-1,3,5-triene
$pK_a = 39$

8 π Electrons make this conjugate base especially unstable (**antiaromatic**).

Because the conjugate base is unstable, the pK_a of cyclohepta-1,3,5-triene is **high**.

cyclopentadiene

6 π electrons
aromatic conjugate base
very stable anion

Because the conjugate base is very stable, the pK_a of cyclopentadiene is much **lower**.

17.16 The compound with the most stable conjugate base is the most acidic.

Conjugate bases:

no resonance
delocalization

most unstable base so
least acidic acid

two resonance
structures

The acid is
intermediate in acidity.

aromatic conjugate base
most stable

The acid is the
most acidic.

17.17

17.18 To be aromatic, the ions must have $4n + 2$ π electrons. Ions in (b) and (c) do not have the right number of π electrons to be aromatic.

a. 2π electrons
$4(0) + 2 = 2$
aromatic

d. 10π electrons
$4(2) + 2 = 10$
aromatic

17.19

A =

The NMR indicates that **A** is aromatic. (a) The C's of the triple bond are *sp* hybridized. (b) Each triple bond has one set of electrons in *p* orbitals that overlap with other *p* orbitals on adjacent atoms in the ring. This overlap allows electrons to delocalize. Each C of the triple bonds also has a *p* orbital in the plane of the ring. The electrons in these *p* orbitals are localized between the C's of the triple bond, not delocalized in the ring. (c) Although **A** has 24 π e$^-$ total, only 18 e$^-$ are delocalized around the ring.

17.20 In using the inscribed polygon method, always draw the vertex pointing down.

2 antibonding MOs

1 bonding MO

2π electrons
All bonding MOs are filled.
aromatic

17.21 Draw the inscribed pentagons with the vertex pointing down. Then draw the molecular orbitals (MOs) and add the electrons.

2 antibonding MOs

3 bonding MOs

Cation:

4 π electrons
Not all bonding MOs are filled.
not aromatic

Radical:

5 π electrons
Not all bonding MOs are filled.
not aromatic

17.22 C_{60} would exhibit only one ^{13}C NMR signal because all the carbons are identical.

17.23

a.

isopropyl group

Name as a derivative of toluene.

Answer: *p*-isopropyltoluene
seven lines in ^{13}C NMR (lettered C_a–C_g)

b.

Name as a derivative of phenol.

Answer: 3-bromo-5-nitrophenol
All C's are different, so there are six lines in the ^{13}C NMR.

17.24

a.

not completely
conjugated
not aromatic

b.

6 π electrons
aromatic

c.

not *sp²* hybridized

aromatic ring

The six-membered ring is aromatic. The bicyclic ring system with the O atom is not completely conjugated, so it is nonaromatic.

17.25

a. If the Kekulé description of benzene was accurate, only one product would form in Reaction [1], but there would be four (not three) dibromobenzenes (**A–D**), because adjacent C–C bonds are different—one is single and one is double. Thus, compounds **A** and **B** would *not* be identical. **A** has two Br's bonded to the same double bond, but **B** has two Br's on different double bonds.

b. In the resonance description, only one product would form in Reaction [1], because all C's are identical, but only three dibromobenzenes (ortho, meta, and para isomers) are possible. **A** and **B** are identical because each C–C bond is identical and intermediate in bond length between a C–C single and C–C double bond.

A   B   C   D

17.26

C₈H₁₀:

C₈H₉Br:

3 isomers   **2 isomers**   **3 isomers**   **1 isomer**

17.27 To name the compounds use the directions from Answer 17.3.

a. *sec*-butylbenzene

b. *m*-chloroethylbenzene

c. toluene
p-chlorotoluene

d. aniline
o-chloroaniline

e. aniline
2,3-dibromoaniline

f. phenol (OH at C1)
2,5-dinitrophenol

g. 1-ethyl-3-isopropyl-5-propylbenzene

h. *cis*-1-bromo-2-phenylcyclohexane

17.28

a. *p*-dichlorobenzene

b. *p*-iodoaniline

c. *o*-bromonitrobenzene

d. 2,6-dimethoxytoluene

e. 2-phenylprop-2-en-1-ol

f. *trans*-1-benzyl-3-phenylcyclopentane

or

17.29

a, b. constitutional isomers of molecular formula C_8H_9Cl and names of the trisubstituted benzenes

stereoisomers
for this isomer

2-chloro-1,3-dimethyl-
benzene

1-chloro-2,3-dimethyl-
benzene

4-chloro-1,2-dimethyl-
benzene

1-chloro-2,4-dimethyl-
benzene

1-chloro-3,5-dimethyl-
benzene

2-chloro-1,4-dimethyl-
benzene

c. stereoisomers

17.30 Count the electrons in the π bonds. Each π bond holds two electrons.

a.   10 π electrons

b.   7 π electrons

c.   10 π electrons

d.   14 π electrons

e.   12 π electrons

17.31 To be aromatic, the compounds must be cyclic, planar, completely conjugated, and have $4n + 2$ π electrons.

a.   Circled C's are not sp^2.
not completely conjugated
not aromatic

c.   12 π electrons
does **not** have $4n + 2$
π electrons
not aromatic

b.   Circled C is not sp^2.
not completely conjugated
not aromatic

d.   12 π electrons
does **not** have $4n + 2$
π electrons
not aromatic

17.32 In determining if a heterocycle is aromatic, count a nonbonded electron pair if it makes the ring aromatic in calculating $4n + 2$. Lone pairs on atoms already part of a multiple bond cannot be delocalized in a ring, so they are never counted in determining aromaticity.

a.

6 π electrons
counting a lone pair from S
4(1) + 2 = 6
aromatic

c.

not aromatic

e.

not aromatic

g.

6 π electrons
counting the lone pair
from N
4(1) + 2 = 6
aromatic

b.

6 π electrons
counting a lone pair from O
4(1) + 2 = 6
aromatic

d.

10 π electrons
4(2) + 2 = 10
aromatic

f.

6 π electrons,
counting a lone pair from O
4(1) + 2 = 6
aromatic

h.

Count
these 2e⁻.

10 π electrons
4(2) + 2 = 10
aromatic

These lone pairs
are on doubly
bonded N atoms,
so they can't be
counted.

17.33

a.

Circled C's are
not sp².
not aromatic

c.

Circled C is
not sp².
not aromatic

b.

10 π electrons
in 10-membered ring
4(2) + 2 = 10
aromatic

d.

4 π electrons
4(1) = 4
antiaromatic

17.34

6 π electrons
in this ring

6 π electrons
in this ring

A

A resonance structure can be drawn for **A** that places a negative charge in the five-membered ring and a positive charge in the seven-membered ring. This resonance structure shows that each ring has six π electrons, making it aromatic. The molecule possesses a dipole such that the seven-membered ring is electron deficient and the five-membered ring is electron rich.

17.35 Each compound is completely conjugated. A compound with $4n + 2$ π electrons is especially stable, whereas a compound with $4n$ π electrons is especially unstable.

pentalene

8 π electrons
4(2) = 8
antiaromatic
unstable

azulene

10 π electrons
4(2) + 2 = 10
aromatic
very stable

heptalene

12 π electrons
6(2) = 12
antiaromatic
unstable

17.36

sp² hybridized but
with lone pair in *p* orbital

purine

a. Each N atom is *sp²* hybridized.
b. The three unlabeled N atoms are *sp²* hybridized with lone pairs in one of the *sp²* hybrid orbitals. The labeled N has its lone pair in a *p* orbital.
c. 10 π electrons
d. Purine is cyclic, planar, completely conjugated, and has 10 π electrons [4(2) + 2], so it is aromatic.

17.37

methotrexate

a. Every N atom is either part of a double bond or is bonded directly to a double bond, so each N atom is *sp²* hybridized.
b. Electron pairs on boxed-in N atoms are located in *p* orbitals to overlap with adjacent π bonds. Lone pairs on all other N atoms are in *sp²* hybrid orbitals.
c. The bicyclic ring system contains 10 π electrons from the five double bonds, making it aromatic. All lone pairs on N atoms in the rings are not delocalized.

17.38

C

a. 16 total π electrons
b. 14 π electrons delocalized in the ring. [Note: Two of the electrons in the triple bond are localized between two C's, perpendicular to the π electrons delocalized in the ring.]
c. By having two of the *p* orbitals of the C–C triple bond co-planar with the *p* orbitals of all the C=C's, the total number of π electrons delocalized in the ring is 14. 4(3) + 2 = 14, so the ring is **aromatic.**

17.39 A second resonance structure can be drawn for the six-membered ring that gives it three π bonds, thus making it aromatic with six π electrons.

6 π electrons

17.40 The rate of an S$_N$1 reaction increases with increasing stability of the intermediate carbocation.

Increasing reactivity

4 π electrons	2° carbocation	6 π electrons
antiaromatic		**aromatic**
very *unstable* intermediate		very *stable* intermediate

The aromatic carbocation is delocalized over the whole ring, making it a very stable intermediate and most easily formed in an S$_N$1 reaction.

Increasing stability

17.41

Two additional resonance structures are not drawn.

17.42 α-Pyrone reacts like benzene because it is aromatic. A second resonance structure can be drawn showing how the ring has six π electrons. Thus, α-pyrone undergoes reactions characteristic of aromatic compounds—that is, substitution rather than addition.

α-pyrone   6 π electrons

17.43

a.

cyclopropenyl radical

b.   and

pyrrole

c.

phenanthrene

17.44

Naphthalene can be drawn as three resonance structures:

In two of the resonance structures bond (a) is a double bond, and
bond (b) is a single bond. Therefore, bond (b) has more single
bond character, making it longer.

17.45

a.

and

pyrrole

Pyrrole is less resonance stabilized than benzene because
four of the resonance structures have charges, making them
less good.

b.

and

furan

Furan is less resonance stabilized than pyrrole because its O atom is less basic,
so it donates electron density less "willingly." Thus, charge-separated resonance
forms are more minor contributors to the hybrid than the charge-separated
resonance forms of pyrrole.

17.46 The compound with the more stable conjugate base is the stronger acid. Draw and compare the
conjugate bases of each pair of compounds.

conjugate bases

a.

more acidic

resonance-stabilized
but not aromatic

6 π electrons, aromatic
more stable conjugate base
Its acid is more acidic.

b.

more acidic

6 π electrons, aromatic
more stable conjugate base
Its acid is more acidic.

antiaromatic
highly destabilized conjugate
base

17.47

indene   + NaNH₂ ⟶   + NH₃

Na⁺

The conjugate base of indene has 10 π electrons, making it aromatic
and very stable. Therefore, indene is more acidic than many hydrocarbons.

17.48

A

H_b is most acidic because its
conjugate base is aromatic (6 π electrons).

B

H_c is least acidic because its
conjugate base has 8 π electrons, making it
antiaromatic.

17.49

pyrrole   conjugate base ⟶

Both pyrrole and the conjugate base of pyrrole have 6 π
electrons in the ring, making them both aromatic. Thus,
deprotonation of pyrrole does not result in a gain of aromaticity
because the starting material is aromatic to begin with.

cyclopentadiene

more acidic ⟶   conjugate base ⟶

Cyclopentadiene is not aromatic, but the conjugate base has 6 π
electrons and is therefore aromatic. This makes the C–H bond in
cyclopentadiene more acidic than the N–H bond in pyrrole, because
deprotonation of cyclopentadiene forms an aromatic conjugate base.

17.50

a.

pyrrole

Protonation at C2 forms conjugate acid **A**
because the positive charge can be delocalized
by resonance. There is no resonance
stabilization of the positive charge in **B.**

b.

A

pK_a = 0.4

Loss of a proton from **A** (which is not aromatic) gives two electrons to N, and forms pyrrole, which has six π electrons that can then delocalize in the five-membered ring, making it aromatic. This makes deprotonation a highly favorable process, and **A** more acidic.

pK_a = 5.2

C

Both **C** and its conjugate base pyridine are aromatic. Since **C** has six π electrons, it is already aromatic to begin with, so there is less to be gained by deprotonation, and **C** is thus less acidic than **A**.

17.51

a.

— 3 antibonding MOs

- 2 nonbonding MOs

— 3 bonding MOs

cyclooctatetraene and its 8 π electrons

cyclooctatetraene →(2 K)→ dianion of cyclooctatetraene

b. Even if cyclooctatetraene were flat, it has two unpaired electrons in its HOMOs (nonbonding MOs), so it cannot be aromatic.

c. The dianion has 10 π electrons.

d. The two additional electrons fill the nonbonding MOs; that is, all the bonding and nonbonding MOs are filled with electrons in the dianion.

e. The dianion is aromatic because its HOMOs are completely filled, and it has no electrons in antibonding MOs.

17.52

4 antibonding MOs

5 bonding MOs

cyclononatetraenyl cation

8 π electrons
antiaromatic

4 antibonding MOs

5 bonding MOs

cyclononatetraenyl radical

9 π electrons
not aromatic

4 antibonding MOs

5 bonding MOs

cyclononatetraenyl anion

10 π electrons
aromatic

All bonding MOs are filled.

17.53 The number of different types of C's = the number of signals.

a.

5 different C's

b.

all unique
9 different C's

c.

3 different C's

d.

4 different C's

17.54 Draw the three isomers and count the different types of carbons in each. Then match the structures with the data.

ortho isomer	meta isomer	para isomer
5 types of C	6 types of C	4 types of C
5 lines in spectrum	6 lines in spectrum	4 lines in spectrum
Spectrum [B]	Spectrum [A]	Spectrum [C]

17.55

a. $C_{10}H_{14}$: IR absorptions at 3150–2850 (sp^2 and sp^3 hybridized C–H), 1600, and 1500 (due to a benzene ring) cm^{-1}

^{1}H NMR data:

Absorption	ppm	# of H's	Explanation
doublet	1.2	6	6 H's adjacent to 1 H
singlet	2.3	3	CH_3
septet	3.1	1	1 H adjacent to 6 H's
multiplet	7–7.4	4	a disubstituted benzene ring

$(CH_3)_2CH$ group

You can't tell from these data where the two groups are on the benzene ring. They are not para, because the para arrangement usually gives two sets of distinct peaks (resembling two doublets), so there are two possible structures—ortho and meta isomers.

or

b. C_9H_{12}: ^{13}C NMR signals at 21, 127, and 138 ppm → means three different types of C's.
^{1}H NMR shows two types of H's: 9 H's probably means 3 CH_3 groups; the other 3 H's are very deshielded so they are bonded to a benzene ring.
Only one possible structure fits:

c. C_8H_{10}: IR absorptions at 3108–2875 (sp^2 and sp^3 hybridized C–H), 1606, and 1496 (due to a benzene ring) cm^{-1}

^{1}H NMR data:

Absorption	ppm	# of H's	Explanation		Structure:
triplet	1.3	3	3 H's adjacent to 2 H's		
quartet	2.7	2	2 H's adjacent to 3 H's		
multiplet	7.3	5	a monosubstituted benzene ring		

17.56

a. Compound **A:** Molecular formula $C_8H_{10}O$

IR absorption at 3150–2850 (sp^2 and sp^3 hybridized C–H) cm^{-1}

^{1}H NMR data:

Absorption	ppm	# of H's	Explanation	Structure:
triplet	1.4	3	3 H's adjacent to 2 H's	
quartet	3.95	2	2 H's adjacent to 3 H's	
multiplet	6.8–7.3	5	a monosubstituted benzene ring	

b. Compound **B:** Molecular formula $C_9H_{10}O_2$

IR absorption at 1669 (C=O) cm^{-1}

^{1}H NMR data:

Absorption	ppm	# of H's	Explanation	Structure:
singlet	2.5	3	CH$_3$ group	
singlet	3.8	3	CH$_3$ group	
doublet	6.9	2	2 H's on a benzene ring	
doublet	7.9	2	2 H's on a benzene ring	

It would be hard to distinguish these two compounds with the given data.

17.57

IR absorptions:
3500–3200 cm^{-1} (O–H)
3150–2850 cm^{-1} (C–H bonds)
1621 and 1585 cm^{-1} (benzene ring)

basic structure of thymol

Thymol must have this basic structure, given the NMR and IR data because it is a trisubstituted benzene ring with one singlet and two doublets in the NMR at ~6.9 ppm. However, which group [OH, CH$_3$, or CH(CH$_3$)$_2$] corresponds to X, Y, and Z is not readily distinguished with the given data. The correct structure for thymol is given.

17.58

^{13}C NMR has four lines that are located in the aromatic region (~110–155 ppm), corresponding to the four different types of carbons in the aromatic ring of the para isomer. The ortho and meta isomers have six different C's, and so six lines would be expected for each of them.

17.59 Because tetrahydrofuran has a higher boiling point and is more water soluble, it must be more polar and have stronger intermolecular forces than furan. There are two contributing factors. One lone pair on furan's O atom is delocalized on the five-membered ring to make it aromatic. This makes it less available for H-bonding with water and other intermolecular interactions. Also, the C–O bonds in furan are less polar than the C–O bonds in tetrahydrofuran because of hybridization. The sp^2 hybridized C's of furan pull a little more electron density towards them than do the sp^3 hybridized C's of tetrahydrofuran. This counteracts the higher electronegativity of O compared to C to a small extent.

17.60

rizatriptan

a. Rizatriptan contains three aromatic rings.
b. The circled N atom is sp^3 hybridized. All other N atoms are either part of a π bond or bonded to a π bond, making them sp^2 hybridized.
c. The lone pair on the circled N atom occupies an sp^3 hybrid orbital. The lone pairs on the boxed-in N atoms are in a p orbital. The lone pairs on the unlabeled N atoms occupy an sp^2 hybrid orbital.

d.

one additional resonance structure

e.

17.61

a, b.

electron pair in an *sp²* hybridized orbital

zolpidem

electron pair in a *p* orbital so it can delocalize

The ring system is aromatic with 10 π electrons, 8 π electrons from the double bonds and 2 π electrons from the N atom common to both rings.

c.

17.62

a.

The enol form is more stable because the enol
double bond makes a highly conjugated system.
The enol OH can also intramolecularly hydrogen
bond to the nearby carbonyl O atom.

b.

The enol O–H proton is more acidic than an alcohol
O–H proton because the conjugate base is
resonance stabilized.

c. Curcumin is colored because it has many conjugated π electrons, which shift absorption of light from the UV to the visible region.

d. Curcumin is an antioxidant because it contains a phenol. Homolytic cleavage affords a resonance-stabilized phenoxy radical, which can inhibit oxidation from occurring, much like vitamin E and BHT in Chapter 15.

phenoxy radical
Resonance delocalizes the radical on the ring and C chain of curcumin.

17.63

a. Pyrazole rings are aromatic because they have six π electrons—two from the lone pair on the N atom that is not part of the double bond and four from the double bonds.

b.

c.

d. The N atom in the NH bond in the pyrazole ring is sp^2 hybridized with 33% s-character, increasing the acidity of the N–H bond. The N–H bond of CH_3NH_2 contains an sp^3 hybridized N atom (25% s-character).

17.64 Both **A** and **B** are cyclic, and if the lone pair of electrons on N is in a p orbital, they are completely conjugated with 10 π electrons, a number that satisfies Hückel's rule. To be aromatic, **A** and **B** must be planar, and the internal bond angles of **A** and **B** would be much larger than 120°, the theoretical bond angle of sp^2 hybridized C's. The fact that **A** is aromatic means that the lone pair on N occupies a p orbital, so it can delocalize onto the nine-membered ring. The stabilization gained by being aromatic is greater than any angle strain. With **B,** the lone pair on N is also delocalized onto the C=O, making it less available for donation to the ring, so the ring is not aromatic.

A
2 electrons in a p orbital

10 π electrons

B
Electron pair is less available so the ring is not aromatic.

17.65 With 14 π electrons in the double bonds, the system is aromatic [4(3) + 2 = 14 π electrons]. The ring current generated by the circulating π electrons deshields the protons on the C=C's, so they absorb downfield (8.14–8.67 ppm). The CH$_3$ groups, however, are very shielded because they lie above and below the plane, so they absorb far upfield (–4.25 ppm). The dianion of **C** now has 16 π electrons, making it antiaromatic, so the position of the absorptions reverses. The C=C protons are now shielded (–3 ppm), and the CH$_3$ protons are now deshielded (21 ppm).

17.66 A second resonance structure for **A** shows that the ring is completely conjugated and has six π electrons, making it aromatic and especially stable. A similar charge-separated resonance structure for **B** makes the ring completely conjugated, but gives the ring four π electrons, making it antiaromatic and especially unstable.

17.67 The conversion of carvone to carvacrol involves acid-catalyzed isomerization of two double bonds and tautomerization of a ketone to an enol tautomer. In this case the enol form is part of an aromatic phenol. Each isomerization of a C=C involves Markovnikov addition of a proton, followed by deprotonation.

17.68 Resonance structures for triphenylene:

Resonance structures **A–H** all keep three double and three single bonds in the three six-membered rings on the periphery of the molecule. This means that each ring behaves like an isolated benzene ring undergoing substitution rather than addition because the π electron density is delocalized within each six-membered ring. Only resonance structure **I** does not have this form. Each C–C bond of triphenylene has four (or five) resonance structures in which it is a single bond and four (or five) resonance structures in which it is a double bond.

Resonance structures for phenanthrene:

With phenanthrene, however, four of the five resonance structures keep a double bond at the labeled C's. (Only **C** does not.) This means that these two C's have more double bond character than other C–C bonds in phenanthrene, making them more susceptible to addition rather than substitution.

17.69

113 ppm

The negative charge and increased electron density make the carbon more shielded and shift the absorption upfield.

130 ppm

The positive charge and decreased electron density make the carbon deshielded and shift the absorption downfield.

Chapter 18 Reactions of Aromatic Compounds

Chapter Review

Mechanism of electrophilic aromatic substitution (18.2)

- Electrophilic aromatic substitution follows a two-step mechanism. Reaction of the aromatic ring with an electrophile forms a carbocation, and loss of a proton regenerates the aromatic ring.
- The first step is rate-determining.
- The intermediate carbocation is stabilized by resonance; a minimum of three resonance structures can be drawn. The positive charge is always located ortho or para to the new C–E bond.

| (+) ortho to E | (+) para to E | (+) ortho to E |

Three rules describing the reactivity and directing effects of common substituents (18.7–18.9)

[1] All ortho, para directors except the halogens activate the benzene ring.
[2] All meta directors deactivate the benzene ring.
[3] The halogens deactivate the benzene ring.

Summary of substituent effects in electrophilic aromatic substitution (18.6–18.9)

	Substituent	Inductive effect	Resonance effect	Reactivity	Directing effect
[1]	R = alkyl	donating	none	activating	ortho, para
[2]	Z = N or O	withdrawing	donating	activating	ortho, para
[3]	X = halogen	withdrawing	donating	deactivating	ortho, para
[4]	Y (δ+ or +)	withdrawing	withdrawing	deactivating	meta

Five examples of electrophilic aromatic substitution
[1] Halogenation—Replacement of H by Cl or Br (18.3)

- Polyhalogenation occurs on benzene rings substituted by OH and NH₂ (and related substituents) (18.10A).

[2] Nitration—Replacement of H by NO₂ (18.4)

[3] Sulfonation—Replacement of H by SO₃H (18.4)

[4] Friedel–Crafts alkylation—Replacement of H by R (18.5)

- Rearrangements can occur.
- Vinyl halides and aryl halides are unreactive.
- The reaction does not occur on benzene rings substituted by meta deactivating groups or NH₂ groups (18.10B).
- Polyalkylation can occur.

Variations:

[1] with alcohols

[2] with alkenes

[5] Friedel–Crafts acylation—Replacement of H by RCO (18.5)

ketone

- The reaction does not occur on benzene rings substituted by meta deactivating groups or NH₂ groups (18.10B).

Nucleophilic aromatic substitution (18.13)
[1] Nucleophilic substitution by an addition–elimination mechanism

X = F, Cl, Br, I
A = electron-withdrawing group

- The mechanism has two steps.
- Strong electron-withdrawing groups at the ortho or para position are required.
- Increasing the number of electron-withdrawing groups increases the rate.
- Increasing the electronegativity of the halogen increases the rate.

[2] Nucleophilic substitution by an elimination–addition mechanism

X = halogen

- Reaction conditions are harsh.
- Benzyne is formed as an intermediate.
- Product mixtures may result.

Other reactions of benzene derivatives
[1] Benzylic halogenation (18.14)

benzylic bromide

[2] Oxidation of alkyl benzenes (18.15A)

benzoic acid

- A benzylic C–H bond is needed for reaction.

[3] Reduction of ketones to alkyl benzenes (18.15B)

[4] Reduction of nitro groups to amino groups (18.15C)

Practice Test on Chapter Review

1.a. Which of the following statements is true about an ethoxy substituent ($-OCH_2CH_3$) on a benzene ring?
1. OCH_2CH_3 increases the rates of both electrophilic substitution and nucleophilic substitution.
2. OCH_2CH_3 decreases the rates of both electrophilic substitution and nucleophilic substitution.
3. OCH_2CH_3 increases the rate of electrophilic substitution and decreases the rate of nucleophilic substitution.
4. OCH_2CH_3 decreases the rate of electrophilic substitution and increases the rate of nucleophilic substitution.
5. None of these statements is true.

b. Which of the following statements is true about a $-CO_2CH_3$ group on a benzene ring?
1. CO_2CH_3 increases the rates of both electrophilic substitution and nucleophilic substitution.
2. CO_2CH_3 decreases the rates of both electrophilic substitution and nucleophilic substitution.
3. CO_2CH_3 increases the rate of electrophilic substitution and decreases the rate of nucleophilic substitution.
4. CO_2CH_3 decreases the rate of electrophilic substitution and increases the rate of nucleophilic substitution.
5. None of these statements is true.

c. Which of the following is *not* a valid resonance structure for the carbocation that results from ortho attack of an electrophile on $C_6H_5C(CH_3)=CH_2$?

2. Draw the organic products formed in the following reactions.

3. (a) Considering the compound drawn below, which ring is *most* reactive in electrophilic aromatic substitution? (b) Which ring is the *least* reactive in electrophilic aromatic substitution?

4. Classify each substituent as [1] ortho, para activating, [2] ortho, para deactivating, or [3] meta deactivating.
   a. –Br
   b. –CH₂CH₂CH₂Br
   c. –COOH
   d. –NHCOCH₂CH₃
   e. –N(CH₂)₆COCH₃
   f. –CCl₃

5. What reagents are needed to convert toluene ($C_6H_5CH_3$) to each compound?
   a. C_6H_5COOH
   b. $C_6H_5CH_2Br$
   c. *p*-bromotoluene
   d. *o*-nitrotoluene
   e. *p*-ethyltoluene
   f.

Answers to Practice Test

1.a. 3
 b. 4
 c. 4

2.

a.

 +

b.

c.

d. CH₃O—

e.

f.

 +

g. O₂N—

h.

 +

3. a. **A**
   b. **C**

4.a. 2
  b. 1
  c. 3
  d. 1
  e. 1
  f. 3

5.a. KMnO₄
  b. Br₂, *hv*
  c. Br₂, FeBr₃
  d. HNO₃, H₂SO₄
  e. CH₃CH₂Cl, AlCl₃
  f. CH₃CH₂COCl, AlCl₃

Answers to Problems

18.1 The π electrons of benzene are delocalized over the six atoms of the ring, increasing benzene's stability and making them less available for electron donation. With an alkene, the two π electrons are localized between the two C's, making them more nucleophilic and thus more reactive with an electrophile than the delocalized electrons in benzene.

18.2

18.3 Reaction with Cl_2 and $FeCl_3$ as the catalyst occurs in two parts. First is the formation of an electrophile, followed by a two-step substitution reaction.

[1]

Lewis base   Lewis acid   electrophile

[2]

resonance-stabilized carbocation

[3]

+ HCl + $FeCl_3$

18.4 There are two parts in the mechanism. The first part is formation of an electrophile. The second part is a two-step substitution reaction.

A

[1]

electrophile

[2]

resonance-stabilized carbocation

[3]

+ H_2SO_4

B

18.5 Friedel–Crafts alkylation results in the transfer of an alkyl group from a halogen to a benzene ring. In Friedel–Crafts acylation an acyl group is transferred from a halogen to a benzene ring.

a.

b.

c.

18.6 Remember that an acyl group is transferred from a Cl atom to a benzene ring. To draw the acid chloride, substitute a Cl for the benzene ring.

a.

c.

b.

18.7 To be reactive in a Friedel–Crafts alkylation reaction, the X must be bonded to an sp^3 hybridized carbon atom.

a.
sp^2
unreactive

b.
sp^3
reactive

c.
sp^2
unreactive

d.
sp^3
reactive

18.8 The product has an "unexpected" carbon skeleton, so rearrangement must have occurred.

[1]

[2]

[3]

18.9 Both alkenes and alcohols can form carbocations for Friedel–Crafts alkylation reactions.

a.

c.

b.

d.

18.10

18.11 In parts (b) and (c), a 1,2-shift occurs to afford a rearrangement product.

18.12

b.

[+ 7 resonance structures]

+ HB⁺   **X**

18.13

a. —CH₂CH₂CH₂CH₃
    alkyl group
    electron donating

b. —Br
    halide
    electron withdrawing

c. —OCH₂CH₃
    electronegative O
    electron withdrawing

18.14   Electron-donating groups place a negative charge in the benzene ring. Draw the resonance structures to show how –OCH₃ puts a negative charge in the ring. Electron-withdrawing groups place a positive charge in the benzene ring. Draw the resonance structures to show how –COCH₃ puts a positive charge in the ring.

a.

b.

18.15   To classify each substituent, look at the atom bonded directly to the benzene ring. All R groups and Z groups (except halogens) are electron donating. All groups with a positive charge, δ+, or halogens are electron withdrawing.

a.
lone pair on O
electron donating

b.
halogen
electron withdrawing

c.
R group
electron donating

18.16 **Electron-donating groups** make the compound *react faster* than benzene in electrophilic aromatic substitution. **Electron-withdrawing groups** make the compound *react more slowly* than benzene in electrophilic aromatic substitution.

a.
electron withdrawing
reacts slower

b.
electron withdrawing
reacts slower

c.
lone pairs on O
electron donating
reacts faster

d.
halogen
electron withdrawing
reacts slower

e.
R group
electron donating
reacts faster

18.17 **Electron-donating groups** make the compound *more reactive* than benzene in electrophilic aromatic substitution. **Electron-withdrawing groups** make the compound *less reactive* than benzene in electrophilic aromatic substitution.

a.
R group
electron donating
more reactive

b.
two OH's
electron donating
more reactive

c.
C with 2 electronegative O's
electron withdrawing
less reactive

d.
electron withdrawing
less reactive

18.18 Chlorine inductively withdraws electron density and decreases the rate of electrophilic aromatic substitution. The closer the Cl is to the ring, the larger the effect it has. The larger the number of Cl's, the larger the effect.

least reactive

intermediate reactivity

most reactive

18.19 Especially stable resonance structures have all atoms with an octet. Carbocations with additional electron donor R groups are also more stable structures. Especially unstable resonance structures have adjacent like charges.

a.

especially stable with additional R group
stabilized carbocation

b.

especially stable
All atoms have an octet.

c.

especially unstable
2 adjacent (+) charges

18.20 Polyhalogenation occurs with highly activated benzene rings containing OH, NH₂, and related groups with a catalyst.

18.21 Friedel–Crafts reactions do not occur with strongly deactivating substituents including NO₂, or with NH₂, NR₂, or NHR groups.

a.

strongly deactivating

b.

Cl is an o,p director.

c.

d.

18.22 To draw the product of a reaction with these disubstituted benzene derivatives and HNO₃, H₂SO₄, remember the following:

- If the two directing effects reinforce each other, the new substituent will be on the position reinforced by both.
- If the directing effects oppose each other, the stronger activator wins.
- No substitution occurs between two meta substituents.

a.

o,p

meta

$$\xrightarrow[\text{H}_2\text{SO}_4]{\text{HNO}_3}$$

c.

o,p

meta

$$\xrightarrow[\text{H}_2\text{SO}_4]{\text{HNO}_3}$$

+

b.

o,p (strong)

o,p

opposing effects
stronger OCH₃ wins out

$$\xrightarrow[\text{H}_2\text{SO}_4]{\text{HNO}_3}$$

+

d.

o,p

o,p

$$\xrightarrow[\text{H}_2\text{SO}_4]{\text{HNO}_3}$$

+

18.23

a.

$$\xrightarrow[]{\text{SO}_3,\ \text{H}_2\text{SO}_4}$$

Put meta director on first.

$$\xrightarrow[]{\text{Cl}_2,\ \text{FeCl}_3}$$

c.

$$\xrightarrow[]{\text{CH}_3\text{Cl},\ \text{AlCl}_3}$$

(+ ortho isomer)

$$\xrightarrow[]{\text{Br}_2}$$

Br goes ortho to
the stronger activator.

b.

$$\xrightarrow[\text{AlCl}_3]{\text{ClCOCH}_3}$$

$$\xrightarrow[]{\text{HNO}_3,\ \text{H}_2\text{SO}_4}$$

18.24

a.

$$\xrightarrow[]{\text{NaOCH}_3}$$

b.

$$\xrightarrow[]{^-\text{OH}}$$

18.25

$$\longrightarrow$$

:OC(CH₃)₃

+ HOC(CH₃)₃

$$\longrightarrow$$

(+ 2 additional
resonance structures)

$$\longrightarrow$$

+ Cl⁻

18.26

a.

$$\xrightarrow[\text{NH}_3]{\text{NaNH}_2}$$

b.

c.

18.27

Two different benzynes are possible.

Ortho, meta, and para products are formed.

18.28 This reaction proceeds via a radical bromination mechanism and two radicals are possible: **A** (2° and benzylic) and **B** (1°). **B** (which leads to $C_6H_5CH_2CH_2Br$) is much less stable, so this radical is not formed and only $C_6H_5CH(Br)CH_3$ is formed as product.

18.29

a.

(+ para isomer)

b.

c.

d.

(from c.)

18.30 First use an acylation reaction, and then reduce the carbonyl group to form the alkyl benzenes.

a.

b.

18.31

p-isobutylacetophenone
(+ ortho isomer)

18.32

a.

b.

c.

(+ para isomer)

18.33

a.

(+ ortho isomer)

b.

Br ← o,p director
o,p director

Both are o,p directors, but they are meta to each other. The alkyl group must be obtained by reduction of a carbonyl.

c.

(+ ortho isomer)

18.34

A

a. Br$_2$ / FeBr$_3$

b. HNO$_3$ / H$_2$SO$_4$

c. CH$_3$CH$_2$COCl / AlCl$_3$

B

a. Br$_2$ / FeBr$_3$

b. HNO$_3$ / H$_2$SO$_4$

c. CH$_3$CH$_2$COCl / AlCl$_3$

18.35 Intramolecular Friedel–Crafts acylation occurs on the more activated aromatic ring.

CH₃O groups activate this ring.

18.36 OH is an ortho, para director.

a.

b.

c.

d.

18.37

a.

No reaction

b.

No Friedel–Crafts reaction

c.

18.38

a.

b.

c.

d.

e.

f.

18.39 Watch out for rearrangements.

a.

2° carbocation

b.

rearrangement

c.

rearrangement

18.40

a.

KMnO₄

b.

(+ cis isomer)

c.

[1] Cl$_2$, FeCl$_3$

[2] Zn(Hg), HCl

d.

CH$_3$NH$_2$

H$_2$ (excess)

Pd-C

18.41

C bonded to 2 H's must use acylation followed by reduction.

a.

AlCl$_3$

Zn(Hg)

HCl

C bonded to 1 H can be added directly by alkylation.

b.

AlCl$_3$

c.

AlCl$_3$

Zn(Hg)

HCl

Method [1]

CH$_3$CH$_2$Cl

AlCl$_3$

Method [2]

Ethyl group can be introduced by two methods.

18.42

KOH

KOH

A

A second resonance structure can be drawn for the **C** ring, showing that it has six π electrons.

Rings **A** and **B** contain 10 π electrons, eight from the double bonds and two from O, making this ring system aromatic.

6 π electrons

18.43

A   B   NaH   C

18.44

D

Path [1]
S$_N$2

Path [2]
nucleophilic aromatic
substitution

+ NaH

+ NaH

18.45

a.

[1] CH$_3$COCl, AlCl$_3$

[2] Cl$_2$, FeCl$_3$

A

Step [1] won't work because a Friedel–Crafts reaction can't be done on a deactivated benzene ring, as is the case with the SO$_3$H substituent. Even if Step [1] did work, the second step would introduce Cl meta to SO$_3$H, not para as drawn.

Alternate synthesis:

Cl$_2$
FeCl$_3$

CH$_3$COCl
AlCl$_3$

(+ para isomer)

SO$_3$
H$_2$SO$_4$

(+ isomer)

b.

[1] CH₃CH₂CH₂CH₂Cl, AlCl₃

[2] HNO₃, H₂SO₄

= **B**

Step [1] involves a Friedel–Crafts alkylation using a 1°
alkyl halide that will undergo rearrangement, so that a
butyl group will not be introduced as a side chain.

Alternate synthesis:

[1] NaH

[2] CH₃Cl

CH₃(CH₂)₂COCl

AlCl₃

(+ ortho isomer)

Zn(Hg), HCl

HNO₃

H₂SO₄

B

18.46 Use the directions from Answer 18.17 to rank the compounds.

a.

least reactive

**intermediate
reactivity**

most reactive

b.

least reactive

**intermediate
reactivity**

most reactive

18.47

[1]

a. withdraw
b. donate
c. less
d. deactivate

[2]

a. withdraw
b. withdraw
c. less
d. deactivate

[3]

a. withdraw
b. donate
c. more
d. activate

18.48

a. **B**   b. **A**   c.

+

18.49

a. more electron rich due to N atom **faster**

b. less electron rich due to (+) charge on N **slower**

c. electron withdrawing — O_2N ... less electron rich due to (+) charge on N and electron-withdrawing NO_2 group **slower**

d. electron withdrawing ... electron donating — Effects cancel out. **similar in reactivity to benzene**

18.50 Electrophilic addition of HBr proceeds by the more stable carbocation.

especially stable resonance structure having all atoms with an octet [+ 4 additional resonance structures]

major product

When the (+) charge is benzylic to the benzene ring with the OCH_3 group, additional resonance stabilization is present, so this pathway is preferred. Such stabilization does not result when the (+) charge is benzylic to the benzene ring with the NO_2 group.

18.51

ortho, para director

Ortho and para products are isolated.

With ortho and para attack there is additional resonance stabilization that delocalizes the positive charge onto the nitroso group. Such additional stabilization is not possible with meta attack. This makes –NO an ortho, para director. Since the N atom bears a partial (+) charge (because it is bonded to a more electronegative O atom), the –NO group inductively withdraws electron density, thus deactivating the benzene ring towards electrophilic attack. In this way, the –NO group resembles the halogens. Thus, the electron-donating resonance effect makes –NO an o,p director, but the electron-withdrawing inductive effect makes it a deactivator.

Ortho attack:

especially stable

Meta attack:

Para attack:

especially stable

18.52

alkyl group on the benzene ring
R stabilizes (+) charges on the o,p positions by an electron-donating inductive effect. This group behaves like any other R group so that ortho and para products are formed in electrophilic aromatic substitution.

(+) charge on atom bonded to the benzene ring
Drawing resonance structures in electrophilic aromatic substitution results in especially unstable structures for attack at the o,p positions—two (+) charges on adjacent atoms. This doesn't happen with meta attack, so meta attack is preferred. This is identical to the situation observed with all meta directors.

18.53 Increasing the number of electron-withdrawing groups (especially at the ortho and para positions to the leaving group) increases the rate of nucleophilic aromatic substitution. Increasing the electronegativity of the halogen increases the rate.

a.

chlorobenzene
least reactive

m-fluoronitro-
benzene

p-fluoronitro-
benzene
most reactive

b.

1-fluoro-3,5-dinitro-
benzene
least reactive

1-fluoro-3,4-
dinitrobenzene

1-fluoro-2,4-dinitro-
benzene
most reactive

c.

4-chloro-3-nitro-
toluene
least reactive

4-fluoro-3-
nitrotoluene

1-fluoro-2,4-dinitro-
benzene
most reactive

18.54

resonance-stabilized carbocation

Use both resonance forms to show how two products are formed.

+ HCl + AlCl$_3$

18.55

18.56

18.57

a. The product has one stereogenic center.

stereogenic center

b. The mechanism for Friedel–Crafts alkylation with this 2° halide involves formation of a trigonal planar carbocation. Because the carbocation is achiral, it reacts with benzene with equal probability from two possible directions (above and below) to afford an optically inactive, racemic mixture of two products.

18.58 The reaction follows the two-step addition–elimination mechanism for nucleophilic aromatic substitution.

Resonance structure **A** is stabilized because the negative charge is located on an electronegative N. This makes nucleophilic aromatic substitution on 2-chloropyridine faster than a similar reaction with chlorobenzene, which has no N atom to stabilize the intermediate negative charge.

18.59 There is no electron-withdrawing group on the benzene ring, so the mechanism likely proceeds via elimination–addition.

18.60

This product is formed.   This product is *not* formed.

Attack to form **A** proceeds via a carbocation for which **7** resonance structures can be drawn. Four resonance structures contain an intact benzene ring.

Attack to form **B** proceeds via a carbocation for which **6** resonance structures can be drawn. Only two resonance structures contain an intact benzene ring.

A reaction that occurs by way of the more stable carbocation is preferred, so product **A** is formed.

18.61

[+ 3 resonance structures]

[+ 3 resonance structures]

18.62 Benzyl bromide forms a resonance-stabilized intermediate that allows it to react rapidly under S_N1 conditions.

Formation of a resonance-stabilized carbocation:

The electron-withdrawing NO_2 group will destabilize the carbocation, so the benzylic halide will be *less* reactive, while the electron-donating OCH_3 group will stabilize the carbocation, so the benzylic halide will be *more* reactive.

18.63

e.

f.

g.

h.

18.64

a.

b.

c.

18.65

a.

b.

18.66

a.

b.

(from a.)   (+ ortho isomer)

c.

(from a.)   (+ ortho isomer)

d.

(from b.)

18.67 One possibility:

(+ ortho isomer)

ibufenac

18.68

18.69

^{1}H NMR data of compound **A** (C$_8$H$_9$Br):

Absorption	ppm	# of H's	Explanation
triplet	1.2	3	3 H's adjacent to 2 H's
quartet	2.6	2	2 H's adjacent to 3 H's
two signals	7.1 and 7.4	4	para disubstituted benzene

Structure:

^{1}H NMR data of compound **B** (C$_8$H$_9$Br):

Absorption	ppm	# of H's	Explanation
triplet	3.1	2	2 H's adjacent to 2 H's
triplet	3.5	2	2 H's adjacent to 2 H's
multiplet	7.1–7.4	5	monosubstituted benzene

Structure:

18.70 IR absorption at 1717 cm^{-1} means compound **C** has a C=O.

^{1}H NMR data of compound **C** (C$_{10}$H$_{12}$O):

Absorption	ppm	# of H's	Explanation
singlet	2.1	3	3 H's
triplet	2.8	2	2 H's adjacent to 2 H's
triplet	2.9	2	2 H's adjacent to 2 H's
multiplet	7.1–7.4	5	monosubstituted benzene

Structure:

18.71

^{1}H NMR data of compound **X** (C$_{10}$H$_{12}$O):

Absorption	ppm	# of H's	Explanation
doublet	1.3	6	6 H's adjacent to 1 H
septet	3.5	1	1 H adjacent to 6 H's
multiplet	7.4–8.1	5	monosubstituted benzene

Structure:

^{1}H NMR data of compound **Y** ($C_{10}H_{14}$):

Absorption	ppm	# of H's	Explanation	Structure:
doublet	0.9	6	6 H's adjacent to 1 H	
multiplet	1.8	1	1 H adjacent to many H's	
doublet	2.5	2	2 H's adjacent to 1 H	
multiplet	7.1–7.3	5	monosubstituted benzene	

18.72

^{1}H NMR spectral data:

1.4 (singlet, 18 H) (a)
2.27 (singlet, 3 H) (b)
5.0 (singlet, 1 H) (c)
7.0 (singlet, 2 H) (d) ppm

p-cresol

2-methyprop-1-ene
(2 equiv)

BHT ($C_{15}H_{24}O$)

Repeat to add the second $C(CH_3)_3$ group.

18.73

18.74 Five resonance structures can be drawn for phenol, three of which place a negative charge on the ortho and para carbons. These illustrate that the electron density at these positions is increased, thus shielding the protons at these positions and shifting the absorptions to lower chemical shift. Similar resonance structures cannot be drawn with a negative charge at the meta position, so it is more deshielded and absorbs farther downfield, at higher chemical shift.

(–) charges on the ortho and para positions

18.75

a. Pyridine: The electron-withdrawing inductive effect of N makes the ring electron poor. Also, electrophiles E^+ can react with N, putting a positive charge on the ring. This makes the ring less reactive with another positively charged species.

To understand why substitution occurs at C3, compare the stability of the carbocation formed by attack at C2 and C3.

Electrophilic attack on N:

Electrophilic attack at C2:

Electrophilic attack at C3:

less reactive than benzene

N does not have an octet.
(+) charge on an electronegative N atom
poor resonance structure
Attack at C2 does not occur.

better resonance structures

Attack at C3 forms a more stable carbocation, so attack at C3 occurs. Attack at C4 generates a carbocation of similar stability to attack at C2, so attack at C4 does *not* occur.

b. Pyrrole is more reactive than benzene because the C's are more electron rich. The lone pair on N has an electron-donating resonance effect.

more reactive than benzene

Attack at C2:

Attack at C3:

2-position
more resonance structures
attack at C2

3-position

fewer resonance structures
Attack at C3 does not occur.

Attack at C2 forms a more stable carbocation, so electrophilic substitution occurs at C2.

18.76

18.77 Draw a stepwise mechanism for the following intramolecular reaction, which was used in the synthesis of the female sex hormone estrone.

18.78

a. In quinoline the lone pair on N occupies an sp^2 hybrid orbital, so it can never be donated to the ring by resonance. The N atom decreases the electron density of the ring in which it is located by an electron-withdrawing inductive effect, so substitution occurs on the other ring. In indole, the N atom donates its electron pair (which is contained in a p orbital) to the five-membered ring, increasing its electron density, so substitution occurs on the five-membered ring with the N atom.

b. In the presence of acid, the N atom is protonated prior to electrophilic attack. For substitution to occur at C8 rather than C7, the carbocation that results from electrophilic addition at C8 must be more stable. Attack at C8 generates a carbocation with more resonance structures, four of which keep one ring aromatic (**1–4**). Attack at C7 generates a carbocation with fewer resonance structures and only two have an intact aromatic ring (**5** and **6**).

c. With indole, attack at C3 forms the more highly resonance-stabilized carbocation.

Attack at C3 forms resonance structures, all of which have an intact aromatic ring, and two of which have all atoms with octets. Attack at C2 forms a cation with more resonance structures, but only two have an intact aromatic ring, and only one has complete octets.

18.79

(+ 3 additional resonance structures)

1,2-shift

Chapter 19 Carboxylic Acids and the Acidity of the O–H Bond

Chapter Review

General facts

- Carboxylic acids contain a carboxy group (COOH). The central carbon is sp^2 hybridized and trigonal planar (19.1).
- Carboxylic acids are identified by the suffixes *-oic acid, carboxylic acid,* or *-ic acid* (19.2).
- Carboxylic acids are polar compounds that exhibit hydrogen bonding interactions (19.3).

Summary of spectroscopic absorptions (19.4)

IR absorptions	C=O	~1710 cm^{-1}
	O–H	3500–2500 cm^{-1} (very broad and strong)
^{1}H NMR absorptions	O–H	10–12 ppm (highly deshielded proton)
	C–H α to COOH	2–2.5 ppm (somewhat deshielded Csp^3–H)
^{13}C NMR absorption	C=O	170–210 ppm (highly deshielded carbon)

General acid–base reaction of carboxylic acids (19.9)

$pK_a \approx 5$ carboxylate anion

- Carboxylic acids are especially acidic because carboxylate anions are resonance stabilized.
- For equilibrium to favor the products, the base must have a conjugate acid with a pK_a > 5. Common bases are listed in Table 19.3.

Factors that affect acidity

Resonance effects. A carboxylic acid is more acidic than an alcohol or phenol because its conjugate base is more effectively stabilized by resonance (19.9).

ROH   OH   O R–C–OH

$pK_a = 16–18$ $pK_a = 10$ $pK_a \approx 5$

Increasing acidity

Inductive effects. Acidity increases with the presence of electron-withdrawing groups (like the electronegative halogens) and decreases with the presence of electron-donating groups (like polarizable alkyl groups) (19.10).

Substituted benzoic acids.

- Electron-donor groups (D) make a substituted benzoic acid *less* acidic than benzoic acid.
- Electron-withdrawing groups (W) make a substituted benzoic acid *more* acidic than benzoic acid.

Other facts

- Extraction is a useful technique for separating compounds having different solubility properties. Carboxylic acids can be separated from other organic compounds by extraction, because aqueous base converts a carboxylic acid into a water-soluble carboxylate anion (19.12).
- A sulfonic acid (RSO_3H) is a strong acid because it forms a weak, resonance-stabilized conjugate base on deprotonation (19.13).
- Amino acids have an amino group on the α carbon to the carboxy group [$RCH(NH_2)COOH$]. Amino acids exist as zwitterions at pH $\approx$ 6. Adding acid forms a species with a net (+1) charge [$RCH(NH_3)COOH$]$^+$. Adding base forms a species with a net (–1) charge [$RCH(NH_2)COO$]$^-$ (19.14).

Practice Test on Chapter Review

1.a. Give the IUPAC name for the following compound.

b. Draw the structure corresponding to the following name: sodium *m*-bromobenzoate.

2.a. Which of the labeled atoms is least acidic?

1. H_a 2. H_b 3. H_c 4. H_d 5. H_e

b. Which of the following carboxylic acids has the lowest pK_a?

c. Which compound(s) can be converted to **A** by an oxidation reaction?

4. Both (1) and (2) can be converted to **A**.

5. Compounds (1), (2), and (3) can all be converted to **A**.

3. Rank the following compounds in order of increasing basicity. Label the *least* basic compound as **1**, the *most* basic compound as **3**, and the compound of *intermediate* basicity as **2**.

A   B   C

4. Draw the organic products formed in each of the following reactions.

a. $\overset{+}{N}H_3$ ... NaOH (1 equiv)

b.   HO— —COOH   NaH (1 equiv)

c.   OH ... [1] NaH   [2] CH_3I

Answers to Practice Test

1.a. *cis*-2-methylcyclo-
pentanecarboxylic acid

b.
COO⁻Na⁺ / Br

2.a. 1
b. 5
c. 5

3. A–2
B–1
C–3

4.a. $\overset{+}{N}H_3$ / CH_3...C—COO⁻ / H

b. HO— —COO⁻

c. OCH₃ ... N

Answers to Problems

19.1 To name a carboxylic acid:
[1] Find the longest chain containing the COOH group and change the *-e* ending to *-oic acid*.
[2] Number the chain to put the COOH carbon at C1, but omit the number from the name.
[3] Follow all other rules of nomenclature.

a.
Number the chain to put COOH at C1.
6 carbon chain = **hexanoic acid**
3,3-dimethylhexanoic acid

c.
Number the chain to put COOH at C1.
6 carbon chain = **hexanoic acid**
2,4-diethylhexanoic acid

b.
Number the chain to put COOH at C1.
5 carbon chain = **pentanoic acid**
3-bromo-4-chloro-2-fluoropentanoic acid

d.
Number the chain to put COOH at C1.
9 carbon chain = **nonanoic acid**
4-isopropyl-6,8-dimethylnonanoic acid

19.2

a. 2-bromo**butanoic acid**

b. 2,3-dimethyl**pentanoic acid**

c. 3,3,4-trimethyl**heptanoic acid**

d. 2-*sec*-butyl-4,4-diethyl**nonanoic acid**

e. 3,4-diethyl**cyclohexanecarboxylic acid**

f. 1-isopropyl**cyclobutane-carboxylic acid**

19.3

a. α-methoxy**valeric acid**

b. β-phenyl**propionic acid**

c. α,β-dimethyl**caproic acid**

d. α-chloro-β-methyl**butyric acid**

19.4

a. **lithium benzoate**

b. **sodium formate**

c. **potassium 2-methylpropanoate**

d. **sodium 4-bromo-6-ethyl-octanoate**

19.5

2-propyl**pentanoic acid**

C2

COOH

sodium 2-propyl**pentanoate**

COO⁻ Na⁺

19.6 More polar molecules have a higher boiling point and are more water soluble.

least polar
lowest boiling point
least H₂O soluble

intermediate polarity
intermediate boiling point

most polar
highest boiling point
most H₂O soluble

19.7 Look for functional group differences to distinguish the compounds by IR. Besides *sp³* hybridized C–H bonds at 3000–2850 cm⁻¹ (which all three compounds have), the following functional group absorptions are seen:

carboxylic acid
2 strong absorptions
~1710 (C=O)
~2500–3500 (OH) cm⁻¹

ester
1 strong absorption
~1700 (C=O) cm⁻¹

alcohol
1 strong absorption
~3600–3200 (OH) cm⁻¹

19.8

s-cis   s-trans

HO

O

Z

Z E   E   E

s-trans s-cis

19.9

HO

COOH

HO   OH

PGF₂ₐ
a prostaglandin

HO

COOH

HO   OH

enantiomer

There are five tetrahedral stereogenic centers labeled with (*). Both double bonds can exhibit cis–trans isomerism. Therefore, there are 2⁷ = 128 stereoisomers.

19.10 1° Alcohols are converted to carboxylic acids by oxidation reactions.

a. ⟹   c. ⟹

b. ⟹

19.11

a. $\xrightarrow[\text{H}_2\text{SO}_4, \text{H}_2\text{O}]{\text{Na}_2\text{Cr}_2\text{O}_7}$ c. $\xrightarrow{\text{KMnO}_4}$

A

C
(Any R group with benzylic H's
can be present para to NO$_2$.)

b. $\xrightarrow[\text{[2] H}_2\text{O}]{\text{[1] O}_3}$ d. $\xrightarrow[\text{H}_2\text{SO}_4, \text{H}_2\text{O}]{\text{CrO}_3}$

B   (2 equiv)

D   1° OH

19.12

not resonance stabilized

Protonation on the hydroxy O gives a
product that is not resonance stabilized,
so this pathway does not occur.

19.13

a. $\xrightarrow{\text{NaOH}}$ c. $\xrightarrow{\text{NaH}}$

b. $\xrightarrow{\text{NaOCH}_3}$ d. $\xrightarrow{\text{NaHCO}_3}$

19.14 CH$_3$COOH has a pK_a of 4.8. Any base having a conjugate acid with a pK_a higher than 4.8 can deprotonate it.

a. F$^-$ pK_a (HF) = 3.2 **not strong enough**
b. (CH$_3$)$_3$CO$^-$ pK_a [(CH$_3$)$_3$COH] = 18 **strong enough**
c. CH$_3^-$ pK_a (CH$_4$) = 50 **strong enough**
d. $^-$NH$_2$ pK_a (NH$_3$) = 38 **strong enough**
e. Cl$^-$ pK_a (HCl) = –7.0 **not strong enough**

19.15

Increasing acidity: H$_a$ < H$_b$ < H$_c$

mandelic acid

– H$_a$ → negative charge on C **unstable conjugate base**

– H$_b$ → negative charge on O **more stable conjugate base**

– H$_c$ → negative charge on O, resonance stabilized **most stable conjugate base**

19.16 Electron-withdrawing groups make an acid more acidic, lowering its pK_a.

least acidic
pK_a = 4.9

one electron-withdrawing group
intermediate acidity
pK_a = 3.2

three electron-withdrawing F's
most acidic
pK_a = 0.2

19.17

a.

least acidic

intermediate acidity

most acidic

b.

least acidic

intermediate acidity

most acidic

19.18

a.

least acidic

intermediate acidity

most acidic

b.

least acidic

intermediate acidity

most acidic

19.19

Phenol **A** has a higher pK_a than phenol because of its substituents. Both OH and $C_{15}H_{31}$ are electron-donating groups, which make the conjugate base less stable. Therefore, the acid is **less acidic.**

19.20 To separate compounds by an extraction procedure, they must have different solubility properties.

a. $CH_3(CH_2)_6COOH$ and $CH_3CH_2CH_2CH_2CH=CH_2$: **YES.** The acid can be extracted into aqueous base, while the alkene will remain in the organic layer.

b. $CH_3CH_2CH_2CH_2CH=CH_2$ and $(CH_3CH_2CH_2)_2O$: **NO.** Both compounds are soluble in organic solvents and insoluble in water. Neither is acidic enough to be extracted into aqueous base.

c. $CH_3(CH_2)_6COOH$ and NaCl: one carboxylic acid, one salt: **YES.** The carboxylic acid is soluble in an organic solvent while the salt is soluble in water.

d. NaCl and KCl: two salts: **NO.**

19.21

weaker conjugate base
better leaving group
↓

$CF_3SO_3H \longrightarrow CF_3SO_3^-$

CF_3 is electron *withdrawing.*
stronger acid
lower pK_a

stronger conjugate base
worse leaving group
↓

$CH_3SO_3H \longrightarrow CH_3SO_3^-$

CH_3 is electron *donating.*
weaker acid
higher pK_a

19.22

a. phenylalanine

COOH · · · C · · · H, H_2N — R

COOH · · · C, H — NH_2 — S

b. methionine

COOH · · · C · · · H, H_2N — $CH_2CH_2SCH_3$ — R

COOH · · · C, H — NH_2, $CH_3SCH_2CH_2$ — S

19.23 Amino acids exist as zwitterions (i.e., salts), so they are too polar to be soluble in organic solvents like diethyl ether. Instead, they are soluble in water.

19.24

$H_3\overset{+}{N}$ — (C=O) — OH
pH = 1

$H_3\overset{+}{N}$ — (C=O) — O^-
glycine
neutral form

H_2N — (C=O) — O^-
pH = 11

19.25

$$pI = \frac{pK_a(COOH) + pK_a(NH_3^+)}{2} = \frac{(2.58) + (9.24)}{2} = 5.91$$

19.26

electron-withdrawing group

The nearby (+) stabilizes the conjugate base by an electron-withdrawing inductive effect, thus making the starting acid more acidic.

19.27

A

a. 2,5-dimethylhexanoic acid

b.

c. sodium 2,5-dimethylhexanoate

d. An alcohol or ether would have a much higher pK_a than a carboxylic acid.

B

a. 3-ethyl-3-methylcyclohexanecarboxylic acid

b.

c. sodium 3-ethyl-3-methylcyclohexanecarboxylate

d.

19.28

least acidic

intermediate acidity

most acidic

19.29 Use the directions from Answer 19.1 to name the compounds.

a.

4,4,5,5-tetramethyloctanoic acid

b.

lithium 2-ethylpentanoate

c.

1-ethyl-3-isobutylcyclopentane-
carboxylic acid

d.

5-bromo-4-ethyl-2-nitrobenzoic acid

e.

sodium 2-methylhexanoate

f.

7-ethyl-5-isopropyl-3-methyldecanoic acid

19.30

a. 3,3-dimethylpentanoic acid

b. 4-chloro-3-phenylheptanoic acid

c. (*R*)-2-chloropropanoic acid

d. *m*-hydroxybenzoic acid

e. potassium acetate

f. sodium α-bromobutyrate

g. 2,2-dichloropentanedioic acid

h. 4-isopropyl-2-methyloctanedioic acid

19.31

a.

OH ——— C2 or α carbon

CO₂H

IUPAC: 2-hydroxy**propanoic acid**
common: α-hydroxy**propionic acid**

b.

C5 or δ carbon

OH

HO   CO₂H

C3 or β carbon

IUPAC: 3,5-dihydroxy-3-methyl**pentanoic acid**
common: β,δ-dihydroxy-β-methyl**valeric acid**

19.32

lowest boiling point

intermediate boiling point

highest boiling point

19.33

a.

b.

c.

d.

19.34

a.

b. HC≡CH $\xrightarrow[\text{[2] CH}_3\text{I}]{\text{[1] NaNH}_2}$ **C** $\xrightarrow[\text{[2] CH}_3\text{CH}_2\text{I}]{\text{[1] NaNH}_2}$ **D** $\xrightarrow[\text{[2] H}_2\text{O}]{\text{[1] O}_3}$ **E** + **F**

c.

19.35

Bases: [1] ⁻OH pK_a (H$_2$O) = 15.7; [2] CH$_3$CH$_2$⁻ pK_a (CH$_3$CH$_3$) = 50; [3] ⁻NH$_2$ pK_a (NH$_3$) = 38; [4] NH$_3$ pK_a (NH$_4$⁺) = 9.4; [5] HC≡C⁻ pK_a (HC≡CH) = 25.

a.

pK_a = 4.3
All of the bases
can deprotonate this RCO$_2$H.

b.

pK_a = 9.4
⁻OH, CH$_3$CH$_2$⁻, ⁻NH$_2$, and HC≡C⁻
can deprotonate this phenol.

c. (CH$_3$)$_3$COH

pK_a = 18
CH$_3$CH$_2$⁻, ⁻NH$_2$, and HC≡C⁻
can deprotonate this ROH.

19.36

a.

| Reaction favors reactants. |

b.

| Reaction favors products. |

c. [benzoic acid with methyl group] + CH$_3^-$Li$^+$ ⇌ [carboxylate] Li$^+$ + CH$_4$

pK$_a$ ≈ 4   pK$_a$ = 50 | Reaction favors products. |

d. [methylphenol] —OH + Na$_2$CO$_3$ ⇌ [phenoxide] —O$^-$Na$^+$ + Na$^+$ HCO$_3^-$

pK$_a$ = 10.2   pK$_a$ = 10.2 | With the same pK$_a$ for the starting acid and the conjugate acid, an equal amount of starting materials and products is present. |

19.37 The stronger acid has a lower pK$_a$ and a weaker conjugate base.

a. [p-methylbenzoic acid] or [p-chlorobenzoic acid]

CH$_3$ is electron donating.
weaker acid
higher pK$_a$
stronger conjugate base

Cl is electron withdrawing.
stronger acid
lower pK$_a$
weaker conjugate base

b. NC—CH$_2$—COOH or CH$_3$—COOH

CN is electron withdrawing.
stronger acid
lower pK$_a$
weaker conjugate base

weaker acid
higher pK$_a$
stronger conjugate base

19.38

a. [butanoic acid] COOH
least acidic

[2-bromo acid] COOH
Br is electronegative.
intermediate acidity

[2-chloro acid] COOH
Cl is more electronegative.
most acidic

b. [p-methylphenol] OH
least acidic

[p-chlorophenol] OH
intermediate acidity

[p-nitrophenol] OH
most acidic

19.39

a. [phenoxide]
weakest base

[anilide N$^-$H]
intermediate basicity

[phenyl anion]
strongest base

b. [p-nitrophenoxide O$_2$N]
weakest base

[phenoxide]
intermediate basicity

[cyclohexoxide]
strongest base

19.40

Increasing acidity →

	ICH$_2$COOH	BrCH$_2$COOH	FCH$_2$COOH	F$_2$CHCOOH	F$_3$CCOOH
pK$_a$ values	least acidic 3.12	2.86	2.66	1.24	most acidic 0.28

19.41 The OH of the phenol group in morphine is more acidic than the OH of the alcohol ($pK_a \approx 10$ versus $pK_a \approx 16$). KOH is basic enough to remove the phenolic OH, the most acidic proton.

most acidic proton
The OH is part of a phenol.
Methylation occurs here.

an alcohol

morphine

[1] KOH

Many resonance structures
stabilize the conjugate base.

[2] CH₃I

codeine

19.42 The closer the electron-withdrawing CH₃CO– group is to the carboxylic acid, the more it will stabilize the conjugate base, making the acid stronger.

pyruvic acid
stronger acid

farther away

acetoacetic acid
weaker acid

19.43

a. The negative charge on the conjugate base of *p*-nitrophenol is delocalized on the NO₂ group, stabilizing the conjugate base, and making *p*-nitrophenol more acidic than phenol (where the negative charge is delocalized only around the benzene ring).

p-nitrophenol
$pK_a = 7.2$

two of the possible resonance structures for the conjugate base
[See part (b) for all the possible resonance structures.]

phenol
$pK_a = 10$

b. In the para isomer, the negative charge of the conjugate base is delocalized over both the benzene ring and onto the NO₂ group, whereas in the meta isomer it cannot be delocalized onto the NO₂ group. This makes the conjugate base from the para isomer more highly resonance stabilized, and the para-substituted phenol more acidic than its meta isomer.

pKa = 7.2
p-nitrophenol

negative charge on
two O atoms
very good resonance structure
more stable conjugate base
stronger acid

pKa = 8.3
m-nitrophenol

19.44 A CH_3O group has an electron-withdrawing inductive effect and an electron-donating resonance effect. In 2-methoxyacetic acid, the OCH_3 group is bonded to an sp^3 hybridized C, so there is no way to donate electron density by resonance. The CH_3O group withdraws electron density because of the electronegative O atom, stabilizing the conjugate base, and making CH_3OCH_2COOH a stronger acid than CH_3COOH.

more acidic acid

– H$^+$

more stable conjugate base

In p-methoxybenzoic acid, the CH_3O group is bonded to an sp^2 hybridized C, so it can donate electron density by a resonance effect. This destabilizes the conjugate base, making the starting material less acidic than C_6H_5COOH.

less acidic acid

– H$^+$

like charges nearby
less stable conjugate base

19.45 Phenol has a pKa of 10, making p-methylthiophenol (pKa = 9.53) the stronger acid. A substituent that increases the acidity of a phenol must withdraw electron density to stabilize the negative charge of the conjugate base. An electron-withdrawing substituent deactivates a benzene ring towards electrophilic aromatic substitution, making p-methylthiophenol less reactive than phenol.

stronger acid
less reactive in electrophilic
substitution

19.46

The O in **A** is more electronegative than the N in **C**, so there is a stronger electron-withdrawing inductive effect. This stabilizes the conjugate base of **A**, making **A** more acidic than **C**.

A	**B**	**C**
$pK_a = 3.2$	$pK_a = 3.9$	$pK_a = 4.4$

Since the O in **A** is closer to the COOH group than the O atom in **B**, there is a stronger electron-withdrawing inductive effect. This makes **A** more acidic than **B**.

19.47

D

– H⁺

Since the benzene ring is bonded to the α carbon (not the carbonyl carbon), this compound is not much different than any alkyl-substituted carboxylic acid.
least acidic

E

– H⁺

The electron-withdrawing inductive effect of the NO_2 group helps stabilize the COO^- group.
intermediate acidity

C

– H⁺

Since the NO_2 group is bonded to a benzene ring that is bonded directly to the carbonyl group, inductive effects and resonance effects stabilize the conjugate base. For example, a resonance structure can be drawn that places a (+) charge close to the COO^- group.
most acidic

Two of the resonance structures for the conjugate base of **C**:

unlike charges nearby
stabilizing

19.48 a. The pK_{a1} of phthalic acid is lower than the pK_{a1} of isophthalic acid because the electron-withdrawing CO_2H is closer to the negatively charged CO_2^- of the conjugate base.

phthalic acid
stronger acid
lower pK_a

stabilizes the conjugate base
by electron withdrawal

isophthalic acid
weaker acid

farther away

b. After loss of the first proton, each compound now has one CO_2H group and one CO_2^-. The CO_2^- is electron donating, so the closer it is located to the CO_2H, the more it destabilizes the resulting conjugate base.

weaker acid

Negative charges are closer—
less stable.

stronger acid
lower pK_{a2}

farther away

19.49

labeled O atom

The resonance-stabilized carboxylate anion can now be protonated on either O atom, the one with the label and the one without the label.

The label is now in two different locations.

19.50

a.

cyclohexane-1,3-dione
increasing acidity: H_b < H_a < H_c

loss of H_b:

The most acidic proton forms the most stable conjugate base.

one Lewis structure
least stable conjugate base

loss of H_a:

2 resonance structures
intermediate stability

loss of H_c:

3 resonance structures
most stable conjugate base

b.

acetanilide
increasing acidity: H_a < H_c < H_b

loss of H_b:

7 resonance structures
most stable conjugate base

loss of H_a:

one Lewis structure
least stable conjugate base

loss of H_c:

2 resonance structures that delocalize the negative charge
intermediate stability

19.51

The conjugate base is resonance stabilized. Two of the structures place a negative charge on an O atom.
weaker conjugate base
stronger acid

The conjugate base has only one Lewis structure.
stronger conjugate base
weaker acid

19.52

The negatively charged C is more nucleophilic than the negatively charged O atom.

strong base (2 equiv)

X

Two equivalents of strong base remove both the O–H and C–H protons.

H_3O^+

hexanoic acid

19.53

acetamide

somewhat less stable with the (–) charge on N

O is more electronegative than N, making the conjugate base of CH_3COOH more stable than the conjugate base of acetamide. Therefore, acetamide is less acidic.

19.54

A   B

- Dissolve both compounds in CH_2Cl_2.
- Add 10% $NaHCO_3$ solution. This makes a carboxylate anion ($C_{10}H_7COO^-$) from **B,** which dissolves in the aqueous layer. The other compound (**A**) remains in the CH_2Cl_2.
- Separate the layers.

19.55

19.56 To separate two compounds in an aqueous extraction, one must be water soluble (or be able to be converted into a water-soluble ionic compound by an acid–base reaction), and the other insoluble. Octan-1-ol has greater than 5 C's, making it insoluble in water. Octane is an alkane, so it, too, is insoluble in water. Neither compound is acidic enough to be deprotonated by a base in aqueous solution. Because their solubility properties are similar, they cannot be separated by an extraction procedure.

19.57

a. Molecular formula: $C_3H_5ClO_2$ ⟶ one double bond or ring
   IR: 3500–2500 cm⁻¹, 1714 cm⁻¹ ⟶ C=O and O–H
   NMR data: 2.87 (triplet, 2 H), 3.76 (triplet, 2 H), and 11.8 (singlet, 1 H) ppm

b. Molecular formula: $C_8H_8O_3$ ⟶ 5 double bonds or rings
   IR: 3500–2500 cm⁻¹, 1688 cm⁻¹ ⟶ C=O and O–H
   NMR data: 3.8 (singlet, 3 H), 7.0 (doublet, 2 H), 7.9 (doublet, 2 H), and 12.7 (singlet, 1 H) ppm

   para disubstituted benzene ring

c. Molecular formula: $C_8H_8O_3$ ⟶ 5 double bonds or rings
   IR: 3500–2500 cm⁻¹, 1710 cm⁻¹ ⟶ C=O and O–H
   NMR data: 4.7 (singlet, 2 H), 6.9–7.3 (multiplet, 5 H), and 11.3 (singlet, 1 H) ppm

   monosubstituted benzene ring

19.58

Compound **A**: Molecular formula $C_4H_8O_2$ (one degree of unsaturation)
IR absorptions at 3600–3200 (O–H), 3000–2800 (C–H), and 1700 (C=O) cm^{-1}
^{1}H NMR data:

Absorption	ppm	# of H's	Explanation	Structure:
singlet	2.2	3	a CH$_3$ group	
singlet	2.55	1	1 H adjacent to none or OH	
triplet	2.7	2	2 H's adjacent to 2 H's	
triplet	3.8	2	2 H's adjacent to 2 H's	**A**

Compound **B**: Molecular formula $C_4H_8O_2$ (one degree of unsaturation)
IR absorptions at 3500–2500 (O–H) and 1700 (C=O) cm^{-1}
^{1}H NMR data:

Absorption	ppm	# of H's	Explanation	Structure:
doublet	1.0	6	6 H's adjacent to 1 H	
septet	2.3	1	1 H adjacent to 6 H's	
singlet (very broad)	10.6	1	OH of RCOOH	**B**

19.59

Compound **C**: Molecular formula $C_4H_8O_3$ (one degree of unsaturation)
IR absorptions at 3600–2500 (O–H) and 1734 (C=O) cm^{-1}
^{1}H NMR data:

Absorption	ppm	# of H's	Explanation	Structure:
triplet	1.3	3	a CH$_3$ group adjacent to 2 H's	
quartet	3.6	2	2 H's adjacent to 3 H's	
singlet	4.2	2	2 H's	
singlet	11.3	1	OH of COOH	**C**

19.60

Compound **D**: Molecular formula $C_9H_9ClO_2$ (five degrees of unsaturation)
^{13}C NMR data: 30, 36, 128, 130, 133, 139, 179 = 7 different types of C's
^{1}H NMR data:

Absorption	ppm	# of H's	Explanation	Structure:
triplet	2.7	2	2 H's adjacent to 2 H's	
triplet	2.9	2	2 H's adjacent to 2 H's	
two signals	7.2	4	on benzene ring	
singlet	11.7	1	OH of COOH	**D**

19.61

A   (structure: tert-butyl COOH)
3 different C's
Spectrum [2]: peaks at 27, 39, 186 ppm

B   (structure: butyl COOH)
5 different C's
Spectrum [1]: peaks at 14, 22, 27, 34, 181 ppm

C   (structure: isovaleric-type acid, OH)
4 different C's
Spectrum [3]: peaks at 22, 26, 43, 180 ppm

19.62

GBL: Molecular formula $C_4H_6O_2$ (two degrees of unsaturation)
IR absorption at 1770 (C=O) cm^{-1}
^1H NMR data:

Absorption	ppm	# of H's	Explanation	Structure:
multiplet	2.28	2	2 H's adjacent to several H's	(structure GBL)
triplet	2.48	2	2 H's adjacent to 2 H's	
triplet	4.35	2	2 H's adjacent to 2 H's	**GBL**

19.63

threonine

2R,3S   2S,3S   2R,3R   2S,3R
naturally
occurring

19.64

proline   enantiomer   zwitterion

19.65

a. methionine

b. serine

pH = 1   pH = 6
form at isoelectric point   pH = 11   pH = 1   pH = 6
form at isoelectric point   pH = 11

19.66

a. cysteine $pI = \dfrac{pK_a(COOH) + pK_a(NH_3^+)}{2} = (2.05) + (10.25) / 2 = \mathbf{6.15}$

b. methionine $pI = \dfrac{pK_a(COOH) + pK_a(NH_3^+)}{2} = (2.28) + (9.21) / 2 = \mathbf{5.75}$

19.67

lysine
This lone pair is localized on the
N atom, making it a base.

tryptophan
This lone pair is delocalized in the π system to give 10 π electrons,
making it aromatic. This is similar to pyrrole (Chapter 17). Since
these electrons are delocalized in the aromatic system, this N atom in
tryptophan is not basic.

19.68

a. At pH = 1, the net charge is (+1).

b. increasing pH: As base is added, the most acidic proton is removed first, then the next most acidic proton, and so forth.

c. monosodium glutamate

19.69   The first equivalent of NaH removes the most acidic proton—that is, the OH proton on the phenol. The resulting phenoxide can then act as a nucleophile to displace I to form a substitution product. With two equivalents, both OH protons are removed. In this case the more nucleophilic O atom is the stronger base—that is, the alkoxide derived from the alcohol (not the phenoxide), so this negatively charged O atom reacts first in a nucleophilic substitution reaction.

19.70

p-hydroxybenzoic acid
less acidic than benzoic acid

The OH group donates electron density by its resonance effect and this destabilizes the conjugate base, making the acid less acidic than benzoic acid.

o-hydroxybenzoic acid
more acidic than benzoic acid

Intramolecular hydrogen bonding stabilizes the conjugate base, making the acid more acidic than benzoic acid.

19.71

2-hydroxybutanedioic acid
increasing acidity:
$H_d < H_c < H_b < H_e < H_a$

H_a and H_e must be the two most acidic protons because they are part of a carboxy group. Loss of a proton forms a resonance-stabilized carboxylate anion that has the negative charge delocalized on two O atoms. H_a is more acidic than H_e because the nearby OH group on the α carbon increases acidity by an electron-withdrawing inductive effect. H_b is the next most acidic proton because the conjugate base places a negative charge on the electronegative O atom, but it is not resonance stabilized.

The least acidic H's are H_c and H_d because these H's are bonded to C atoms. The electronegative O atom further acidifies H_c by an electron-withdrawing inductive effect.

19.72

The conjugate base has three resonance structures, two of which place a negative charge on the oxygens. In this way the conjugate base resembles a carboxylate anion. In addition, the C=C's in **A** and **C** are conjugated.

Chapter 20 Introduction to Carbonyl Chemistry; Organometallic Reagents; Oxidation and Reduction

Chapter Review

Reduction reactions

[1] Reduction of aldehydes and ketones to 1° and 2° alcohols (20.4)

$$\text{R} \overset{O}{\underset{}{\text{C}}} \text{R'} \quad \xrightarrow[\substack{\text{or} \\ [1]\ LiAlH_4\ [2]\ H_2O \\ \text{or} \\ H_2,\ Pd\text{-}C}]{NaBH_4,\ CH_3OH} \quad \text{R} \overset{OH}{\underset{R'}{\text{C}}} \text{H}$$

R' = H or alkyl

1° or 2° alcohol

[2] Reduction of α,β-unsaturated aldehydes and ketones (20.4C)

$\xrightarrow[CH_3OH]{NaBH_4}$ • reduction of the C=O only

$\xrightarrow[Pd\text{-}C]{H_2\ (1\ equiv)}$ • reduction of the C=C only

$\xrightarrow[Pd\text{-}C]{H_2\ (excess)}$ • reduction of both π bonds

[3] Enantioselective ketone reduction (20.6)

$\xrightarrow[\text{[2] }H_2O]{\text{[1] }(S)\text{- or }(R)\text{-}\ CBS\ reagent}$

(R) 2° alcohol   or   (S) 2° alcohol

• A single enantiomer is formed.

[4] Reduction of acid chlorides (20.7A)

$\xrightarrow[\text{[2] }H_2O]{\text{[1] }LiAlH_4}$ R⌒OH   1° alcohol

• LiAlH₄, a strong reducing agent, reduces an acid chloride to a 1° alcohol.

$\xrightarrow[\text{[2] }H_2O]{\text{[1] }LiAlH[OC(CH_3)_3]_3}$ aldehyde

• With LiAlH[OC(CH₃)₃]₃, a milder reducing agent, reduction stops at the aldehyde stage.

[5] Reduction of esters (20.7A)

- LiAlH$_4$, a strong reducing agent, reduces an ester to a 1° alcohol.

- With DIBAL-H, a milder reducing agent, reduction stops at the aldehyde stage.

[6] Reduction of carboxylic acids to 1° alcohols (20.7B)

$$R-COOH \xrightarrow[\text{[2] H}_2\text{O}]{\text{[1] LiAlH}_4} R\text{-}CH_2OH$$
1° alcohol

[7] Reduction of amides to amines (20.7B)

$$R-CO-N \xrightarrow[\text{[2] H}_2\text{O}]{\text{[1] LiAlH}_4} R-CH_2-N$$
amine

Oxidation reactions
Oxidation of aldehydes to carboxylic acids (20.8)

$$R-CHO \xrightarrow[\substack{\text{or} \\ \text{Ag}_2\text{O, NH}_4\text{OH}}]{\text{CrO}_3,\ \text{Na}_2\text{Cr}_2\text{O}_7,\ \text{K}_2\text{Cr}_2\text{O}_7,\ \text{KMnO}_4} R-COOH$$
carboxylic acid

- All Cr^{6+} reagents except PCC oxidize RCHO to RCOOH.
- Tollens reagent (Ag$_2$O + NH$_4$OH) oxidizes RCHO only. Primary (1°) and secondary (2°) alcohols do not react with Tollens reagent.

Preparation of organometallic reagents (20.9)

[1] Organolithium reagents:

$$R-X\ +\ 2\,Li \longrightarrow R-Li\ +\ LiX$$

[2] Grignard reagents:

$$R-X\ +\ Mg \xrightarrow{(\text{CH}_3\text{CH}_2)_2\text{O}} R-Mg-X$$

[3] Organocuprate reagents:

$$R-X\ +\ 2\,Li \longrightarrow R-Li\ +\ LiX$$

$$2\,R-Li\ +\ CuI \longrightarrow R_2Cu^-\ Li^+\ +\ LiI$$

[4] Lithium and sodium acetylides:

$$R-C \equiv C-H \xrightarrow{Na^+ \ ^-NH_2} \boxed{\begin{array}{c} R-C \equiv C^- \ Na^+ \\ \text{a sodium acetylide} \end{array}} + \ NH_3$$

$$R-C \equiv C-H \xrightarrow{R-Li} \boxed{\begin{array}{c} R-C \equiv C-Li \\ \text{a lithium acetylide} \end{array}} + \ R-H$$

Reactions with organometallic reagents

[1] Reaction as a base (20.9C)

$$R-M \ + \ H-\overset{..}{\underset{..}{O}}-R \longrightarrow R-H \ + \ M^+ \ + \ \overset{..}{\underset{..}{:}}\overset{..}{O}-R$$

- $RM = RLi,\ RMgX,\ R_2CuLi$
- This acid–base reaction occurs with H_2O, ROH, RNH_2, R_2NH, RSH, RCOOH, $RCONH_2$, and RCONHR.

[2] Reaction with aldehydes and ketones to form 1°, 2°, and 3° alcohols (20.10)

$$\underset{\substack{R \quad R' \\ R' = H \text{ or alkyl}}}{\overset{O}{\|}}\ \xrightarrow[{[2] \ H_2O}]{[1] \ R''MgX \ \text{or} \ R''Li}\ \boxed{\underset{\substack{R \ R'' \\ R' \\ \text{1°, 2°, or 3° alcohol}}}{\overset{OH}{|}}}$$

[3] Reaction with esters to form 3° alcohols (20.13A)

$$\underset{R \quad OR'}{\overset{O}{\|}}\ \xrightarrow[{[2] \ H_2O}]{\substack{[1] \ R''Li \ \text{or} \ R''MgX \\ (2 \ \text{equiv})}}\ \boxed{\underset{\substack{R \ R'' \\ R'' \\ \text{3° alcohol}}}{\overset{OH}{|}}}$$

[4] Reaction with acid chlorides (20.13)

- More reactive organometallic reagents—R"Li and R"MgX—add two equivalents of R" to an acid chloride to form a 3° alcohol with two identical R" groups.

- Less reactive organometallic reagents— R'$_2$CuLi—add only one equivalent of R' to an acid chloride to form a ketone.

[5] Reaction with carbon dioxide—Carboxylation (20.14A)

$$R-MgX \xrightarrow[\text{[2] } H_3O^+]{\text{[1] } CO_2} \text{ carboxylic acid}$$

[6] Reaction with epoxides (20.14B)

[1] RLi, RMgX, or R_2CuLi

[2] H_2O

alcohol

[7] Reaction with α,β-unsaturated aldehydes and ketones (20.15B)

[1] R'Li or R'MgX

[2] H_2O

allylic alcohol

[1] R'_2CuLi

[2] H_2O

ketone

- More reactive organometallic reagents—R'Li and R'MgX—react with α,β-unsaturated carbonyls by 1,2-addition.

- Less reactive organometallic reagents—R'$_2$CuLi— react with α,β-unsaturated carbonyls by 1,4-addition.

Protecting groups (20.12)

[1] Protecting an alcohol as a *tert*-butyldimethylsilyl ether

$R-O-H$ + $[Cl-TBDMS]$ → $[R-O-TBDMS]$

tert-butyldimethylsilyl ether

[2] Deprotecting a *tert*-butyldimethylsilyl ether to re-form an alcohol

$[R-O-TBDMS] \xrightarrow{Bu_4N^+F^-} R-O-H$ + $[F-TBDMS]$

Practice Test on Chapter Review

1. Which compounds undergo nucleophilic addition and which undergo substitution?

a. [structure] b. [structure with Cl] c. [structure with O] d. [benzaldehyde structure with H]

2. What product is formed when $CH_3CH_2CH_2Li$ reacts with each compound, followed by quenching with water and acid?

a. CH_3CH_2CHO
b. $(CH_3)_2CO$
c. $CH_3CH_2CO_2CH_3$
d. CH_3CH_2COCl

e. CO_2
f. $CH_2=CHCOCH_3$
g. ethylene oxide
h. CH_3COOH

3. What product is formed when $HO(CH_2)_4CHO$ is treated with each reagent?

a. $NaBH_4$, CH_3OH
b. PCC

c. Ag_2O, NH_4OH
d. $Na_2Cr_2O_7$, H_2SO_4, H_2O

4. What reagent is needed to convert $(CH_3CH_2)_2CHCOCl$ into each compound?

a. $(CH_3CH_2)_2CHCOCH_2CH_3$
b. $(CH_3CH_2)_2CHCHO$

c. $(CH_3CH_2)_2CHC(OH)(CH_2CH_3)_2$
d. $(CH_3CH_2)_2CHCH_2OH$

5. Draw the organic products formed in the following reactions.

6. What starting materials are needed to synthesize each compound using the indicated reagent or functional group?

a. Synthesize:

from an ester

c. Synthesize:

using a Grignard reagent

b. Synthesize:

using an organocuprate reagent

Answers to Practice Test

1.a. addition
  b. substitution
  c. substitution
  d. addition

2. a. $CH_3CH_2CH(OH)CH_2CH_2CH_3$
  b. $(CH_3)_2C(OH)CH_2CH_2CH_3$
  c. $CH_3CH_2C(OH)(CH_2CH_2CH_3)_2$
  d. $CH_3CH_2C(OH)(CH_2CH_2CH_3)_2$
  e. $CH_3CH_2CH_2CO_2H$
  f. $CH_2{=}CHC(OH)(CH_3)CH_2CH_2CH_3$
  g. $CH_3(CH_2)_4OH$
  h. $CH_3CH_2CH_3 + CH_3CO_2H$

3.a. $HO(CH_2)_5OH$
  b. $OHC(CH_2)_3CHO$
  c. $HO(CH_2)_4CO_2H$
  d. $HO_2C(CH_2)_3CO_2H$

4.a. $(CH_3CH_2)_2CuLi$
  b. $LiAlH[OC(CH_3)_3]_3$
  c. CH_3CH_2MgBr
  d. $LiAlH_4$

5.

a.

b.

c.

d.

6.

a.

+ CH_3MgBr

b.

+ $\left(\right)_2 CuLi$

c.

+

Answers to Problems

20.1

a.

[3] $C_{sp^2}{-}C_{sp^2}$

[1] $C_{sp^2}{-}C_{sp^2}$

[2] σ: $C_{sp^2}{-}O_{sp^2}$
    π: $C_p{-}O_p$

α-sinensal

b. The O is sp^2 hybridized.
Both lone pairs occupy sp^2 hybrid orbitals.

20.2 A carbonyl compound with a leaving group (NR$_2$ or OR bonded to the C=O) undergoes substitution reactions. Those without leaving groups undergo addition.

no good leaving group
addition reactions

All other C=O's have leaving groups.
substitution reactions

20.3 Aldehydes are more reactive than ketones. In carbonyl compounds with leaving groups, the better the leaving group, the more reactive the carbonyl compound.

a. or

less hindered carbonyl
more reactive

c. or

better leaving group
more reactive

b. or

less hindered carbonyl
more reactive

d. or

better leaving group
more reactive

20.4 NaBH$_4$ reduces aldehydes to 1° alcohols and ketones to 2° alcohols.

a.

c.

b.

20.5 1° Alcohols are prepared from aldehydes and 2° alcohols are from ketones.

a. b. c.

20.6

3° Alcohols cannot be made by reduction of a carbonyl group, because they do not contain a H on the C with the OH.

1-methylcyclohexanol

20.7

a.

b.

c.

d.

e.

f.

20.8

a.

b.

c.

20.9   The 2° alcohol comes from a carbonyl group. Since hydride was delivered from the back side, the (*R*)-CBS reagent must be used.

20.10

Part [1]: Nucleophilic substitution of H for Cl

Part [2]: Nucleophilic addition of H⁻ to form an alcohol

20.11   Acid chlorides and esters can be reduced to 1° alcohols. Keep the carbon skeleton the same in drawing an ester and acid chloride precursor.

a.

OH ⟹ (cyclopentanecarbonyl chloride) Cl or (cyclopentanecarboxylate methyl ester) O—

c.

CH₃O— —OH ⟹ CH₃O— —Cl

or

CH₃O— —C(=O)—O—

b.

OH ⟹ Cl or O—

20.12

a. [1] LiAlH₄ / [2] H₂O

b. [1] LiAlH₄ / [2] H₂O

c. [1] LiAlH₄ / [2] H₂O

d. [1] LiAlH₄ / [2] H₂O

20.13

a. benzylamine ⟹ benzamide

b. ⟹ or

c. ⟹

20.14

a. [1] LiAlH₄ / [2] H₂O → + HOCH₃

NaBH₄ / CH₃OH →

b. CH₃O... OH [1] LiAlH₄ / [2] H₂O → HO...OH + HOCH₃

NaBH₄ / CH₃OH → **Neither functional group reduced**

c. CH₃O... [1] LiAlH₄ / [2] H₂O →

NaBH₄ / CH₃OH →

20.15 Tollens reagent reacts only with aldehydes.

a. —OH

Ag₂O, NH₄OH → **No reaction**

Na₂Cr₂O₇ / H₂SO₄, H₂O →

b.

Ag₂O, NH₄OH →

Na₂Cr₂O₇ / H₂SO₄, H₂O →

20.16

a. **B** $\xrightarrow[\text{CH}_3\text{OH}]{\text{NaBH}_4}$

b. **B** $\xrightarrow[\text{[2] H}_2\text{O}]{\text{[1] LiAlH}_4}$

c. **B** $\xrightarrow{\text{PCC}}$

d. **B** $\xrightarrow{\text{Ag}_2\text{O, NH}_4\text{OH}}$

e. **B** $\xrightarrow[\text{H}_2\text{SO}_4,\ \text{H}_2\text{O}]{\text{CrO}_3}$

20.17

a. $\text{CH}_3\text{CH}_2\text{Br} + 2\ \text{Li} \longrightarrow \text{CH}_3\text{CH}_2\text{Li} + \text{LiBr}$

b. $\text{CH}_3\text{CH}_2\text{Br} + \text{Mg} \longrightarrow \text{CH}_3\text{CH}_2\text{MgBr}$

c. $\text{CH}_3\text{CH}_2\text{Br} + 2\ \text{Li} \longrightarrow \text{CH}_3\text{CH}_2\text{Li} + \text{LiBr}$

$\quad 2\ \text{CH}_3\text{CH}_2\text{Li} + \text{CuI} \longrightarrow \text{LiCu(CH}_2\text{CH}_3)_2 + \text{LiI}$

20.18

$+ \text{NaH} \longrightarrow$ $+ \text{H}_2 \longleftarrow$ hydrogen gas

$+ \text{CH}_3\text{MgBr} \longrightarrow$ $+ \text{CH}_4 \longleftarrow$ methane gas

20.19 Organometallic compounds have a carbon–metal bond.

a. $\text{BrMgC}{\equiv}\text{CCH}_2\text{CH}_3$

organometallic
compound

b. $\text{NaOCH}_2\text{CH}_3$

NOT an
organometallic
compound

c. $\text{KOC(CH}_3)_3$

NOT an
organometallic
compound

d. PhLi =

organometallic
compound

20.20

a. $+\ \text{H}_2\text{O} \longrightarrow$ $+\ \text{LiOH}$

b. $+\ \text{H}_2\text{O} \longrightarrow$ $+\ \text{HOMgBr}$

c. $+\ \text{H}_2\text{O} \longrightarrow$ $+\ \text{HOMgBr}$

d. $+\ \text{H}_2\text{O} \longrightarrow$ $+\ \text{LiOH}$

20.21 To draw the products, add the alkyl or phenyl group to the carbonyl carbon and protonate the oxygen.

a.

b.

c.

d.

20.22 Addition of RM always occurs from above and below the plane of the molecule.

a.

b.

20.23

a.

b.

c.

d.

20.24

a.

linalool
(three methods)

lavandulol

b.

NaBH$_4$
CH$_3$OH

c. Linalool is a 3° ROH. Therefore, it has no H on the carbon with the OH group, and cannot be prepared by reduction of a carbonyl compound.

20.25

venlafaxine

+   BrMg

20.26

estrone

TBDMS–Cl
imidazole

TBDMSO

[1] Li−C≡CH
[2] H$_2$O

OH

TBDMSO

(Bu)$_4$NF

OH

HO

ethynylestradiol

20.27

a.

[1]   MgBr
(2 equiv)
[2] H$_2$O

HO

b.

[1]   MgBr
(2 equiv)
[2] H$_2$O

HO

c.

[1]   MgBr
(2 equiv)
[2] H$_2$O

HO

20.28

a.

OH

+   MgBr
(2 equiv)

b.

c.

20.29 The R group of the organocuprate has replaced the Cl on the acid chloride.

a.   [1] (CH$_3$)$_2$CuLi   [2] H$_2$O

c.   [1] (Ph)$_2$CuLi   [2] H$_2$O

b.   [1] [CH$_3$CH$_2$CH(CH$_3$)]$_2$CuLi   [2] H$_2$O

20.30

a.   [1] LiAlH[OC(CH$_3$)$_3$]$_3$ [2] H$_2$O

c.   [1] (CH$_3$CH$_2$)$_2$CuLi   [2] H$_2$O

b.   [1] CH$_3$CH$_2$Li (2 equiv)   [2] H$_2$O

d.   [1] LiAlH$_4$ [2] H$_2$O

20.31

a.   ⇒ ( )$_2$CuLi + Cl

or   ⇒ + (CH$_3$)$_2$CuLi

b.   ⇒ Cl + [(CH$_3$)$_3$C]$_2$CuLi

or   ⇒ Cl + [(CH$_3$)$_2$CH]$_2$CuLi

20.32

a.   Br [1] Mg → MgBr [2] CO$_2$ → O$^-$ [3] H$_3$O$^+$ → OH

b.   Cl [1] Mg → MgCl [2] CO$_2$ → O$^-$ [3] H$_3$O$^+$ → OH

c.   Br [1] Mg → MgBr [2] CO$_2$ → O$^-$ [3] H$_3$O$^+$ → OH

20.33

a. (+ enantiomer)

b.

c.

d.

20.34 The characteristic reaction of α,β-unsaturated carbonyl compounds is nucleophilic addition. Grignard and organolithium reagents react by 1,2-addition and organocuprate reagents react by 1,4-addition.

a.

[1] (CH₃)₂CuLi
[2] H₂O

[1] H−C≡C−Li
[2] H₂O

b.

[1] (CH₃)₂CuLi
[2] H₂O

[1] H−C≡C−Li
[2] H₂O

c.

[1] (CH₃)₂CuLi
[2] H₂O

[1] H−C≡C−Li
[2] H₂O

20.35

a.
OH → HBr or PBr₃ → Br → Mg → MgBr

OH → PCC → O → [1] MgBr [2] H₂O → OH

b.
OH → HBr → Br
(from a.)

c.
OH → H₂SO₄ → + → H₂ Pd-C →
(from a.)

d.
OH → HBr or PBr₃ → Br → Mg → MgBr

OH → PCC → (aldehyde)

→ H₂O → OH → PCC → O

e.

(from d.)

20.36

A

a. NaBH$_4$ / CH$_3$OH

b. [1] LiAlH$_4$ / [2] H$_2$O

c. [1] CH$_3$MgBr (excess) / [2] H$_2$O

d. [1] C$_6$H$_5$Li (excess) / [2] H$_2$O

e. Na$_2$Cr$_2$O$_7$ / H$_2$SO$_4$ / H$_2$O → No reaction

B

a. NaBH$_4$ / CH$_3$OH → No reaction

b. [1] LiAlH$_4$ / [2] H$_2$O

c. [1] CH$_3$MgBr (excess) / [2] H$_2$O

d. [1] C$_6$H$_5$Li (excess) / [2] H$_2$O

e. Na$_2$Cr$_2$O$_7$ / H$_2$SO$_4$ / H$_2$O → No reaction

20.37

a.

PBr$_3$

PCC

Mg

[1] BrMg—CH$_2$CH$_3$ / [2] H$_2$O

b.

CH₃OH
↓ PBr₃
CH₃Br
↓ Mg

(from a.) →[[1] CH₃MgBr / [2] H₂O]→ OH →[PCC]→ O →[[1] CH₃MgBr / [2] H₂O]→ OH

c.

OH (from b.) →[PBr₃]→ Br →[[1] Mg / [2] H₂C=O / [3] H₂O]→ OH

CH₃OH →[PCC]→

d.

OH →[H₂SO₄]→ CH₂=CH₂ →[mCPBA]→ (epoxide) →[[1] BrMg—CH₂CH₃ (from a.) / [2] H₂O]→ OH

20.38

a. →[NaBH₄ / CH₃OH]→ OH

b. →[[1] LiAlH₄ / [2] H₂O]→ OH

c. →[H₂ / Pd-C]→ OH

d. →[PCC]→ **No reaction**

e. →[Na₂Cr₂O₇ / H₂SO₄, H₂O]→ (carboxylic acid) OH

f. →[Ag₂O / NH₄OH]→ (carboxylic acid) OH

g. →[[1] CH₃MgBr / [2] H₂O]→ OH

h. →[[1] C₆H₅Li / [2] H₂O]→ OH (with phenyl)

i. →[[1] (CH₃)₂CuLi / [2] H₂O]→ **No reaction**

j. →[[1] HC≡CNa / [2] H₂O]→ OH (with alkyne)

k. →[[1] CH₃C≡CLi / [2] H₂O]→ OH (with alkyne)

l. OH →[TBDMSCl / imidazole]→ O–TBDMS

20.39

a. (benzoyl chloride) →[(CH₃CH₂CH₂CH₂)₂CuLi]→ (phenyl ketone)

b. (methyl benzoate) →[(CH₃CH₂CH₂CH₂)₂CuLi]→ **No reaction**

c.

[1] $(CH_3CH_2CH_2CH_2)_2CuLi$

[2] H_2O

d.

[1] $(CH_3CH_2CH_2CH_2)_2CuLi$

[2] H_2O

20.40 Arrange the larger group [$(CH_3)_3C–$] on the left side of the carbonyl.

a. NaBH$_4$, CH$_3$OH

H   OH   ...   HO   H

S   +   R

b. [1] (S)-CBS reagent; [2] H$_2$O

HO   H

R

c. [1] (R)-CBS reagent; [2] H$_2$O

H   OH

S

20.41

A

a. NaBH$_4$, CH$_3$OH   → OH

b. H$_2$ (1 equiv), Pd-C →

c. H$_2$ (excess), Pd-C →   OH

d. [1] CH$_3$Li; [2] H$_2$O → HO

e. [1] CH$_3$CH$_2$MgBr; [2] H$_2$O → HO

f. [1] (CH$_2$=CH)$_2$CuLi; [2] H$_2$O →

20.42

a.

NaBH$_4$

CH$_3$OH

OH

b.

[1] LiAlH$_4$

[2] H$_2$O

HO

OH

c.

[1] LiAlH$_4$

[2] H$_2$O

N   OH

d.

Cl   OH

[1] LiAlH[OC(CH$_3$)$_3$]$_3$

[2] H$_2$O

H   OH

20.43

a.

b.

c.

d.

e.

f.

20.44

a.

b.

c.

d.

20.45 Three carbons bear a δ+ because they are bonded to electronegative O atoms.

C1 is an sp^3 hybridized C bonded to an O so it bears a δ+. There are no additional resonance structures that affect C1. C2 is part of a carbonyl that has three resonance structures, only one of which places a (+) charge on C. The O atom of the ester donates its electron pair, making the carbonyl C less electrophilic than C3. C3 is most electrophilic. Its carbonyl is stabilized by two resonance structures, one of which places a (+) charge on carbon.

O donates electron density.

full (+) charge
most electrophilic site

20.46 The organolithium reagent is a nucleophile and a base. As a base it can remove the most acidic proton (between the benzene ring and C=O) to form an enolate that is protonated by D_3O^+.

A

B
$C_{12}H_{13}DO_3$

20.47 Both ketones are chiral molecules with carbonyl groups that have one side more sterically hindered than the other. In both reductions, hydride approaches from the less hindered side.

2 H's
less hindered
Attack comes from above.

[1] LiAlH₄
[2] H₂O

H ← **from above**

endo OH group

The concave shape of the six-membered ring makes the bottom face of the C=O more sterically hindered. Addition of the H⁻ occurs from above to place the new C–H bond exo, making the OH endo.

The CH₃ groups on the bridgehead carbon make the top more hindered. H⁻ attacks from below to afford an exo OH group.
Attack comes from below.

[1] LiAlH₄
[2] H₂O

from below

exo OH group

20.48 A Grignard reagent contains a carbon atom with a partial negative charge, so it acts as a base and reacts with the OH of the starting halide, $BrCH_2CH_2CH_2CH_2OH$. This acid–base reaction destroys the Grignard reagent, so that addition cannot occur. To get around this problem, the OH group can be protected as a *tert*-butyldimethylsilyl ether, from which a Grignard reagent can be made.

basic site acidic site

These will react.

INSTEAD: Use a protecting group.

20.49 Compounds **F**, **G**, and **K** are all alcohols with aromatic rings so there will be many similarities in their proton NMR spectra. These compounds, however, will show differences in absorptions due to the CH protons on the carbon bearing the OH group. **F** has a CH$_2$OH group, which will give a singlet in the 3–4 ppm region of the spectrum. **G** is a 3° alcohol that has no protons on the C bonded to the OH group so it will have no peak in the 3–4 ppm region of the spectrum. **K** is a 2° alcohol that will give a doublet in the 3–4 ppm region of the spectrum for the CH proton on the carbon with the OH group.

20.50

20.51

20.52

20.53

20.54

Protonate four alkoxides.

20.55

a.

or

b.

or

c.

20.56

(a)

(b)

(c)

(a) + MgBr

(b) + BrMg

(c) + MgBr

20.57

a. OH ⟹ + CH₃–MgBr (2 equiv)   b. ⟹ + BrMg (2 equiv)

20.58

a. OH ⟹ Li +   (two ways)

or ⟹ + Li

b. OH ⟹ + Li (three ways)

or ⟹ + Li (2 equiv)

or ⟹ + Li

20.59

a. OH ⟹ MgBr +

b. C₆H₅ OH ⟹ O + BrMg—C₆H₅

c. HO ⟹ + MgBr

20.60

C₆H₅ Br →(Mg) C₆H₅ MgBr →(H₂O) C₆H₅

C₆H₅ Br →(KOC(CH₃)₃) C₆H₅ →(H₂/Pd-C) C₆H₅

C₆H₅ Br →(LiAlH₄) C₆H₅

20.61

20.62

a.

b.

c.

d.

(from a.)

20.63

a.

b.

20.64

a.

b.

OH $\xrightarrow{\text{PCC}}$ (aldehyde) + (PhMgBr) $\xrightarrow{\text{H}_2\text{O}}$ (OH) $\xrightarrow{\text{PCC}}$ (ketone) $\xrightarrow[\text{FeBr}_3]{\text{Br}_2}$ (Br-aryl ketone)

(from a.)

20.65

a. OH $\xrightarrow{\text{PCC}}$ (aldehyde) $\xrightarrow[\text{[2] H}_2\text{O}]{\text{[1]} \quad \text{MgBr}}$ (OH) $\xrightarrow{\text{PCC}}$ (ketone)

OH $\xrightarrow{\text{PBr}_3}$ Br $\xrightarrow{\text{Mg}}$ MgBr

b. (aldehyde) $\xrightarrow[\text{[2] H}_2\text{O}]{\text{[1]} \quad \text{MgBr}}$ (OH) $\xrightarrow{\text{PBr}_3}$ Br $\xrightarrow{\text{Mg}}$ MgBr

(from a.)

OH $\xrightarrow{\text{PBr}_3}$ Br $\xrightarrow{\text{Mg}}$ MgBr

$\xrightarrow[\text{[2] H}_3\text{O}^+]{\text{[1] CO}_2}$ CO$_2$H

c. OH $\xrightarrow{\text{HBr}}$ Br $\xrightarrow{\text{Mg}}$ MgBr

+ (aldehyde) $\xrightarrow{\text{H}_2\text{O}}$ OH $\xrightarrow{\text{PBr}_3}$ Br $\xrightarrow{\text{K}^+ \ ^-\text{OC(CH}_3)_3}$ (alkene)

(from a.)

20.66

a. (cyclohexenone) $\xrightarrow[\text{[2] H}_2\text{O}]{\text{[1] (CH}_3)_2\text{CuLi}}$ (ketone) $\xrightarrow[\text{[2] H}_2\text{O}]{\text{[1] CH}_3\text{MgBr}}$ OH $\xrightarrow{\text{H}_2\text{SO}_4}$ (alkene) $\xrightarrow{\text{H}_2, \text{Pd-C}}$ (dimethylcyclohexane)

+

(alkene isomer)

b. HC≡CH $\xrightarrow[\text{[3] H}_2\text{O}]{\begin{array}{l}\text{[1] NaH}\\\text{[2] CH}_3\text{CH=O}\end{array}}$ HO (alkyne) $\xrightarrow[\text{imidazole}]{\text{TBDMS–Cl}}$ TBDMSO (alkyne) $\xrightarrow{\text{NaH}}$ TBDMSO (alkyne)$^-$

$\xrightarrow[\text{[2] H}_2\text{O}]{\text{[1]}\quad \triangle\!\!O}$

HO (alkyne) OH $\xleftarrow{\text{(Bu)}_4\text{NF}}$ TBDMSO (alkyne) OH

20.67

20.68

(E)-tetradec-11-enal

20.69

IR peak: 1716 cm⁻¹ (C=O)
¹H NMR: 2 signals (ppm)
doublet 1.2 (H_b)
septet 2.7 (H_a)

C₇H₁₄O
A

IR peak: 3600–3200 cm⁻¹ (OH)
¹H NMR: 4 signals (ppm)
doublet 0.9 (H_d)
singlet 1.5 (H_a)
multiplet 1.7 (H_c)
triplet 3.0 (H_b)

C₇H₁₆O
B

20.70

C (C₄H₈O)

$H_b \rightarrow H$, O, $CH_3 \leftarrow H_a$
$H_b \rightarrow H$, $CH_3 \leftarrow H_a$

C_4H_8O
C

[1] C₆H₅MgBr
[2] H₂O

H_c H_c
H H OH_b
CH_3 CH_3
H_a H_a

$C_{10}H_{14}O$
D

¹H NMR: 2 signals (ppm)
  singlet (6 H) 1.3 (H$_a$)
  singlet (2 H) 2.4 (H$_b$)

IR peak 3600–3200 cm⁻¹ (OH)
¹H NMR: 4 signals (ppm)
  singlet (6 H) 1.2 (H$_a$)
  singlet (1 H) 1.6 (H$_b$)
  singlet (2 H) 2.7 (H$_c$)
  multiplet (5 H) 7.2 (benzene ring)

20.71

O
CH_3 C OCH_2CH_3
H_a H_b H_c

$C_4H_8O_2$
E

IR peak 1743 cm⁻¹ (C=O)
¹H NMR: 3 signals (ppm)
  triplet (3 H) 1.2 (H$_c$)
  singlet (3 H) 2.0 (H$_a$)
  quartet (2 H) 4.1 (H$_b$)

[1] CH₃CH₂MgBr (excess)
[2] H₂O

H_b OH_d H_b
$CH_3CH_2 - C - CH_2CH_3$
H_a CH_3 H_a
H_c

$C_6H_{14}O$
F

IR peak 3600–3200 cm⁻¹ (OH)
¹H NMR: 4 signals (ppm)
  triplet (6 H) 0.9 (H$_a$)
  singlet (3 H) 1.1 (H$_c$)
  quartet (4 H) 1.5 (H$_b$)
  singlet (1 H) 1.55 (H$_d$)

20.72 Molecular ion at $m/z = 86$: $C_5H_{10}O$ (possible molecular formula).

≡N

[1] CH₃MgBr
[2] H₃O⁺

H_b H_b
O H H
CH_3 CH_3
H_c H H H_a
H_d H_d

G

IR peak 1721 cm⁻¹ (C=O)
¹H NMR: 4 signals (ppm)
  triplet (3 H) 0.9 (H$_a$)
  sextet (2 H) 1.6 (H$_b$)
  singlet (3 H) 2.1 (H$_c$)
  triplet (2 H) 2.4 (H$_d$)

20.73 Molecular ion at $m/z = 86$: $C_5H_{10}O$ (possible molecular formula).

[1] (CH₃)₃CLi
[2] CH₂=O
[3] H₂O

H_b H_d H_d
CH_3 H H
H OH
H_e H H H H_a
H_f H_c H_c

H

IR peaks: 3600–3200 cm⁻¹ (OH)
        1651 cm⁻¹ (C=C)
¹H NMR: 6 signals (ppm)
  singlet (1 H) 1.7 (H$_a$)
  singlet (3 H) 1.8 (H$_b$)
  triplet (2 H) 2.2 (H$_c$)
  triplet (2 H) 3.8 (H$_d$)
  two signals at 4.8 and 4.9 due
  to 2 H's: H$_e$ and H$_f$

20.74

20.75

20.76

4-*tert*-butylcyclohexanone → *cis*-4-*tert*-butylcyclohexanol (major product)

L-Selectride adds H⁻ to a C=O group. There are two possible reduction products—cis and trans isomers—but the cis isomer is favored. The key element is that the three *sec*-butyl groups make L-selectride a large, bulky reducing agent that attacks the carbonyl group from the less hindered direction.

When H⁻ adds from the equatorial direction, the product has an axial OH and a new equatorial H. Since the equatorial direction is less hindered, this mode of attack is favored with large bulky reducing agents like L-selectride. In this case, the product is cis.

The axial H's hinder H⁻ attack from the axial direction. As a result, this mode of attack is more difficult with larger reducing agents. In this case the product is trans. This product is not formed to any appreciable extent.

20.77 The β carbon of an α,β-unsaturated carbonyl compound absorbs farther downfield in the ^{13}C NMR spectrum than the α carbon, because the β carbon is deshielded and bears a partial positive charge as a result of resonance. Because three resonance structures can be drawn for an α,β-unsaturated carbonyl compound, one of which places a positive charge on the β carbon, the decrease of electron density at this carbon deshields it, shifting the ^{13}C absorption downfield. This is not the case for the α carbon.

20.78

20.79

20.80

(+ 1 resonance structure)

20.81

Chapter 21 Aldehydes and Ketones—Nucleophilic Addition

Chapter Review

General facts

- Aldehydes and ketones contain a carbonyl group bonded to only H atoms or R groups. The carbonyl carbon is sp^2 hybridized and trigonal planar (21.1).
- Aldehydes are identified by the suffix -al, whereas ketones are identified by the suffix -one (21.2).
- Aldehydes and ketones are polar compounds that exhibit dipole–dipole interactions (21.3).

Summary of spectroscopic absorptions of RCHO and R$_2$CO (21.4)

IR absorptions	C=O	~1715 cm^{-1} for ketones
		• increasing frequency with decreasing ring size
		~1730 cm^{-1} for aldehydes
		• For both RCHO and R$_2$CO, the frequency decreases with conjugation.
	C$_{sp^2}$–H of CHO	~2700–2830 cm^{-1} (one or two peaks)
^{1}H NMR absorptions	CHO	9–10 ppm (highly deshielded proton)
	C–H α to C=O	2–2.5 ppm (somewhat deshielded C$_{sp^3}$–H)
^{13}C NMR absorption	C=O	190–215 ppm

Nucleophilic addition reactions
[1] Addition of hydride (H$^-$) (21.8)

- The mechanism has two steps.
- H:$^-$ adds to the planar C=O from both sides.

[2] Addition of organometallic reagents (R$^-$) (21.8)

- The mechanism has two steps.
- R:$^-$ adds to the planar C=O from both sides.

[3] Addition of cyanide ($^-$CN) (21.9)

- The mechanism has two steps.
- $^-$CN adds to the planar C=O from both sides.

[4] Wittig reaction (21.10)

- The reaction forms a new C–C σ bond and a new C–C π bond.
- $Ph_3P=O$ is formed as by-product.

[5] Addition of 1° amines (21.11)

- The reaction is fastest at pH 4–5.
- The intermediate carbinolamine is unstable, and loses H_2O to form the C=N.

[6] Addition of 2° amines (21.12)

- The reaction is fastest at pH 4–5.
- The intermediate carbinolamine is unstable, and loses H_2O to form the C=C.

[7] Addition of H₂O—Hydration (21.13)

- The reaction is reversible. Equilibrium favors the product only with less stable carbonyl compounds (e.g., H_2CO and Cl_3CCHO).
- The reaction is catalyzed with either H^+ or ^-OH.

[8] Addition of alcohols (21.14)

- The reaction is reversible.
- The reaction is catalyzed with acid.
- Removal of H_2O drives the equilibrium to favor the products.

Other reactions

[1] Synthesis of Wittig reagents (21.10A)

- Step [1] is best with CH_3X and RCH_2X because the reaction follows an S_N2 mechanism.
- A strong base is needed for proton removal in Step [2].

[2] Conversion of cyanohydrins to aldehydes and ketones (21.9)

- This reaction is the reverse of cyanohydrin formation.

[3] Hydrolysis of nitriles (21.9)

R' = H or alkyl

[4] Hydrolysis of imines and enamines (21.12)

[5] Hydrolysis of acetals (21.14)

R' = H or alkyl

- The reaction is acid catalyzed and is the reverse of acetal synthesis.
- A large excess of H_2O drives the equilibrium to favor the products.

Practice Test on Chapter Review

1. Give the IUPAC name for the following compounds.

a.

b.

c.

2. (a) Considering compounds **A–D,** which compound forms the smallest amount of hydrate? (b) Which compound forms the largest amount of hydrate?

CH_3O—⟨ ⟩—CHO

A

Cl—⟨ ⟩—

B

O_2N—⟨ ⟩—CHO

C

CH_3O—⟨ ⟩—

D

3. (a) Considering compounds **A–D,** which compound absorbs at the *lowest* wavenumber in its IR spectrum? (b) Which compound absorbs at the *highest* wavenumber in its IR spectrum?

A

B

C

D

4. Fill in the lettered reagents (**A–G**) in the following reaction scheme.

5. Draw the organic products formed in the following reactions.

a. [1] Ph₃P / [2] BuLi / [3] (cyclohexanone)

b. (pentanal) CHO + cyclohexyl-NH₂, mild acid

c. (cyclopentanone) [1] NaCN, HCl; [2] H₂O, H⁺, Δ

d. (3-pentanone) + piperidine NH, mild acid

e. (cyclopentanone) + HO–CH₂CH₂–OH, TsOH

f. HO–(pyranose)–OH, CH₃CH₂OH, H⁺ [Indicate stereochemistry.]

Answers to Practice Test

1.a. 5-isopropyl-2,4-dimethyl-cyclohexanone
   b. 3,3-dimethyl-5-phenylpentan-2-one
   c. 4-ethyl-2-methyl-cyclohexane-carbaldehyde

4.
   A = [1] LiC≡CH; [2] H_2O
   B = [1] R_2BH; [2] H_2O_2, ⁻OH
   C = Ag_2O, NH_4OH
   D = H_2O, H_2SO_4, $HgSO_4$
   E = TBDMS–Cl, imidazole
   F = [1] CH_3Li; [2] H_2O
   G = $Bu_4N^+F^-$

2.a. D
   b. C
3.a. B
   b. C

5.
   a. (methylenecyclohexane / ethylidene)
   b. (imine product) N–cyclohexyl
   c. (cyclopentane OH, CO₂H)
   d. (enamine with piperidine N) (E + Z)
   e. (cyclic ketal, spiro dioxolane)
   f. (pyran ring –OH, OEt) +

Answers to Problems

21.1 As the number of R groups bonded to the carbonyl C increases, reactivity towards nucleophilic attack decreases. Steric hindrance decreases reactivity as well.

(2,6-dimethylcyclohexanone) (2-methylcyclohexanone) (cyclohexanone)

Increasing reactivity
decreasing steric hindrance

21.2   More stable aldehydes are less reactive towards nucleophilic attack.

benzaldehyde
Several resonance structures delocalize the partial
positive charge on the carbonyl carbon, making
it more stable and less reactive towards
nucleophilic attack.

cyclohexanecarbaldehyde
This aldehyde has no added resonance
stabilization.

21.3   • To name an aldehyde with a chain of atoms: [1] Find the longest chain with the CHO group and change the -*e* ending to -*al*. [2] Number the carbon chain to put the CHO at C1, but omit this number from the name. Apply all other nomenclature rules.
   • To name an aldehyde with the CHO bonded to a ring: [1] Name the ring and add the suffix -*carbaldehyde*. [2] Number the ring to put the CHO group at C1, but omit this number from the name. Apply all other nomenclature rules.

a.

5 C chain = pentanal   **3,3,4,4-tetramethylpentanal**

c.

4 C ring =
cyclobutanecarbaldehyde

**3,3-dichlorocyclobutane-
carbaldehyde**

b.

8 C chain = octanal   **2,5,6-trimethyloctanal**

21.4   Work backwards from the name to the structure, referring to the nomenclature rules in Answer 21.3.

a. 2-isobutyl-3-isopropyl**hexanal**

6 C chain

b. *trans*-3-methyl**cyclopentanecarbaldehyde**

5 carbon ring

or

c. 1-methyl**cyclopropanecarbaldehyde**

3 carbon ring

d. 3,6-diethyl**nonanal**

9 C chain

21.5 • To name an acyclic ketone: [1] Find the longest chain with the carbonyl group and change the -e ending to -one. [2] Number the carbon chain to give the carbonyl C the lower number. Apply all other nomenclature rules.
• To name a cyclic ketone: [1] Name the ring and change the -e ending to -one. [2] Number the C's to put the carbonyl C at C1 and give the next substituent the lower number. Apply all other nomenclature rules.

a.

8 C chain = octanone

5-ethyl-4-methyloctan-3-one

c.

5 C chain = pentanone

2,2,4,4-tetramethylpentan-3-one

b.

5 C ring = cyclopentanone

3-*tert*-butyl-2-methylcyclopentanone

21.6 Most common names are formed by naming both alkyl groups on the carbonyl C, arranging them alphabetically, and adding the word ketone.

a. *sec*-butyl ethyl ketone

b. methyl vinyl ketone

c. *p*-ethylacetophenone

d. 3-benzoyl-2-benzylcyclopentanone

benzyl group: benzoyl group: 5 C ketone

e. 6,6-dimethylcyclohex-2-enone

6 C ketone

f. 3-ethylhex-5-enal

CHO

21.7 Compounds with both a C–C double bond and an aldehyde are named as enals.

a. (Z)-3,7-dimethylocta-2,6-dienal

7   3   2
  6
    1 CHO

neral

b. (2E,6Z)-nona-2,6-dienal

7   3   1
            CHO
  6   2

cucumber aldehyde

21.8   Even though both compounds have polar C–O bonds, the electron pairs around the sp^3 hybridized O atom of diethyl ether are more crowded and less able to interact with electron-deficient sites in other diethyl ether molecules. The O atom of the carbonyl group of butan-2-one extends out from the carbon chain, making it less crowded. The lone pairs of electrons on the O atom can more readily interact with the electron-deficient sites in the other molecules, resulting in stronger forces and a higher boiling point.

butan-2-one   diethyl ether

21.9   For cyclic ketones, the carbonyl absorption shifts to higher wavenumber as the size of the ring decreases and the ring strain increases. Conjugation of the carbonyl group with a C=C or a benzene ring shifts the absorption to lower wavenumber.

a. [structure] CHO   or   [structure] CHO   b. [structure] =O   or   [structure] =O

conjugated C=O   **higher wavenumber**   smaller ring
lower wavenumber   **higher wavenumber**

21.10 The number of lines in their ^{13}C NMR spectra can distinguish the constitutional isomers.

pentan-2-one   pentan-3-one   3-methylbutan-2-one
5 lines   3 lines   4 lines

21.11

a. [structure] $\xrightarrow{\text{[1] DIBAL-H}}_{\text{[2] H}_2\text{O}}$ [structure] H

b. [structure] OH $\xrightarrow{\text{PCC}}$ [structure] H

c. [structure] $\xrightarrow{\text{[1] R}_2\text{BH}}_{\text{[2] H}_2\text{O}_2,\ \text{HO}^-}$ [structure] H

d. [structure] $\xrightarrow{\text{[1] O}_3}_{\text{[2] Zn, H}_2\text{O}}$ [structure] H

21.12

a. [benzene] $\xrightarrow{\text{Cl—C(=O)CH}_3}_{\text{AlCl}_3}$ [acetophenone]

b. [benzoyl chloride] $\xrightarrow{\text{[1] (CH}_3)_2\text{CuLi}}_{\text{[2] H}_2\text{O}}$ [acetophenone]

c. [phenylacetylene] $\xrightarrow{\text{H}_2\text{O}}_{\substack{\text{H}_2\text{SO}_4 \\ \text{HgSO}_4}}$ [acetophenone]

21.13

Cleave this C=C with O_3.

21.14 Addition of hydride or R–M occurs at a planar carbonyl C, so two different configurations at a new stereogenic center are possible.

a.

new stereogenic center

Add stereochemistry:

b.

Add stereochemistry:

21.15 Treatment of an aldehyde or ketone with NaCN, HCl adds HCN across the double bond. Cyano groups are hydrolyzed by H_3O^+ to replace the three C–N bonds with three C–O bonds.

a.

b.

21.16

amygdalin

enzyme

enzyme

toxic by-product

+ HCN

21.17

a.

$\quad + \quad Ph_3P=CH_2 \quad \longrightarrow$

b.

$\quad + \quad Ph_3P=$

21.18

a.

b.

c.

21.19

a.

b.

c.

21.20 To draw the starting materials of the Wittig reactions, find the C=C and cleave it. Replace it with a C=O in one half of the molecule and a C=PPh₃ in the other half. The preferred pathway uses a Wittig reagent derived from a less hindered alkyl halide.

a.

2° halide precursor
(CH₃)₂CHX

1° halide precursor
XCH₂CH₂CH₃
preferred pathway

b.

(only one route possible)

c.

1° halide precursor
C₆H₅CH₂X

(both routes possible)

1° halide precursor
XCH₂CH₃

21.21

a.   Two-step sequence:

One-step sequence:

minor product   tetrasubstituted **major product**   (*E* and *Z* isomers)

only product

b.   Two-step sequence:

trisubstituted conjugated C=C   trisubstituted

One-step sequence:

only product

21.22 When a 1° amine reacts with an aldehyde or ketone, the C=O is replaced by C=NR.

a.

b.

c.

21.23 Remember that the C=NR is formed from a C=O and an NH₂ group of a 1° amine.

a.

b.

21.24

21.25 • Imines are hydrolyzed to 1° amines and a carbonyl compound.
• Enamines are hydrolyzed to 2° amines and a carbonyl compound.

a.

imine → (H₂O, H⁺) → benzaldehyde + cyclopentyl amine (1° amine)

b.

enamine → (H₂O, H⁺) → 2° amine + cyclopentanone

c.

enamine → (H₂O, H⁺) → $(CH_3)_2NH$ (2° amine) + carbonyl compound

21.26

proton transfer

$H_2\ddot{O}$: + ← ←

21.27

• A substituent that **donates** electron density to the carbonyl C stabilizes it, **decreasing** the percentage of hydrate at equilibrium.
• A substituent that **withdraws** electron density from the carbonyl C destabilizes it, **increasing** the percentage of hydrate at equilibrium.

a.

one R group on C=O
higher percentage of hydrate

2 R groups
on C=O

b.

F atoms are electron withdrawing.
higher percentage of hydrate

21.28

+ $H_2\ddot{O}$

(+ 1 resonance structure)

21.29 Treatment of an aldehyde or ketone with two equivalents of alcohol results in the formation of an acetal (a C bonded to two OR groups).

a. [cyclopentanone] =O + 2 CH₃OH →(TsOH)→ [cyclopentane with OCH₃, OCH₃]

b. [ketone] + HO—CH₂CH₂—OH →(TsOH)→ [cyclic acetal with O, O]

21.30

a. [cyclohexane with OCH₃ and OCH₃ on different C's]
   2 OR groups
   on different C's
   2 ethers

b. [cyclohexane with OCH₃, OCH₃ on same C]
   2 OR groups
   on same C
   acetal

c. [bicyclic dioxolane with O, O]
   2 OR groups
   on same C
   acetal

d. [chain with OCH₃ and OH on same C]
   1 OR group and
   1 OH group
   on same C
   hemiacetal

21.31 The mechanism has two parts: [1] nucleophilic addition of ROH to form a hemiacetal; [2] conversion of the hemiacetal to an acetal.

overall reaction

hemiacetal

carbocation
re-drawn

acetal

21.32

a. [CH₃O OCH₃ acetal] + H₂O →(H₂SO₄)→ [acetone] + 2 CH₃OH

b. [bicyclic acetal] + H₂O →(H₂SO₄)→ [acetone] + [cyclohexane-1,2-diol with OH, OH]

21.33

safrole → formaldehyde +

H_2O / H_2SO_4

21.34

acetal   oleandrin H_3O^+ →

21.35 Use an acetal protecting group to carry out the reaction.

HO—OH / TsOH

[1] CH_3Li (2 equiv)
[2] H_2O

H_3O^+

21.36

a.

HO — C5   C1   →   OH — C1   C5

b.

C4   OH   C1   →   OH — C1   C4

21.37

acetal

HO

OCH$_3$

COOH

HO   OH

monensin

hemiacetal

acetal   acetal

HO,

HO,

HO,

OH

acetal

digoxin

OH   OH

Ether **O** atoms are indicated in **bold.**

21.38 The hemiacetal OH is replaced by an OR group to form an acetal.

a.

b.

21.39

a.

5 stereogenic centers
(labeled with *)

b.

hemiacetal C

α-D-galactose

c.

β-D-galactose

d.

e. +

21.40

a.

3,3-dimethylbutanal
A

cis-5-isopropyl-2-
methylcyclohexanone
B

b. [1] **A** $\xrightarrow[\text{CH}_3\text{OH}]{\text{NaBH}_4}$

[2] **A** $\xrightarrow[]{\text{CH}_3\text{MgBr}} \xrightarrow[]{\text{H}_2\text{O}}$ +

[3] **A** $\xrightarrow[]{\text{Ph}_3\text{P=CHOCH}_3}$ OCH$_3$ (+ cis isomer)

[4] **A** $\xrightarrow[\text{mild H}^+]{\qquad \text{NH}_2}$

[5] **A** $\xrightarrow[\text{H}^+]{\text{HOCH}_2\text{CH}_2\text{CH}_2\text{OH}}$

[1] **B** $\xrightarrow{\text{NaBH}_4}$ $\xrightarrow{\text{CH}_3\text{OH}}$ [structure with HO] + [structure with HO]

[2] **B** $\xrightarrow{\text{CH}_3\text{MgBr}}$ $\xrightarrow{\text{H}_2\text{O}}$ [structure with HO] + [structure with HO]

[3] **B** $\xrightarrow{\text{Ph}_3\text{P=CHOCH}_3}$ [structure with OCH$_3$] + [structure with CH$_3$O]

[4] **B** $\xrightarrow[\text{mild H}^+]{\text{NH}_2}$ [structure with N]

[5] **B** $\xrightarrow[\text{H}^+]{\text{HOCH}_2\text{CH}_2\text{CH}_2\text{OH}}$ [structure with O, O]

21.41 The least hindered carbonyl group is the most reactive.

least reactive   intermediate reactivity   most reactive

21.42

a. [structure] ⟹ [CHO structure] + HO⌒OH

b. [structure] ⟹ [structure with OH, OH] + [acetone]

21.43 Use the rules from Answers 21.3 and 21.5 to name the aldehydes and ketones.

a. [structure]
6 C ring = cyclohexanone
5-ethyl-2-methyl-cyclohexanone

b. [CHO structure]
***trans*-2-benzylcyclohexane-carbaldehyde**

c. [structure with NO$_2$]
***o*-nitroacetophenone**

d. [CHO structure]
6 C = hexanal
3,4-diethylhexanal

e. [structure, E]
8 C = octenone
(*E*)-2,5-dimethyloct-5-en-4-one

f. [CHO structure]
6 C = hexenal
3,4-diethyl-2-methylhex-3-enal

21.44

a. 2-methyl-3-phenylbutanal

b. 3,3-dimethylcyclohexanecarbaldehyde

c. 3-benzoylcyclopentanone

d. 2-formylcyclopentanone

e. (*R*)-3-methylheptan-2-one

f. *m*-acetylbenzaldehyde

g. 2-*sec*-butylcyclopent-3-enone

h. 5,6-dimethylcyclohex-1-enecarbaldehyde

21.45

a.
$$\text{Br} \quad [1]\ Ph_3P \quad\longrightarrow\quad (+\ Z\text{ isomer})$$
$$[2]\ BuLi$$
$$[3]\ C_6H_5CH_2CH_2CHO$$

b.
$$\text{Cl} \quad [1]\ Ph_3P \quad\longrightarrow\quad (+\ Z\text{ isomer})$$
$$[2]\ BuLi$$
$$[3]\ CH_3CH_2CH_2CHO$$

21.46

a.
$$\quad + \quad H_2N-\quad \xrightarrow[\text{acid}]{\text{mild}} \quad$$

b.
$$\xrightarrow[H^+]{HO\quad OH}$$

c.
$$\xrightarrow{H_3O^+} \quad + \quad H_2N$$

d.
$$C_6H_5 \quad + \quad \xrightarrow[\text{acid}]{\text{mild}} \quad (E\ \text{and}\ Z\ \text{isomers})$$

e.
$$\xrightarrow{H_3O^+,\ \Delta}$$

f.

$$\text{(tetrahydrofuran-2-ol)} \xrightarrow[\text{H}^+]{\text{CH}_3\text{CH}_2\text{OH}} \text{(2-ethoxytetrahydrofuran)}$$

g.

$$\text{(enamine)} \xrightarrow{\text{H}_3\text{O}^+} \text{(3,3-dimethylcyclopentanone)} + \text{HN (piperidine)}$$

h.

$$\text{(cyclopentanone)} + \text{Ph}_3\text{P}=\text{(ylide)}\text{OCH}_3 \longrightarrow \text{(alkene product)}\text{OCH}_3$$

21.47

a.

$$\text{(acetal)} \longrightarrow \text{(3-methyl-2-butanone)} + \text{HO (ethanol)}$$

c.

$$\text{(2-ethoxytetrahydropyran)} \longrightarrow \text{HO} \text{(5-hydroxypentanal)}\text{H} + \text{HOCH}_2\text{CH}_3$$

b.

$$\text{(orthoester)} \longrightarrow \text{(benzophenone deriv)} + \text{HOCH}_3$$

21.48

a.

$$\text{(butanal)}\text{H} \xrightarrow{\text{Ph}_3\text{P}=\text{(propylide)}} \text{(E-alkene)} + \text{(Z-alkene)}$$

b.

$$\text{(butanone)} \xrightarrow[\text{HCl}]{\text{NaCN}} \text{HO,CN (cyanohydrin)} + \text{NC,OH (cyanohydrin)}$$

c.

$$\text{(3-ethylcyclohexanone)} \xrightarrow[\text{CH}_3\text{OH}]{\text{NaBH}_4} \text{(cis-ol)}\text{OH} + \text{(trans-ol)}\text{OH}$$

d.

$$\text{HO (dihydroxypyran)}\text{OH} \xrightarrow[\text{HCl}]{\text{CH}_3\text{OH}} \text{HO (product)}\text{OCH}_3 + \text{HO (product)}\text{OCH}_3$$

21.49

A achiral

An equal mixture of enantiomers results, so the product is optically inactive.

B chiral

A mixture of diastereomers results. Both compounds are chiral and they are not enantiomers, so the mixture is optically active.

21.50

a.

b.

21.51

acetal / enamine / imine

21.52

a.

etoposide

b. Lines of cleavage are drawn in.

21.53 The less stable carbonyl compound forms the higher percentage of hydrate.

The aldehyde is destabilized because there is a δ+ on its adjacent carbon, which is part of a C=O. Thus, PhCOCHO has the higher concentration of hydrate.

21.54 Electron-donating groups decrease the amount of hydrate at equilibrium by stabilizing the carbonyl starting material. Electron-withdrawing groups increase the amount of hydrate at equilibrium by destabilizing the carbonyl starting material. Electron-donating groups make the IR absorption of the C=O shift to lower wavenumber because they stabilize the charge-separated resonance form, giving the C=O more single bond character.

	p-nitroacetophenone	*p*-methoxyacetophenone
a.	NO_2 withdrawing group **less stable**	CH_3O donating group **more stable**
b.	**higher percentage of hydrate**	**lower percentage of hydrate**
c.	**higher wavenumber**	**lower wavenumber**

21.55 Use the principles from Answer 21.20.

a.

1° alkyl halide precursor
($XCH_2CH_2CH_2CH_3$)
preferred pathway

2° alkyl halide precursor
[$(CH_3CH_2)_2CHX$]

b.

methyl halide precursor
(CH_3X)
preferred pathway

2° alkyl halide precursor
($CH_3CH_2CH_2CHXCH_3$)

c.

1° alkyl halide precursor
($C_6H_5CH_2X$)
preferred pathway

2° alkyl halide precursor

21.56

a.

b.

c.

d.

21.57

One-step sequence:

preferred route
only one product formed

or

Two-step sequence:

+ other alkenes that result from
carbocation rearrangement

21.58

a.

One possibility:

(+ Z isomer)

b.

21.59

a.

b.

(from a.)

21.60

a.

(+ ortho isomer)

(+ Z isomer)

(+ ortho isomer)

b.

c.

21.61

a.

b.

21.62

21.63

a.

methoxy methyl ether

b.

acetal

c.

(The three organic products are boxed in.)

21.64

21.65

celecoxib

21.66 The OH groups react with the C=O in an intramolecular reaction, first to form a hemiacetal, and then to form an acetal.

OH adds here to form a hemiacetal.
Then, the acetal is formed by a second
intramolecular reaction.

hemiacetal

acetal
$C_9H_{16}O_2$

21.67

a.

b.

21.68

21.69

21.70

dopamine

salsolinol

+ H₃O⁺ → $+ \ H_3\overset{..}{O}{}^+$

(+ 3 more resonance structures)

$+ \ H_2\overset{..}{O}$:

21.71 Hemiacetal **A** is in equilibrium with its acyclic hydroxy aldehyde. The aldehyde can undergo hydride reduction to form butane-1,4-diol and a Wittig reaction to form an alkene.

a.
A

This can now be reduced with $NaBH_4$.

butane-1,4-diol

b.
A

reacts with the Wittig reagent

(+ Z isomer)

c.

Y

$NaOCH_2CH_3$

X

aldehyde for Wittig reaction

New C=C could be cis or trans. The more stable trans C=C is drawn.

isotretinoin

21.72

CH₃–MgI

5,5-dimethoxypentan-2-one

(+ 1 resonance structure)

+ H₂Ö:

(+ 1 resonance
structure)

+ H–Ö–CH₃

+ H₂Ö:

H₂Ö:

Y

+ H₃Ö⁺

(+ 1 resonance structure)

+ H–Ö–CH₃

21.73

cyclopropenone
(1640 cm⁻¹)

cyclohex-2-enone
(1685 cm⁻¹)

These three resonance structures include an aromatic ring; $4n + 2 = 2$ π electrons. Although they are charge separated, the stabilized aromatic ring makes these three structures contribute to the hybrid more than usual. Because these three resonance contributors have a C–O single bond, the absorption is shifted to a lower wavenumber.

There are three resonance structures for cyclohex-2-enone, but the charge-separated resonance structures are not aromatic, so they contribute less to the resonance hybrid. The C=O absorbs in the usual region for a conjugated carbonyl.

21.74

A. Molecular formula $C_5H_{10}O$ ⟶ 1 degree of unsaturation
IR absorptions at 1728, 2791, 2700 cm⁻¹ ⟶ C=O, CHO
NMR data: singlet at 1.08 (9 H) ⟶ 3 CH₃ groups
singlet at 9.48 (1 H) ppm ⟶ CHO

B. Molecular formula $C_5H_{10}O$ ⟶ 1 degree of unsaturation
IR absorption at 1718 cm⁻¹ ⟶ C=O
NMR data: doublet at 1.10 (6 H) ⟶ 2 CH₃'s adjacent to H
singlet at 2.14 (3 H) ⟶ CH₃
septet at 2.58 (1 H) ppm ⟶ CH adjacent to 2 CH₃'s

C. Molecular formula $C_{10}H_{12}O$ ⟶   5 degrees of unsaturation (4 due to a benzene ring)
IR absorption at 1686 cm^{-1} ⟶   C=O
NMR data:

    triplet at 1.21 (3 H)   ⟶ CH_3 adjacent to 2 H's

    singlet at 2.39 (3 H)   ⟶ CH_3

    quartet at 2.95 (2 H)   ⟶ CH_2 adjacent to 3 H's

    doublet at 7.24 (2 H)   ⟶   2 H's on benzene ring

    doublet at 7.85 (2 H) ppm   ⟶   2 H's on benzene ring

D. Molecular formula $C_{10}H_{12}O$ ⟶   5 degrees of unsaturation (4 due to a benzene ring)
IR absorption at 1719 cm^{-1} ⟶   C=O
NMR data:

    triplet at 1.02 (3 H)   ⟶ CH_3 adjacent to 2 H's

    quartet at 2.45 (2 H)   ⟶   2 H's adjacent to 3 H's

    singlet at 3.67 (2 H)   ⟶ CH_2

    multiplet at 7.06–7.48 (5 H) ppm   ⟶   a monosubstituted benzene ring

21.75

$C_7H_{16}O_2$: 0 degrees of unsaturation
IR: 3000 cm^{-1}: C–H bonds
NMR data (ppm):
  H_a: quartet at 3.8 (**4 H**), split by 3 H's
  H_b: singlet at 1.5 (**6 H**)
  H_c: triplet at 1.2 (**6 H**), split by 2 H's

$$CH_3CH_2-O-\underset{\underset{H_b}{\overset{\displaystyle CH_3}{|}}}{\overset{\overset{\displaystyle CH_3 \leftarrow H_b}{|}}{C}}-O-CH_2CH_3$$

21.76

A. Molecular formula $C_9H_{10}O$
5 degrees of unsaturation
IR absorption at 1700 cm^{-1} → C=O
IR absorption at ~2700 cm^{-1} → CH of RCHO
NMR data (ppm):

    triplet at 1.2 (2 H's adjacent)
    quartet at 2.7 (3 H's adjacent)
    doublet at 7.3 (2 H's on benzene)
    doublet at 7.7 (2 H's on benzene)
    singlet at 9.9 (CHO)

B. Molecular formula $C_9H_{10}O$
5 degrees of unsaturation
IR absorption at 1720 cm^{-1} → C=O
IR absorption at ~2700 cm^{-1} → CH of RCHO
NMR data (ppm):

    2 triplets at 2.85 and 2.95 (suggests $-CH_2CH_2-$)
    multiplet at 7.2 (benzene H's)
    signal at 9.8 (CHO)

21.77

C. Molecular formula $C_6H_{12}O_3$
1 degree of unsaturation
IR absorption at 1718 cm^{-1} → C=O
NMR data (ppm):
    singlet at 2.1 (3 H's)
    doublet at 2.7 (2 H's)
    singlet at 3.3 (6 H's – 2 OCH_3 groups)
    triplet at 4.8 (1 H)

21.78

D. Molecular ion at $m/z = 150$: $C_9H_{10}O_2$ (possible molecular formula)
5 degrees of unsaturation
IR absorption at 1692 cm^{-1} → C=O
NMR data (ppm):

triplet at 1.5 (3 H's – CH_3CH_2)
quartet at 4.1 (2 H's – CH_3CH_2)
doublet at 7.0 (2 H's – on benzene ring)
doublet at 7.8 (2 H's – on benzene ring)
singlet at 9.9 (1 H – on aldehyde)

21.79

a.

b.

21.80

β-D-glucose

$+$ Cl$^-$

$+$ H$_2$Ö:

CH$_3$ÖH
above

$+$ HCl
acetal

CH$_3$ÖH
below

acetal $+$ HCl

:ÖCH$_3$

The carbocation is trigonal planar, so CH$_3$OH attacks from two different
directions, and two different acetals are formed.

21.81

$+$ H$_3$O$^+$

21.82

a.

brevicomin

b.

21.83

21.84

a.

acetal carbon

hemiacetal carbon

c.

(OH can be up or down in both products.)

b.

[1] H₃O⁺ →

← (OH can be up or down.)

[2] CH₃OH, HCl →

← (OCH₃ can be up or down.)

[3] NaH (excess) / CH₃I (excess) →

21.85

[1] O₃
[2] (CH₃)₂S

NaBH₄
CH₃OH

RSO₃H

R

S
C₆H₁₀O₃

Mechanism of **R → S:**

21.86 The mechanism involves S$_N$2 displacement of Br, followed by intramolecular enamine formation.

conivaptan

Chapter 22 Carboxylic Acids and Their Derivatives—Nucleophilic Acyl Substitution

Chapter Review

Summary of spectroscopic absorptions of RCOZ (22.5)

IR absorptions
- All RCOZ compounds have a C=O absorption in the region $1600–1850$ cm^{-1}.
  - RCOCl: 1800 cm^{-1}
  - (RCO)$_2$O: 1820 and 1760 cm^{-1} (two peaks)
  - RCOOR': $1735–1745$ cm^{-1}
  - RCONR'$_2$: $1630–1680$ cm^{-1}
- Additional amide absorptions occur at $3200–3400$ cm^{-1} (N–H stretch) and 1640 cm^{-1} (N–H bending).
- Decreasing the ring size of a cyclic lactone, lactam, or anhydride increases the frequency of the C=O absorption.
- Conjugation shifts the C=O to lower wavenumber.

^{1}H NMR absorptions
- C–H α to the C=O absorbs at $2–2.5$ ppm.
- N–H of an amide absorbs at $7.5–8.5$ ppm.

^{13}C NMR absorption
- C=O absorbs at $160–180$ ppm.

Summary of spectroscopic absorptions of RCN (22.5)

IR absorption
- C≡N absorption at 2250 cm^{-1}

^{13}C NMR absorption
- C≡N absorbs at $115–120$ ppm.

Summary: The relationship between the basicity of Z⁻ and the properties of RCOZ

- **Increasing basicity of the leaving group** (22.2)
- **Increasing resonance stabilization** (22.2)

acid chloride   anhydride   carboxylic acid   ester   amide

- **Increasing leaving group ability** (22.7B)
- **Increasing reactivity** (22.7B)
- **Increasing frequency of the C=O absorption in the IR** (22.5)

General features of nucleophilic acyl substitution
- The characteristic reaction of compounds having the general structure RCOZ is nucleophilic acyl substitution (22.1).
- The mechanism consists of two steps (22.7A):
  [1] Addition of a nucleophile to form a tetrahedral intermediate
  [2] Elimination of a leaving group
- More reactive acyl compounds can be used to prepare less reactive acyl compounds. The reverse is not necessarily true (22.7B).

Nucleophilic acyl substitution reactions

[1] Reaction that synthesizes acid chlorides (RCOCl)

**From RCOOH
(22.10A):**

$$R-COOH + SOCl_2 \longrightarrow R-COCl + SO_2 + HCl$$

[2] Reactions that synthesize anhydrides [(RCO)$_2$O]

**a. From RCOCl
(22.8):**

$$R-COCl + {}^-O-CO-R' \longrightarrow R-CO-O-CO-R' + Cl^-$$

**b. From dicarboxylic
acids (22.10B):**

$$\xrightarrow{\Delta} \text{cyclic anhydride} + H_2O$$

cyclic anhydride

[3] Reactions that synthesize carboxylic acids (RCOOH)

**a. From RCOCl
(22.8):**

$$R-COCl + H_2O \xrightarrow{pyridine} R-COOH + \text{pyridinium } Cl^-$$

**b. From (RCO)$_2$O
(22.9):**

$$R-CO-O-CO-R + H_2O \longrightarrow 2 \; R-COOH$$

**c. From RCOOR'
(22.11):**

$$R-COOR' + H_2O \xrightarrow{(H^+ \text{ or } {}^-OH)} R-COOH \; or \; R-CO-O^- + R'OH$$

(with acid)   (with base)

**d. From RCONR'$_2$
(R' = H or alkyl,
22.13):**

R-CO-NR'$_2$
R' = H or alkyl

$$\xrightarrow{H_2O, H^+} R-COOH + R'_2 \overset{+}{N}H_2$$

$$\xrightarrow{H_2O, {}^-OH} R-CO-O^- + R'_2NH$$

[4] Reactions that synthesize esters (RCOOR')

**a. From RCOCl
(22.8):**

$$R-COCl + R'OH \xrightarrow{pyridine} R-CO-OR' + \text{pyridinium } Cl^-$$

b. From (RCO)₂O
(22.9):

$$\underset{R}{\overset{O}{\|}}C\overset{O}{\underset{O}{\|}}CR + R'OH \longrightarrow \underset{R}{\overset{O}{\|}}C\underset{OR'}{} + RCOOH$$

c. From RCOOH
(22.10C):

$$\underset{R}{\overset{O}{\|}}C\underset{OH}{} + R'OH \xrightarrow{H_2SO_4} \underset{R}{\overset{O}{\|}}C\underset{OR'}{} + H_2O$$

[5] Reactions that synthesize amides (RCONH₂) [The reactions are written with NH₃ as the nucleophile to form RCONH₂. Similar reactions occur with R'NH₂ to form RCONHR', and with R'₂NH to form RCONR'₂.]

a. From RCOCl
(22.8):

$$\underset{R}{\overset{O}{\|}}C\underset{Cl}{} + \underset{(2\ equiv)}{NH_3} \longrightarrow \underset{R}{\overset{O}{\|}}C\underset{NH_2}{} + NH_4^+Cl^-$$

b. From (RCO)₂O
(22.9):

$$\underset{R}{\overset{O}{\|}}C\overset{O}{\underset{O}{\|}}CR + \underset{(2\ equiv)}{NH_3} \longrightarrow \underset{R}{\overset{O}{\|}}C\underset{NH_2}{} + RCOO^-NH_4^+$$

c. From RCOOH
(22.10D):

$$\underset{R}{\overset{O}{\|}}C\underset{OH}{} \xrightarrow[[2]\ \Delta]{[1]\ NH_3} \underset{R}{\overset{O}{\|}}C\underset{NH_2}{} + H_2O$$

$$\underset{R}{\overset{O}{\|}}C\underset{OH}{} + R'NH_2 \xrightarrow{DCC} \underset{R}{\overset{O}{\|}}C\underset{NHR'}{} + H_2O$$

d. From RCOOR'
(22.11):

$$\underset{R}{\overset{O}{\|}}C\underset{OR'}{} + NH_3 \longrightarrow \underset{R}{\overset{O}{\|}}C\underset{NH_2}{} + R'OH$$

Nitrile synthesis (22.18)

Nitriles are prepared by S$_N$2 substitution using unhindered alkyl halides as starting materials.

$$R-X + {}^-CN \xrightarrow{S_N2} R-C{\equiv}N + X^-$$
$$R = CH_3, 1°$$

Reactions of nitriles
[1] Hydrolysis (22.18A)

$$R-C{\equiv}N \xrightarrow[(H^+\ or\ ^-OH)]{H_2O} \underset{R}{\overset{O}{\|}}C\underset{OH}{} \ or \ \underset{R}{\overset{O}{\|}}C\underset{O^-}{}$$
$$\qquad\qquad\quad (with\ acid)\quad (with\ base)$$

[2] Reduction (22.18B)

R–C≡N

[1] LiAlH₄ → [2] H₂O → R–NH₂ 1° amine

[1] DIBAL-H → [2] H₂O → R–CHO (aldehyde)

---

[3] Reaction with organometallic reagents (22.18C)

R–C≡N $\xrightarrow[\text{[2] H}_2\text{O}]{\text{[1] R'MgX or R'Li}}$ ketone (R–CO–R')

Practice Test on Chapter Review

1. Give the IUPAC name for each of the following compounds.

a. b. c.

2. (a) Which compound absorbs at the *lowest* wavenumber in the IR? (b) Which compound absorbs at the *highest* wavenumber?

A   B   C   D

3.a. Which of the following reaction conditions can be used to synthesize an ester?

1. RCOCl + R'OH + pyridine
2. RCOOH + R'OH + H₂SO₄
3. RCOOH + R'OH + NaOH
4. Both methods (1) and (2) can be used to synthesize an ester.
5. Methods (1), (2), and (3) can all be used to synthesize an ester.

b. Which of the following compounds is most reactive in nucleophilic acyl substitution?

1. CH₃COCl
2. CH₃COOCH₃
3. CH₃CON(CH₃)₂
4. (CH₃CO)₂O
5. CH₃COOH

4. What reagent is needed to convert $(CH_3CH_2)_2CHCOOH$ into each compound?
   a. $(CH_3CH_2)_2CHCOO^-Na^+$
   b. $(CH_3CH_2)_2CHCOCl$
   c. $(CH_3CH_2)_2CHCON(CH_3)_2$
   d. $(CH_3CH_2)_2CHCO_2CH_2CH_3$
   e. $[(CH_3CH_2)_2CHCO]_2O$

5. What reagent is needed to convert $CH_3CH_2CH_2CN$ to each compound?
   a. $CH_3CH_2CH_2COOH$
   b. $CH_3CH_2CH_2CH_2NH_2$
   c. $CH_3CH_2CH_2COCH_2CH_3$
   d. $CH_3CH_2CH_2CHO$

6. Draw the organic products formed in the following reactions.

Answers to Practice Test

1.a. 5-ethyl-2-methyl-
    heptanenitrile
  b. cyclohexyl 2-
    methylbutanoate
  c. *N*-cyclohexyl-*N*-
    methylbenzamide
2.a. **C**
  b. **A**
3.a. 4
  b. 1
4.a. NaOH
  b. $SOCl_2$
  c. $HN(CH_3)_2$, DCC
  d. CH_3CH_2OH, H_2SO_4
  e. heat

5.a. H_3O^+
  b. [1] $LiAlH_4$; [2] H_2O
  c. [1] CH_3CH_2Li; [2] H_2O
  d. [1] DIBAL-H; [2] H_2O
6.

Answers to Problems

22.1 The number of C–N bonds determines the classification as a 1°, 2°, or 3° amide.

oxytocin
All seven others are 2° amides.

22.2 As the basicity of Z increases, the stability of RCOZ increases because of added resonance stabilization.

The **basicity of Z** determines how much
this structure contributes to the hybrid.
Br⁻ is less basic than ⁻OH, so RCOBr
is less stable than RCOOH.

22.3

CH_3—Cl

This resonance structure contributes little to the hybrid
because Cl⁻ is a weak base. Thus, the C–Cl
bond has little double bond character, making it similar in
length to the C–Cl bond in CH_3Cl.

CH_3—NH_2

This resonance structure contributes more to the hybrid
because ⁻NH_2 is more basic. Thus, the C–N bond
in $HCONH_2$ has more double bond character, making
it shorter than the C–N bond in CH_3NH_2.

22.4

a. **2-ethylbutanoyl chloride**
← 2-ethyl

b. **methyl benzoate**
alkyl group = methyl
acyl group = benzoate

c. *N*-ethyl-*N*-methyl

acyl group =
propanamide **N-ethyl-N-methylpropanamide**

d.   H   alkyl group = ethyl

acyl group =   **ethyl formate**
formate

e. **benzoic propanoic anhydride**

acyl group =   acyl group =
propanoic   benzoic

f. **3-ethylhexanenitrile**

6 carbon chain =
hexanenitrile

22.5

a. 5-methylheptanoyl chloride

b. isopropyl propanoate

c. acetic formic anhydride

d. *N*-isobutyl-*N*-methylbutanamide

e. 3-methylpentanenitrile

f. *o*-cyanobenzoic acid

g. *sec*-butyl 2-methylhexanoate

h. *N*-ethylhexanamide

22.6

CH$_3$CONH$_2$ has a higher boiling point than CH$_3$CO$_2$H because it has more opportunities for hydrogen bonding. Each CH$_3$CONH$_2$ can hydrogen bond to three other molecules (at its O atom and two N–H bonds), whereas CH$_3$CO$_2$H has only two intermolecular hydrogen bonds possible (at its carbonyl O atom and OH bond).

22.7

a.   and

amide: C=O at
lower wavenumber

b.   and

smaller ring:
C=O at a higher
wavenumber

c.   and

2° amide: 1 N–H
absorption at
3200–3400 cm^{-1}

1° amide: 2
N–H absorptions

d.   and

anhydride:
2 C=O peaks

22.8

H_b → H H ... O
H_c signal from the 5 H's on the aromatic ring

A

Molecular formula $C_9H_{10}O_2$
5 degrees of unsaturation
IR: 1743 cm⁻¹ from ester C=O
    3091–2895 cm⁻¹ from sp^2 and sp^3 C–H
¹H NMR:
    H_a = 2.06 ppm (singlet, 3 H) – CH₃
    H_b = 5.08 ppm (singlet, 2 H) – CH₂
    H_c = 7.33 ppm (broad singlet, 5 H)

H_a → H H ... O
H_b →

B

H_b signal from the 10 H's on the two aromatic rings

Molecular formula $C_{14}H_{12}O_2$
9 degrees of unsaturation
IR: 1718 cm⁻¹ from conjugated ester C=O
    3091–2953 cm⁻¹ from sp^2 and sp^3 C–H
¹H NMR:
    H_a = 5.35 ppm (singlet, 2 H)
    H_b = 7.26–8.15 ppm (multiplets, 10 H)

22.9   More reactive acyl compounds can be converted to less reactive acyl compounds.

a.   CH₃COCl   ——→   CH₃COOH
    more reactive   **YES**   less reactive

c.   CH₃COOCH₃   ——→   CH₃COCl
    less reactive   **NO**   more reactive

b.   CH₃CONHCH₃   ——→   CH₃COOCH₃
    less reactive   **NO**   more reactive

d.   (CH₃CO)₂O   ——→   CH₃CONH₂
    more reactive   **YES**   less reactive

22.10   The better the leaving group is, the more reactive the carboxylic acid derivative. The weakest base is the best leaving group.

a.

–NH₂ strongest base
least reactive

–OCH₃
intermediate

–Cl weakest base
most reactive

b.

–NHCH₃ strongest base
least reactive

–OH
intermediate

–O₂CCH₂CH₃ weakest base
most reactive

22.11

acetic anhydride

trichloroacetic anhydride

The Cl atoms are electron withdrawing, which makes the conjugate base (the leaving group, CCl₃COO⁻) weaker and more stable.

22.12

22.13 The mechanism has three steps: [1] nucleophilic attack by O; [2] proton transfer; and [3] elimination of the Cl⁻ leaving group to form the product.

22.14

22.15 Reaction of a carboxylic acid with thionyl chloride converts it to an acid chloride.

a.

b.

22.16

a.

b.

c.

d.

22.17

22.18

22.19

a.

b.

c.

22.20

fenofibrate → hydrolysis → fenofibric acid

22.21 The three bonds broken during hydrolysis are indicated.

ginkgolide B

22.22

sucrose → RCOOH, H$_2$SO$_4$ (a long-chain fatty acid) → olestra

22.23

hydrolysis → glycerol + soap

22.24

22.25 Aspirin has an ester, a more reactive acyl group, whereas acetaminophen has an amide, a less reactive acyl group. The ester makes aspirin more easily hydrolyzed with water from the air than acetaminophen. Therefore, Tylenol can be kept for many years, whereas aspirin decomposes.

22.26

22.27

22.28

1,4-dihydroxymethylcyclohexane

terephthalic acid

Kodel

In the polyester Kodel, most of the bonds in the polymer backbone are part of a ring, so there are fewer degrees of freedom. Fabrics made from Kodel are stiff and crease resistant, due to these less flexible polyester fibers.

22.29

Reaction occurs here (–H₂O).

PLA
polyl(lactic acid)

22.30 Acetyl CoA acetylates the NH_2 group of glucosamine, because the NH_2 group is the most nucleophilic site.

glucosamine

NAG

22.31

a.

c.

b.

22.32

a.

c.

b.

22.33

a. [structure] Br [1] NaCN [2] LiAlH₄ [3] H₂O → [structure] NH₂

b. [structure] C≡N [1] DIBAL-H [2] H₂O → [structure]

22.34

a. [structure] OCH₃, CN [1] ethyl MgCl [2] H₂O → [structure] OCH₃, ketone

b. [structure] CN [1] C₆H₅Li [2] H₂O → [structure] phenyl ketone

22.35

a. [structure] CN [1] CH₃MgBr [2] H₂O → [structure] ketone

b. [structure] CN [1] (CH₃)₃CMgBr [2] H₂O → [structure] ketone

c. [structure] CN [1] DIBAL-H [2] H₂O → [structure] aldehyde

d. [structure] CN H₃O⁺ → [structure] carboxylic acid OH

22.36

CH₃—C≡N [1] ethyl MgBr [2] H₂O → [structure ketone]

[structure] N [1] CH₃MgBr [2] H₂O → [structure ketone]

22.37 The better the leaving group, the more reactive the acyl compound.

[structure with OH]
worst leaving group
least reactive

[structure with SH]
**intermediate
reactivity**

[structure with Cl]
best leaving group
most reactive

22.38

a. 3 [structure] 2 1 O isobutyl

A
isobutyl 2,2-dimethyl-
propanoate

1 CN
6 5 4 3 2 [structure]

B
2-ethylhexanenitrile

b. A $\xrightarrow{\text{[1] } H_3O^+}$ (CH₃)₂C(CH₃)COOH + HO—CH(CH₃)₂

A $\xrightarrow[H_2O]{\text{[2] } ^-OH}$ (CH₃)₂C(CH₃)COO⁻ + HO—CH(CH₃)₂

A $\xrightarrow{\text{[3] } \text{CH}_2\text{CH}_2\text{CH}_2\text{MgBr}}$ $\xrightarrow{H_2O}$ (tertiary alcohol product) + HO—CH(CH₃)₂

A $\xrightarrow{\text{[4] } LiAlH_4}$ $\xrightarrow{H_2O}$ (CH₃)₃C—CH₂OH + HO—CH(CH₃)₂

B $\xrightarrow{\text{[1] } H_3O^+}$ (COOH substituted chain)

B $\xrightarrow[H_2O]{\text{[2] } ^-OH}$ (COO⁻ substituted chain)

B $\xrightarrow{\text{[3] } \text{CH}_2\text{CH}_2\text{CH}_2\text{MgBr}}$ $\xrightarrow{H_2O}$ (ketone product)

B $\xrightarrow{\text{[4] } LiAlH_4}$ $\xrightarrow{H_2O}$ (NH₂ substituted chain)

22.39 Better leaving groups make acyl compounds more reactive. **C** has an electron-withdrawing NO_2 group, which stabilizes the negative charge of the leaving group, whereas **D** has an electron-donating OCH_3 group, which destabilizes the leaving group.

C

an electron-withdrawing substituent

D

an electron-donating substituent

leaving group from **C**

one possible resonance structure

Delocalizing the negative charge on the NO_2 stabilizes the leaving group, making **C** more reactive than **D**.

leaving group from **D**

one possible resonance structure

Adjacent negative charges destabilize the leaving group.

22.40

a.

cyclohexanecarboxylic anhydride

b.

phenyl phenylacetate

c.

N-cyclohexylbenzamide

d.

m-chlorobenzonitrile

e.

cis-2-bromocyclohexane-carbonyl chloride

f.

N,N-diethylcyclohexanecarboxamide

22.41

a. cyclohexyl propanoate

d. vinyl acetate

g. octyl butanoate

b. cyclohexanecarboxamide

e. benzoic propanoic anhydride

h. N,N-dibenzylformamide

c. 4-methylheptanenitrile

f. 3-methylhexanoyl chloride

22.42

imidazolide

resonance structures for the leaving group

The leaving group is both resonance stabilized and aromatic (6 π electrons), making it a much better leaving group than exists in a regular amide.

22.43

Reaction as an acid:

These two resonance structures
make the conjugate base more stable,
and therefore CH_3CONH_2 a stronger acid.

no resonance stabilization
of the conjugate base

Reaction as a base:

This electron pair is delocalized by resonance,
making it less available for electron donation.
Thus, CH_3CONH_2 is a much weaker base.

This electron pair is
localized on N.

22.44 a. The electron pair on the O atom in phenyl acetate is delocalized on both the carbonyl group and the benzene ring. Thus, the C=O has less single bond character than the C=O of cyclohexyl acetate, so the absorption occurs at higher wavenumbers.

+ other forms with a C=O

This form contributes
less to the hybrid.

b. Because the lone pair on the O atom in cyclohexyl acetate can only be delocalized on the C=O, the C=O of cyclohexyl acetate is more effectively stabilized by resonance.

only two resonance forms

c. Phenyl acetate has a better leaving group and is more reactive.

weaker base
better leaving group

+ 3 more resonance
structures

stronger base

22.45

C₆H₅CH₂COOH → $C_6H_5CH_2COOH$

a. NaHCO₃ → [product] + H₂CO₃

b. NaOH → [product] + H₂O

c. SOCl₂ → [product]

d. NaCl → no reaction

e. NH₃ (1 equiv) → [product]

f. [1] NH₃ [2] Δ → [product]

g. CH₃OH / H₂SO₄ → [product]

h. CH₃OH / ⁻OH → [product]

i. [1] NaOH [2] CH₃COCl → [product]

j. CH₃NH₂ / DCC → [product]

k. [1] SOCl₂ [2] CH₃CH₂CH₂NH₂ → [product]

l. [1] SOCl₂ [2] (CH₃)₂CHOH → [product]

22.46

a. H₃O⁺ → [product]

b. H₂O, ⁻OH → [product]

c. [1] CH₃MgBr [2] H₂O → [product]

d. [1] CH₃CH₂Li [2] H₂O → [product]

e. [1] DIBAL-H [2] H₂O → [product]

f. [1] LiAlH₄ [2] H₂O → [product]

22.47

a.

b.

c.

d.

e.

f.

g.

h.

22.48

cinnamoylcocaine
Hydrolyze both esters.

22.49

a.

b.

c.

d.

22.50

22.51

a.

b.

c.

d. (2 equiv)

22.52 Hydrolyze the amide and ester bonds in both starting materials to draw the products.

a.

oseltamivir

b.

aspartame → phenylalanine + CH₃OH

22.53

CH_3SO_2Cl / $(CH_3CH_2)_3N$

(–H⁺)

intramolecular amide formation

F
$C_{18}H_{18}FNO$

22.54

22.55

Two possibilities for **A**:

22.56

γ-butyrolactone

4-hydroxybutanoic acid
GHB

22.57

aspirin

salicylic acid

inactive enzyme

22.58

22.59

22.60

22.61

22.62 The mechanism is composed of two parts: hydrolysis of the acetal and intramolecular Fischer esterification of the hydroxy carboxylic acid.

22.63

A

$C_6H_{10}O_2$: 2 degrees of unsaturation

IR: 1770 cm^{-1} from ester C=O in a five-membered ring

^{1}H NMR:

H_a = 1.27 ppm (singlet, 6 H) – 2 CH$_3$ groups
H_b = 2.12 ppm (triplet, 2 H) – CH$_2$ bonded to CH$_2$
H_c = 4.26 ppm (triplet, 2 H) – CH$_2$ bonded to CH$_2$

22.64 Fischer esterification is the treatment of a carboxylic acid with an alcohol in the presence of an acid catalyst to form an ester.

22.65

22.66

a. CH_3Cl + NaCN $\longrightarrow$ CH_3-CN $\xrightarrow{H_3O^+}$ CH_3-COOH

CH_3-Cl + Mg $\longrightarrow$ CH_3-MgCl $\xrightarrow[\text{[2] } H_3O^+]{\text{[1] } CO_2}$ CH_3-COOH

b.

+ NaCN $\longrightarrow$ This method can't be used because an S_N2 reaction can't be done on an sp^2 hybridized C.

+ Mg $\longrightarrow$ $\xrightarrow[\text{[2] } H_3O^+]{\text{[1] } CO_2}$

c. + NaCN $\longrightarrow$ $\xrightarrow{H_3O^+}$

+ Mg $\longrightarrow$ This method can't be used because you can't make a Grignard reagent with an acidic OH group.

22.67

(+ ortho isomer)

22.68

a.

(+ para isomer)

salicylamide

b.

(+ ortho isomer)   (More nucleophilic NH_2 reacts first.)   acetaminophen

c.

acetaminophen
(from b.)

p-acetophenetidin

22.69

a.

b.

22.70

a.

b.

22.71

a.

b.

22.72

a. Docetaxel has fewer C's and one more OH group than paclitaxel. This makes docetaxel more water soluble than paclitaxel.

b.

carbamate

docetaxel

carbamate

1
most stable

4
least stable

3

2
More basic N atom allows N to donate electron density more than O, so this structure contributes more than **3** to the hybrid.

Increasing stability: **4 < 3 < 2 < 1**

c.

$+ H_2O$

$+ H_2\ddot{O}:$

$+ H_3\overset{+}{O}$

$H_2\ddot{O}:$

$H - \overset{+}{OH_2}$

$H - \overset{+}{OH_2}$

$:O:$
$\|$
C $+ \overset{\cdot\cdot}{N}H_2R$
$\|$
$:O:$

$H_3\overset{+}{N}R$ $+$ H_2O

d.

docetaxel

H_3O^+ → + CO_2 + ... + CH_3CO_2H

22.73

a. and

Acid chloride CO absorbs at much higher wavenumber.

ketone

b. and

$C=O$ at <1700 cm^{-1} due to the stabilized amide

2 NH absorptions at 3200–3400 cm^{-1} $C=O$ absorption at higher wavenumber

22.74

Increasing wavenumber

22.75

a. $C_6H_{12}O_2 \rightarrow$ one degree of unsaturation
IR: 1738 cm^{-1} → $C=O$
NMR: 1.12 (triplet, 3 H), 1.23 (doublet, 6 H), 2.28 (quartet, 2 H), 5.00 (septet, 1 H) ppm

b. C_4H_7N
IR: 2250 cm^{-1} → triple bond
NMR: 1.08 (triplet, 3 H), 1.70 (multiplet, 2 H), 2.34 (triplet, 2 H) ppm

c. C_8H_9NO
IR: 3328 (NH), 1639 (conjugated amide $C=O$) cm^{-1}
NMR: 2.95 (singlet, 3 H), 6.95 (singlet, 1 H), 7.3–7.7 (multiplet, 5 H) ppm

d. $C_4H_7ClO \rightarrow$ one degree of unsaturation
IR: 1802 cm^{-1} → $C=O$ (high wavenumber, RCOCl)
NMR: 0.95 (triplet, 3 H), 1.07 (multiplet, 2 H), 2.90 (triplet, 2 H) ppm

e. $C_{10}H_{12}O_2 \rightarrow$ five degrees of unsaturation
   IR: 1740 cm$^{-1} \rightarrow$ C=O
   NMR: 1.2 (triplet, 3 H), 2.4 (quartet, 2 H),
   5.1 (singlet, 2 H), 7.1–7.5 (multiplet, 5 H) ppm

22.76

A. Molecular formula $C_{10}H_{12}O_2 \rightarrow$ five degrees of unsaturation
   IR absorption at 1718 cm$^{-1} \rightarrow$ C=O
   NMR data (ppm):

          triplet at 1.4 (CH$_3$ adjacent to 2 H's)
          singlet at 2.4 (CH$_3$)
          quartet at 4.4 (CH$_2$ adjacent to CH$_3$)
          doublet at 7.2 (2 H's on benzene ring)
          doublet at 7.9 (2 H's on benzene ring)

B. IR absorption at 1740 cm$^{-1} \rightarrow$ C=O
   NMR data (ppm):

          singlet at 2.1 (CH$_3$)
          triplet at 2.9 (CH$_2$ adjacent to CH$_2$)
          triplet at 4.3 (CH$_2$ adjacent to CH$_2$)
          multiplet at 7.3 (5 H's, monosubstituted benzene)

22.77

Molecular formula $C_{10}H_{13}NO_2 \rightarrow$ five degrees of unsaturation
IR absorptions at 3300 (NH) and 1680 (C=O, amide or conjugated) cm^{-1}
NMR data (ppm):

          triplet at 1.4 (CH$_3$ adjacent to CH$_2$)
          singlet at 2.2 (CH$_3$C=O)
          quartet at 3.9 (CH$_2$ adjacent to CH$_3$)
          doublet at 6.8 (2 H's on benzene ring)
          singlet at 7.1 (NH)
          doublet at 7.3 (2 H's on benzene ring)

phenacetin

22.78

Molecular formula $C_{11}H_{15}NO_2 \rightarrow$ five degrees of unsaturation
IR absorption 1699 (C=O, amide or conjugated) cm^{-1}
NMR data (ppm):

          triplet at 1.3 (3 H) (CH$_3$ adjacent to CH$_2$)
          singlet at 3.0 (6 H) (2 CH$_3$ groups on N)
          quartet at 4.3 (2 H) (CH$_2$ adjacent to CH$_3$)
          doublet at 6.6 (2 H) (2 H's on benzene ring)
          doublet at 7.9 (2 H) (2 H's on benzene ring)

C

22.79

a. Molecular formula $C_6H_{12}O_2 \rightarrow$ one degree of unsaturation
   IR absorption at 1743 cm$^{-1} \rightarrow$ C=O
   ^{1}H NMR data (ppm):

          triplet at 0.9 (3 H) – CH$_3$ adjacent to CH$_2$
          multiplet at 1.35 (2 H) – CH$_2$
          multiplet at 1.60 (2 H) – CH$_2$
          singlet at 2.1 (3 H – from CH$_3$ bonded to C=O)
          triplet at 4.1 (2 H) – CH$_2$ adjacent to the electronegative O atom and another CH$_2$

D

b. Molecular formula $C_6H_{12}O_2 \rightarrow$ one degree of unsaturation
IR absorption at 1746 cm–1 $\rightarrow$ C=O
^{1}H NMR data (ppm):

> doublet at 0.9 (6 H) – 2 CH$_3$'s adjacent to CH
> multiplet at 1.9 (1 H)
> singlet at 2.1 (3 H) – CH$_3$ bonded to C=O
> doublet at 3.85 (2 H) – CH$_2$ bonded to electronegative O and CH

E

22.80 The extent of resonance stabilization affects the position of the C=O absorption in the IR of an amide.

A

This resonance structure does not contribute significantly to the hybrid because it places a double bond at the bridgehead N, an impossible geometry in small rings. The C=O has more double bond character, so the absorption is at a higher wavenumber.

B

This resonance structure contributes significantly to the hybrid, so the C=O has more single bond character. The absorption shifts to a lower wavenumber.

22.81

4.02 ppm

7.35 and 7.60 ppm

different environments

This resonance structure makes a significant contribution to the resonance hybrid.

There is restricted rotation around the amide C–N bond. The 2 H's are in different environments (one is cis to an O atom, and one is cis to CH$_2$Cl), so they give different NMR signals.

22.82

ethyl benzoate

Two OH groups are now equivalent and either can lose H$_2$O to form labeled or unlabeled ethyl benzoate.

Unlabeled starting material was recovered.

22.83

22.84 Both acetals are hydrolyzed by the usual mechanism for acetal hydrolysis (Steps [1 –[6]), forming four new OH groups. Intramolecular esterification forms a lactone (Steps [10]–[15]), followed by conversion of a carbonyl tautomer to an enol (Steps [16]–[17]). Both acetals are hydrolyzed at once in the given mechanism. For ease in drawing this long mechanism, the stereochemistry of intermediates is omitted.

+ 2 Cl⁻

H₂O:

H₂O:

[1]   [2]   [3]

[4]

H—Cl   :OH   OH

+ Cl⁻   + 2 Cl⁻

[7]   [6]   [5]

[8]

+   + 2 HCl
(2 equiv)   :O:
(2 equiv)

[9]

H—Cl   :OH   :Cl:

[10]   [11]

+ Cl⁻   [12]

HO   HO   HO

+ H₂O   + Cl⁻

[14]   [13]

[15]

HO   HO   HO

[16]   [17]   + HCl

H—Cl   vitamin C

:Cl: + Cl⁻

22.85

Chapter 23 Substitution Reactions of Carbonyl Compounds at the α Carbon

Chapter Review

Kinetic versus thermodynamic enolates (23.4)

kinetic enolate

thermodynamic enolate

Kinetic enolate
- The less substituted enolate
- Favored by strong base, polar aprotic solvent, low temperature: LDA, THF, –78 °C

Thermodynamic enolate
- The more substituted enolate
- Favored by strong base, protic solvent, higher temperature: $NaOCH_2CH_3$, CH_3CH_2OH, room temperature

Halogenation at the α carbon
[1] Halogenation in acid (23.7A)

$$R\text{-}CO\text{-}CH(CH_3)H \xrightarrow[CH_3COOH]{X_2} R\text{-}CO\text{-}CH(CH_3)X$$

$X_2 = Cl_2, Br_2, or\ I_2$

α-halo aldehyde or ketone

- The reaction occurs via enol intermediates.
- Monosubstitution of X for H occurs on the α carbon.

[2] Halogenation in base (23.7B)

$$R\text{-}CO\text{-}CRH_2 \xrightarrow[^-OH]{X_2\ (excess)} R\text{-}CO\text{-}CRX_2$$

$X_2 = Cl_2, Br_2, or\ I_2$

- The reaction occurs via enolate intermediates.
- Polysubstitution of X for H occurs on the α carbon.

[3] Halogenation of *methyl* ketones in base—The haloform reaction (23.7B)

$$R\text{-}CO\text{-}CH_3 \xrightarrow[^-OH]{X_2\ (excess)} R\text{-}CO\text{-}O^- + HCX_3$$

$X_2 = Cl_2, Br_2, or\ I_2$

haloform

- The reaction occurs with methyl ketones and results in cleavage of a carbon–carbon σ bond.

Reactions of α-halo carbonyl compounds (23.7C)
[1] Elimination to form α,β-unsaturated carbonyl compounds

- Elimination of the elements of Br and H forms a new π bond, giving an α,β-unsaturated carbonyl compound.

[2] Nucleophilic substitution

- The reaction follows an S$_N$2 mechanism, generating an α-substituted carbonyl compound.

Alkylation reactions at the α carbon
[1] Direct alkylation at the α carbon (23.8)

- The reaction forms a new C–C bond to the α carbon.
- LDA is a common base used to form an intermediate enolate.
- The alkylation in Step [2] follows an S$_N$2 mechanism.

[2] Malonic ester synthesis (23.9)

diethyl malonate

- The reaction is used to prepare carboxylic acids with one or two alkyl groups on the α carbon.
- The alkylation in Step [2] follows an S$_N$2 mechanism.

[3] Acetoacetic ester synthesis (23.10)

ethyl acetoacetate

- The reaction is used to prepare ketones with one or two alkyl groups on the α carbon.
- The alkylation in Step [2] follows an S$_N$2 mechanism.

Practice Test on Chapter Review

1.a. Which of the following compounds can be prepared using either the acetoacetic ester synthesis or the malonic ester synthesis?

4. Both (1) and (2) can be prepared by one of these routes.
5. Compounds (1), (2), and (3) can all be prepared by these routes.

b. Which of the following compounds is **not** an enol form of dicarbonyl compound **A?**

5. Compounds (1)–(4) are all enols of **A.**

2.a. Which proton in compound **A** has the *lowest* pKa?
b. Which proton in compound **A** has the *highest* pKa?

c. Which proton in compound **B** is the least acidic?
d. Which proton in compound **B** is the most acidic?

3. What reagents are needed to convert heptan-4-one to each compound?
   a. 3-bromoheptan-4-one
   b. 3-methylheptan-4-one
   c. 3,5-dimethylheptan-4-one
   d. hept-2-en-4-one
   e. 2-methylheptan-4-one

4. Draw the organic products formed in each of the following reactions.

a. I₂ (excess) / ⁻OH

d. [1] NaOEt / [2] CH₃CH₂CH₂CH₂Br / [3] H₃O⁺, Δ

b. [1] LDA / [2] CH₃CH₂Br / [3] H₃O⁺, Δ

e. [1] LDA / [2] CH₃CH₂Br

c. $CH_2(CO_2Et)_2$ — [1] NaOEt / [2] CH₃CH₂Br → [1] NaOEt / [2] C₆H₅CH₂Cl / [3] H₃O⁺, Δ

5.a. What starting materials are needed to synthesize carboxylic acid **A** by a malonic ester synthesis?

A

b. What starting materials are needed to synthesize ketone **B** by the acetoacetic ester synthesis?

B

Answers to Practice Test

1.a. 1
  b. 4
2.a. H_b
  b. H_d
  c. H_b
  d. H_c
3.a. Br₂, CH₃CO₂H
  b. [1] LDA; [2] CH₃I
  c. []1 LDA; [2] CH₃I; [3] LDA; [4] CH₃I
  d. [1] Br₂, CH₃CO₂H; [2] Li₂CO₃, LiBr, DMF
  e. [1] Br₂, CH₃CO₂H; [2] Li₂CO₃, LiBr, DMF; [3] (CH₃)₂CuLi; [4] H₂O

4.
a. + HCI₃
b.
c.
d.
e.

5.
a.
$CH_2(CO_2Et)_2$
b.

Answers to Problems

23.1 • To convert a ketone to its enol tautomer, change the C=O to C–OH, make a new double bond to an α carbon, and remove a proton at the other end of the C=C.

• To convert an enol to its keto form, find the C=C bonded to the OH. Change the C–OH to a C=O, add a proton to the other end of the C=C, and delete the double bond.

[In cases where *E* and *Z* isomers are possible, only one stereoisomer is drawn.]

a.

d.

b.

e.

c.

f.

[Draw mono enol tautomers only.]

(Conjugated enols are preferred.)

23.2   The mechanism has two steps for each part: protonation followed by deprotonation.

23.3

23.4

a.

b.

c.

23.5 The indicated H's are α to a C=O or C≡N group, making them more acidic because their removal forms conjugate bases that are resonance stabilized.

a. b. c. d.

23.6

no resonance stabilization
least acidic

Two resonance structures stabilize
the conjugate base.
intermediate acidity

Three resonance structures stabilize
the conjugate base.
most acidic

23.7 In each of the reactions, the LDA pulls off the most acidic proton.

a.

b.

c.

d.

23.8 In addition to being strong bases, organolithiums are good nucleophiles that can add to a carbonyl group instead of pulling off a proton to generate an enolate.

23.9 • LDA, THF forms the kinetic enolate by removing a proton from the less substituted C.
   • Treatment with NaOCH₃, CH₃OH forms the thermodynamic enolate by removing a proton from the more substituted C.

a.

b.

c.

23.10

a. This acidic H is removed with base to form an achiral enolate.

(R)-2-methylcyclohexanone   achiral

Protonation of the planar achiral enolate occurs with equal probability from two sides, so a racemic mixture is formed. The racemic mixture is optically inactive.

b.

(R)-3-methylcyclohexanone

This stereogenic center is not located at the α carbon, so it is not deprotonated with base. Its configuration is retained in the product, and the product remains optically active.

23.11

a.

b.

c.

23.12

a.

b.

c.

23.13

a.

b.

c.

23.14 Bromination takes place on the α carbon to the carbonyl, followed by an S$_N$2 reaction with the nitrogen nucleophile.

LSD

(Section 18.5)

M

23.15

a.

b.

c.

d.

23.16

a.
[1] LDA, THF
[2] CH₃I

b.
[1] LDA, THF
[2] CH₃I

c.
[1] LDA, THF
[2] CH₃I

23.17 Three steps are needed: [1] formation of an enolate; [2] alkylation; [3] hydrolysis of the ester.

A

[1] LDA

[2] CH₃I

[3] H₃O⁺

The product is racemic because the new stereogenic center is formed by alkylation of a planar enolate with equal probability from above and below.

naproxen

23.18

a.
LDA

THF

CH₃CH₂Br

b.
KOC(CH₃)₃

(CH₃)₃COH

CH₃Br

KOC(CH₃)₃

(CH₃)₃COH

CH₃Br

c.
(from a.)
LDA

THF

CH₃I

d.
LDA

THF

CH₃Br

(from b.)

23.19

α-methylene-γ-butyrolactone

23.20 Decarboxylation occurs only when a carboxy group is bonded to the α C of another carbonyl group.

a. YES   b. NO   c. YES   d. NO

23.21

a. $CH_2(CO_2Et)_2$ $\xrightarrow{\text{[1] NaOEt}}$ $\xrightarrow[\Delta]{H_3O^+}$

b. $CH_2(CO_2Et)_2$ $\xrightarrow{\text{[1] NaOEt}}$ $\xrightarrow{\text{[1] NaOEt}}$ $\xrightarrow[\Delta]{H_3O^+}$

23.22

23.23 Locate the α C to the COOH group, and identify all of the alkyl groups bonded to it. These groups are from alkyl halides, and the remainder of the molecule is from diethyl malonate.

c.

$CH_2(CO_2Et)_2$ → [1] NaOEt / [2] (alkyl bromide) → [1] NaOEt / [2] (alkyl bromide) → H_3O^+, Δ → (product carboxylic acid)

23.24 The reaction works best when the alkyl halide is 1° or CH_3X, because this is an S_N2 reaction.

a.

(CH₃)₃CX
3° alkyl halide
(too crowded)

b.

X
aryl halide
(leaving group on an
sp^2 hybridized C)

Aryl halides are unreactive
in S_N2 reactions.

c.

α

This compound has 3 CH₃ groups on the α
carbon to the COOH. The malonic ester
synthesis can be used to prepare mono- and
disubstituted carboxylic acids only: RCH_2COOH
and $R_2CHCOOH$, but not R_3CCOOH.

23.25

a. (ethyl acetoacetate) → [1] NaOEt / [2] CH₃I / [3] H₃O⁺, Δ → (ketone product)

b. (ethyl acetoacetate) → [1] NaOEt / [2] CH₃CH₂CH₂Br / [3] NaOEt / [4] C₆H₅CH₂I / [5] H₃O⁺, Δ → (ketone product)

23.26 Locate the α C. All alkyl groups on the α C come from alkyl halides, and the remainder of the molecule comes from ethyl acetoacetate.

23.27

(ethyl acetoacetate) + Br—CH₂CH₂—Br → NaOEt (2 equiv) → (product)

X

23.28

a.

b.

23.29 Use the directions from Answer 23.1 to draw the enol tautomer(s). In cases where E and Z isomers can form, only one isomer is drawn.

a.

b.

unconjugated enol
(less stable)

+

23.30

a.

b.

one axial and one equatorial
group, **less stable**

Both groups are equatorial.
more stable

Both groups are equatorial = cis.

This isomerization will occur because it makes
a more stable compound.

Compound **C** will not isomerize because it
already has the more stable arrangement
of substituents.

23.31 Use the directions from Answer 23.1 to draw the enol tautomer(s). In cases where E and Z isomers can form, only one isomer is drawn.

a.

conjugated enol
(more stable)

c.

(mono enol form)

conjugated enol
(more stable)

b.

23.32

ethyl acetoacetate

The ester C=O is resonance stabilized, and is therefore less available for tautomerization. Since the carbonyl form of the ester group is stabilized by electron delocalization, less enol is present at equilibrium.

23.33

a.

Conjugation stabilizes this enol.
higher percentage of enol

not conjugated

b.

not conjugated

conjugated C=C
higher percentage of enol

23.34

a.

b.

c.

23.35

a.

H_c is bonded to an sp^2 hybridized C = **least acidic.**
H_a is bonded to an α C = **intermediate acidity.**
H_b is bonded to an α C, and is adjacent
  to a benzene ring = **most acidic.**

b.

H_b is bonded to an sp^3 hybridized
  C = **least acidic.**
H_c is bonded to an α C =
  intermediate acidity.
H_a is bonded to O = **most acidic.**

23.36

a.

LDA
THF

b.

CN   LDA
THF

c.

LDA
THF

d.

LDA
THF

23.37

Removal of H_a gives two resonance structures. The negative charge is never on O.

Removal of H_b gives three resonance structures. The negative charge is on O in one resonance structure, making the conjugate base more stable and H_b more acidic (lower pK_a).

23.38 a.

keto form
acyclovir

enol form

In the enol form, the bicyclic ring system has four π bonds (eight π electrons) and a lone pair on N, for a total of 10 π electrons. This makes it aromatic by Hückel's rule (Section 17.7).

b. The keto form of acyclovir can also be drawn in a resonance form that gives it 10 π electrons, making it aromatic as well.

aromatic
completely conjugated
10 π electrons

23.39

pentane-2,4-dione

base (1 equiv)

[1] CH$_3$I
[2] H$_2$O

A

base
(2nd equiv)

more nucleophilic site

[1] CH$_3$I
[2] H$_2$O

B

One equivalent of base removes the most acidic proton between the two C=O's, to form **A** on alkylation with CH$_3$I.

With a second equivalent of base a dianion is formed. Since the second enolate is less resonance stabilized, it is more nucleophilic and reacts first in an alkylation with CH$_3$I, forming **B** after protonation with H$_2$O.

23.40

enediol   keto tautomer **A**   keto tautomer **B**

The enediol is more stable than either of the two keto tautomers because it is more highly conjugated. More resonance structures can be drawn, which delocalize the lone pairs of the two OH groups bonded to the C=C. Such delocalization is not possible with either keto tautomer.

23.41 The mechanism of acid-catalyzed halogenation consists of two parts: **tautomerization** of the carbonyl compound to the enol form and **reaction of the enol with halogen.**

A higher percentage of the more stable enol is present.

CH$_3$COOH

Br$_2$

pentan-2-one

C=C has 1 bond to C

C=C has 2 bonds to C
more stable
(E and Z isomers)

A

B
major product formed from the more stable enol

23.42 In the haloform reaction, the three H's of the CH$_3$ group are successively replaced by X, to form an intermediate that is oxidatively cleaved with base.

23.43 Use the directions from Answer 23.23.

a. $\Longrightarrow$ CH$_3$OCH$_2$Br

c. $\Longrightarrow$

b. $\Longrightarrow$

23.44

a.

b.

23.45

valproic acid

23.46

a. $CH_2(CO_2Et)_2$ $\xrightarrow{\text{[1] NaOEt} \\ \text{[2] BrCH}_2\text{CH}_2\text{CH}_2\text{CH}_2\text{CH}_2\text{Br} \\ \text{[3] NaOEt}}$ $\xrightarrow[\Delta]{H_3O^+}$ [cyclohexane]—COOH $\xrightarrow{\text{[1] LiAlH}_4 \\ \text{[2] H}_2\text{O}}$ [cyclohexane]—OH

b. [cyclohexane]—COOH $\xrightarrow[H_2SO_4]{CH_3OH}$ [cyclohexane]—CO_2CH_3 $\xrightarrow{\text{[1] CH}_3\text{MgBr (2 equiv)} \\ \text{[2] H}_2\text{O}}$ [cyclohexane]—C(OH)(CH₃)₂

(from a.)

23.47

a. [epoxide] $\xrightarrow{\text{[1] Na}^+ \text{ }^-\text{CH(COOEt)}_2 \\ \text{[2] H}_2\text{O}}$ EtO–C(=O)–CH(–C(=O)OEt)–CH₂CH(OH)CH₃

nucleophilic attack here

c. CH_3COCl $\xrightarrow{\text{[1] Na}^+ \text{ }^-\text{CH(COOEt)}_2 \\ \text{[2] H}_2\text{O}}$ EtO–C(=O)–CH(–C(=O)OEt)–C(=O)CH₃

b. HCHO $\xrightarrow{\text{[1] Na}^+ \text{ }^-\text{CH(COOEt)}_2 \\ \text{[2] H}_2\text{O}}$ EtO–C(=O)–CH(–C(=O)OEt)–CH₂OH

d. (CH₃CO)₂O $\xrightarrow{\text{[1] Na}^+ \text{ }^-\text{CH(COOEt)}_2 \\ \text{[2] H}_2\text{O}}$ EtO–C(=O)–CH(–C(=O)OEt)–C(=O)CH₃

+ CH_3COOH

23.48 Use the directions from Answer 23.26.

a. [ketone with α marked] | [CH₃COCH₂CO₂Et] $\xrightarrow{\text{[1] NaOEt} \\ \text{[2] Br-CH}_2\text{CH}_2\text{CH(CH}_3)\text{...}}$ $\xrightarrow[\Delta]{H_3O^+}$ [ketone product]

b. [ketone with α marked] | [CH₃COCH₂CO₂Et] $\xrightarrow{\text{[1] NaOEt} \\ \text{[2] CH}_3\text{CH}_2\text{Br}}$ $\xrightarrow{\text{[1] NaOEt} \\ \text{[2] Br-CH}_2\text{-cyclohexane}}$ $\xrightarrow[\Delta]{H_3O^+}$ [ketone product]

c. [ketone with α marked] | [CH₃COCH₂CO₂Et] $\xrightarrow{\text{[1] NaOEt} \\ \text{[2] Br-CH}_2\text{CH}_2\text{CH(CH}_3)\text{CH}_2\text{-Br} \\ \text{[3] NaOEt}}$ $\xrightarrow[\Delta]{H_3O^+}$ [ketone product]

23.49

a. [CH₃COCH₂CO₂Et] $\xrightarrow{\text{[1] NaOEt} \\ \text{[2] CH}_3\text{Br}}$ $\xrightarrow{\text{[1] NaOEt} \\ \text{[2] CH}_3\text{Br}}$ $\xrightarrow[\Delta]{H_3O^+}$ [ketone product]

b. [methyl isopropyl ketone] $\xrightarrow[\text{THF}]{\text{LDA}}$ [enolate] $\xrightarrow{CH_3I}$ [ketone product]

(from a.)

23.50

a.

b.

c.

d.

e.

f.

g.

h.

23.51

a.

b.

c.

23.52

a.

p-isobutylbenzaldehyde

A

B

C

ibuprofen

b.

Removal of the most acidic proton with LDA forms a carboxylate anion that reacts as a nucleophile with CH₃I to form an ester as substitution product.

23.53

a.

b. **X** can be converted to meperidine by hydrolysis of the nitrile and esterification.

23.54

23.55

23.56 Protonation in Step [3] can occur from below (to re-form the *R* isomer) or from above to form the *S* isomer as shown.

R isomer
inactive enantiomer

achiral enol

S isomer
active enantiomer

(+ one more resonance structure)

23.57

23.58

23.59 Protons on the γ carbon of an α,β-unsaturated carbonyl compound are acidic because of resonance.

There is no H on this C, so a planar enolate cannot form and this stereogenic center cannot change.

Remove the H on this γ C.

Removal of this proton forms a resonance-stabilized anion. One resonance structure places a negative charge on O.

Protonation of the planar enolate can occur from below (to re-form starting material **X**), or from above to form **Y**.

23.60

LDA = B:

This reaction occurs with both bases [LDA and KOC(CH$_3$)$_3$].

+ Br$^-$

23.61

1st new C–C bond

23.62

a. Because there are two C's bonded to the α carbon, there are two possible intramolecular alkylation reactions.

Path [1]

two different halo
ketones possible

Path [2]

b.

Form both bonds to the α carbon during
acetoacetic ester synthesis.

23.63

a.

[1] LDA, THF
[2] CH₃Br

[1] LDA, THF
[2] CH₃CH₂Br

b.

c.

23.64

bupropion

23.65

anastrozole

23.66

23.67

most acidic H

To synthesize the desired product, a protecting group is needed:

23.68

23.69

a.

$Y =$ (CH$_3$)$_2$CH—C (structure)

$C_7H_{14}O \rightarrow$ one degree of unsaturation

IR peak at 1713 cm^{-1} $\rightarrow$ C=O

^{1}H NMR signals at (ppm)

H_e: triplet at 0.7 (3 H)

H_a: doublet at 0.9 (6 H)

H_d: sextet at 1.3 (2 H)

H_c: triplet at 1.9 (2 H)

H_b: septet at 2.1 (1 H)

b.

23.70 Removal of H_a with base does not generate an anion that can delocalize onto the carbonyl O atom, whereas removal of H_b generates an enolate that is delocalized on O.

Delocalization of this sort can't occur by removal of H_a, making H_a less acidic.

Removal of H_b gives an anion that is resonance stabilized, so H_b is more acidic.

Mechanism:

+ CH₃OH

+ Br⁻

+ HB⁺

23.71

23.72

23.73

23.74 a. In the presence of base an achiral enolate that can be protonated from both sides is formed.

(−)-hyoscyamine

conjugated

achiral
A

racemic mixture
optically inactive

b. The enolate **A** formed from (−)-hyoscyamine is conjugated with the benzene ring, making it easier to form. The enolate **B** formed from (−)-littorine is not conjugated, so it is less readily formed.

(−)-littorine

NOT conjugated

B

Chapter 24 Carbonyl Condensation Reactions

Chapter Review

The four major carbonyl condensation reactions

Reaction type	Reaction
[1] Aldol reaction (24.1)	
[2] Claisen reaction (24.5)	
[3] Michael reaction (24.8)	
[4] Robinson annulation (24.9)	

Useful variations

[1] Directed aldol reaction (24.3)

[2] Intramolecular aldol reaction (24.4)

a. With 1,4-dicarbonyl compounds:

b. With 1,5-dicarbonyl compounds:

[3] Dieckmann reaction (24.7)

a. With 1,6-diesters:

b. With 1,7-diesters:

Practice Test on Chapter Review

1.a. Which compounds are possible Michael acceptors?

1.

2.

3.

4. Both compounds (1) and (2) are Michael acceptors.

5. Compounds (1), (2), and (3) are all Michael acceptors.

b. Which of the following compounds can be formed by an aldol reaction?

1.

2.

3.

4. Both (1) and (2) can be formed by aldol reactions.

5. Compounds (1), (2), and (3) can all be formed by aldol reactions.

c. Which compounds can be formed in a Robinson annulation?

1.

3.

2.

4. Compounds (1) and (2) can be formed by Robinson annulations.

5. Compounds (1), (2), and (3) can all be formed by Robinson annulations.

d. What compounds can be used to form **A** by a condensation reaction?

A

1. and

2.

3. and

4. Compounds (1) and (2) can be used to form **A**.

5. Compounds (1), (2), and (3) can all be used to form **A**.

2. Give the reagents required for each step.

3. Draw the organic products formed in the following reactions.

a.
[1] LDA
[2] CH$_3$CH$_2$CHO
[3] H$_2$O

b. + HCO$_2$Et →
[1] NaOEt, EtOH
[2] H$_3$O$^+$

c. + →
NaOEt
EtOH

d. + →
$^-$OH, H$_2$O

4.a. What organic starting materials are needed to synthesize **D** by a Robinson annulation reaction?

D

b. What organic starting materials are needed to synthesize β-keto ester **B** by a Dieckmann reaction?

B

c. What starting materials are needed to synthesize **A** by an aldol reaction?

A

Answers to Practice Test

1.a. 4
 b. 2
 c. 1
 d. 4
2.a. [1] O_3; [2] $(CH_3)_2S$
 b. CrO_3, H_2SO_4, H_2O
 c. CH_3CH_2OH, H_2SO_4
 d. [1] NaOEt, EtOH; [2] H_3O^+
 e. [1] LDA; [2] CH_3I

3.

a.

b.

c.

d.

(*E* and *Z* isomers)

4.

a. +

b.

c.

Chapter 24: Answers to Problems

24.1

a.

b.

c.

d.

24.2

a.

no α H
no aldol reaction

b.

α H

yes

c.

no α H
no aldol reaction

d.

α H

yes

24.3

a. base

c. base

b. base

(*E* and *Z* isomers)

24.4

+ $H_2\ddot{O}$: HSO_4^- + H_2SO_4

24.5

Locate the α and β C's to the carbonyl group, and break the molecule into two halves at this bond. The α C and all of the atoms bonded to it belong to one carbonyl component. The β C and all of the atoms bonded to it belong to the other carbonyl component.

a.

b.

c.

24.6

+ $H_2\ddot{O}$:

24.7

a. and → or   b. and → (*E* and *Z* isomers)

24.8

a. $CH_2(CO_2Et)_2$ c. CH_3COCH_2CN (*E* and *Z* isomers)

b. $CH_2(COCH_3)_2$

24.9

24.10 Find the α and β C's to the carbonyl group and break the bond between them.

24.11

24.12

24.13 All enolates have a second resonance structure with a negative charge on O.

24.14

a.

b.

24.15

24.16 Join the α C of one ester to the carbonyl C of the other ester to form the β-keto ester.

a.

b.

24.17 In a crossed Claisen reaction between an ester and a ketone, the enolate is formed from the ketone, and the product is a β-dicarbonyl compound.

a.

Only this compound
can form an enolate.

b.

The ketone
forms the enolate.

24.18 A β-dicarbonyl compound like avobenzone is prepared by a crossed Claisen reaction between a ketone and an ester.

Break the molecule into
two components at either
dashed line.

avobenzone

24.19

a.

b.

24.20

ibuprofen

24.21

24.22 A Michael acceptor is an α,β-unsaturated carbonyl compound.

a. α,β-unsaturated
yes — Michael acceptor

b. not α,β-unsaturated

c. not α,β-unsaturated

d. α,β-unsaturated
yes — Michael acceptor

24.23

a.

b.

24.24

a.

b.

24.25 The Robinson annulation forms a six-membered ring and three new carbon–carbon bonds: two σ bonds and one π bond.

a.

new C–C bond

new σ and π bonds

b.

new C–C bond

c.

new C–C bond

new σ and π bonds

d.

new C–C bond

new σ and π bonds

24.26 A Robinson annulation forms a conjugated cyclohex-2-enone. In a bicyclic product, one carbon of the C=C must be shared by both rings.

A
yes

C
yes

B
no

← wrong position

D
no

wrong position

24.27

a.

c.

b.

24.28

a. ^-OH / H_2O or

b.   + C_6H_5CHO ^-OH / H_2O (*E* and *Z* isomers)

24.29

A   [1] O_3 / [2] $(CH_3)_2S$ CHO CHO   NaOEt / EtOH   CHO   B

24.30 The product of an aldol reaction is a β-hydroxy carbonyl compound or an α,β-unsaturated carbonyl compound. The α,β-unsaturated carbonyl compound is drawn as product unless elimination of H_2O cannot form a conjugated system.

a. ^-OH / H_2O OH

b.   + ^-OH / H_2O HO

c.   + ^-OH / H_2O CHO   (*E* and *Z* isomers)

24.31

a.   [1] LDA / [2] $CH_3CH_2CH_2CHO$ / [3] H_2O OH O

b.   OEt   [1] LDA / [2] CHO / [3] H_2O OH OEt

24.32

a.   CHO   →   O

b.   OHC   CHO   →   CHO

c.   →

24.33 Locate the α and β C's to the carbonyl group, and break the molecule into two halves at this bond. The α C and all of the atoms bonded to it belong to one carbonyl component. The β C and all the atoms bonded to it belong to the other carbonyl component.

Carbonyl Condensation Reactions 24–13/

a.   b.   c.   d.

24.34 Base removes the most acidic proton between the two C=O's in **B.** This enolate reacts with the aldehyde in **A** to form a product that loses H_2O.

Form enolate here.

A   electrophilic C   B   base $-H_2O$

(*E* and *Z* isomers can form.)

24.35

a.   b.   c.   d.

24.36 Ozonolysis cleaves the C=C, and base catalyzes an intramolecular aldol reaction.

[1] O_3
[2] $(CH_3)_2S$

NaOH
H_2O

C   D

$C_{10}H_{14}O$

24.37 Aldol reactions proceed via resonance-stabilized enolates. **K** can form an enolate that allows for delocalization of the negative charge on O. Delocalization is not possible in **J,** because a double bond would be placed at a bridgehead carbon, which is geometrically impossible.

from **K**

from **J**

Bond angles don't allow
this double bond.

24.38

a.

OEt → OEt

b.

OEt → OEt

24.39

a. OEt + → c. O + HCO₂Et → H

b. OEt + EtO OEt → EtO OEt d. + Cl OEt → EtO

24.40

a. CH₃O ═══⟹ CH₃O + (EtO)₂C=O

or

CH₃O OEt + EtO

b. C₆H₅ ═══⟹ OEt + C₆H₅

c. CHO ═══⟹ + H OEt

or

+ H

d. EtO OEt ═══⟹ OEt + EtO OEt

24.41 Only esters with two H's or three H's on the α carbon form enolates that undergo Claisen reaction to form resonance-stabilized enolates of the product β-keto ester. Thus, the enolate forms on the CH₂ α to one ester carbonyl, and cyclization yields a five-membered ring.

This is the only α carbon with 2 H's.

24.42

a.

b.

24.43

a.   b.   c.

24.44

a.

(*E* or *Z* isomer can be used.)

b.

+ H$_2$O

+ H$_2$O

24.45

a.

b.

c.

d.

24.46

a.

c.

b.

24.47

a.

b.

c.

d.

e.

f.

24.48

24.49

a. This reaction is a crossed Claisen between an ester and ketone. Cyclization forms the more stable six-membered ring.

b.

24.50

24.51 Enolate **A** is more substituted (and more stable) than either of the other two possible enolates and attacks an aldehyde carbonyl group, which is sterically less hindered than a ketone carbonyl. The resulting ring size (five-membered) is also quite stable. That is why 1-acetylcyclopentene is the major product.

24.52 Removal of a proton from CH_3NO_2 forms an anion for which three resonance structures can be drawn.

24.53 All enolates have a second resonance structure with a negative charge on O.

24.54

24.55 Et₃N reacts with phenylacetic acid to form a carboxylate anion that acts as a nucleophile to displace Br, forming **Y**. Then an intramolecular crossed Claisen reaction yields rofecoxib.

24.56

coumarin

24.57 The mechanism consists of an intramolecular aldol reaction using an enolate formed by removal of a γ proton of an α,β-unsaturated carbonyl. The β-hydroxy carbonyl compound loses H₂O by the two-step E1cB mechanism to generate the final product.

A

(+ 2 resonance structures)

+ H₂O

β-hydroxy carbonyl

B

24.58 Because the reaction is carried out in acid, enols (not enolates) must be intermediates.

24.59

ethyl hexa-2,4-dienoate

diethyl oxalate

b. The protons on C6 are more acidic than other sp^3 hybridized C–H bonds because a highly resonance-stabilized carbanion is formed when a proton is removed. One resonance structure places a negative charge on the carbonyl O atom. This makes the protons on C6 similar in acidity to the α H's to a carbonyl.

c. This is a crossed Claisen because it involves the enolate of a conjugated ester reacting with the carbonyl group of a second ester.

24.60

24.61

24.62

a.

b.

24.63

24.64

a.

octinoxate

b.

24.65

a.

b.

c.

d.

Two products are possible.

e.

conjugated tautomers favored

24.66

24.67 Rearrangement generates a highly resonance-stabilized enolate between two carbonyl groups.

(+ 2 resonance structures)

This product is a highly resonance-stabilized enolate. This drives the reaction.

24.68 All enolates have a second resonance structure with a negative charge on O.

isophorone

+ HB+
(+ 2 resonance structures)

24.69

β-alkoxy carbonyl

1,4-dicarbonyl compound

intramolecular aldol

24.70 All enolates have a second resonance structure with a negative charge on O.

new bond

new bond

H₂O

+ ⁻OH

24.71

a.

(+ other resonance structures)

one possible resonance structure
The negative charge is delocalized on the electronegative N
atom. This factor is what makes the CH₃ group bonded to the
pyridine ring more acidic, and allows the condensation to occur.

+ H₂Ö

A

+ ⁻ÖH

b. The condensation reaction can occur only if the CH₃ group bonded to the pyridine ring has acidic
hydrogens that can be removed with ⁻OH.

2-methylpyridine

(+ other resonance structures)

Since the negative charge is delocalized on the N, the CH₃ contains
acidic H's and reaction will occur.

3-methylpyridine

No resonance structure places the negative charge on the N, so the CH₃
is not acidic and condensation does not occur.

24.72 The reaction takes place in acid, so enols are involved. After the initial condensation reaction, the NH$_2$ and C=O groups form an imine by an intramolecular reaction.

24.73

The enolate
opens the
epoxide ring.

Chapter 25 Amines

Chapter Review

General facts

- Amines are organic nitrogen compounds having the general structure RNH₂, R₂NH, or R₃N, with a lone pair of electrons on N (25.1).
- Amines are named using the suffix -*amine* (25.3).
- All amines have polar C–N bonds. Primary (1°) and 2° amines have polar N–H bonds and are capable of intermolecular hydrogen bonding (25.4).
- The lone pair on N makes amines strong organic bases and nucleophiles (25.8).

Summary of spectroscopic absorptions (25.5)

Mass spectra	Molecular ion	Amines with an odd number of N atoms give an odd molecular ion.
IR absorptions	N–H	3300–3500 cm^{-1} (two peaks for RNH₂, one peak for R₂NH)
¹H NMR absorptions	NH	0.5–5 ppm (no splitting with adjacent protons)
	CH–N	2.3–3.0 ppm (deshielded Csp^3–H)
¹³C NMR absorption	C–N	30–50 ppm

Comparing the basicity of amines and other compounds (25.10)

- Alkylamines (RNH₂, R₂NH, and R₃N) are more basic than NH₃ because of the electron-donating R groups (25.10A).
- Alkylamines (RNH₂) are more basic than arylamines (C₆H₅NH₂), which have a delocalized lone pair from the N atom (25.10B).
- Arylamines with electron-donor groups are more basic than arylamines with electron-withdrawing groups (25.10B).
- Alkylamines (RNH₂) are more basic than amides (RCONH₂), which have a delocalized lone pair from the N atom (25.10C).
- Aromatic heterocycles with a localized electron pair on N are more basic than those with a delocalized lone pair from the N atom (25.10D).
- Alkylamines with a lone pair in an sp^3 hybrid orbital are more basic than those with a lone pair in an sp^2 hybrid orbital (25.10E).

Preparation of amines (25.7)

[1] Direct nucleophilic substitution with NH₃ and amines (25.7A)

- The mechanism is S$_N$2.
- The reaction works best for CH₃X or RCH₂X.
- The reaction works best to prepare 1° amines and ammonium salts.

[2] Gabriel synthesis (25.7A)

- The mechanism is S_N2.
- The reaction works best for CH_3X or RCH_2X.
- Only 1° amines can be prepared.

[3] Reduction methods (25.7B)

a. From nitro compounds

$$R-NO_2 \xrightarrow[\substack{Fe, HCl \text{ or} \\ Sn, HCl}]{H_2, Pd\text{-}C \text{ or}} R-NH_2$$

1° amine

b. From nitriles

$$R-C\equiv N \xrightarrow[\text{[2] } H_2O]{\text{[1] } LiAlH_4} R\diagup NH_2$$

1° amine

c. From amides

R' = H or alkyl

1°, 2°, and 3° amines

[4] Reductive amination (25.7C)

R', R" = H or alkyl

1°, 2°, and 3° amines

- Reductive amination adds one alkyl group (from an aldehyde or ketone) to a nitrogen nucleophile.
- Primary (1°), 2°, and 3° amines can be prepared.

Reactions of amines

[1] Reaction as a base (25.9)

[2] Nucleophilic addition to aldehydes and ketones (25.11)

With 1° amines:

R = H or alkyl

imine

With 2° amines:

R = H or alkyl

enamine

[3] Nucleophilic substitution with acid chlorides and anhydrides (25.11)

[4] Hofmann elimination (25.12)

- The less substituted alkene is the major product.

[5] Reaction with nitrous acid (25.13)

With 1° amines:

alkyl diazonium salt

With 2° amines:

N-nitrosamine

Reactions of diazonium salts
[1] Substitution reactions (25.14)

With H$_2$O:

phenol

With CuX:

aryl chloride or
aryl bromide
X = Cl or Br

With HBF$_4$:

aryl fluoride

With NaI or KI:

aryl iodide

With CuCN:

benzonitrile

With H$_3$PO$_2$:

benzene

[2] Coupling to form azo compounds (25.15)

Y = NH₂, NHR, NR₂, OH

(a strong electron-donor group)

azo compound

Practice Test on Chapter Review

1. Give a systematic name for each of the following compounds.

a.

b.

2. (a) Which compound is the weakest base? (b) Which compound is the strongest base?

A

B

C

D

3. (a) Which compound is the weakest base? (b) Which compound is the strongest base?

A

B

C

D

4. Draw the organic products formed in each of the following reactions.

a.
[1] NaNO₂, HCl
[2] CuCN

b.
[1] KOH
[2] PhCH₂CH₂Br
[3] ⁻OH, H₂O

c.
[1] NaNO₂, HCl
[2] NaI

d.
+
NaBH₃CN

e.
[1] CH₃I (excess)
[2] Ag₂O
[3] heat

5. Draw the products formed when the given amine is treated with [1] CH₃I (excess); [2] Ag₂O; [3] Δ, and indicate the major product. You need not consider any stereoisomers formed in the reaction.

6. What organic starting materials are needed to synthesize **B** by reductive amination?

B

Answers to Practice Test

1.a. *N*-ethyl-2,4-dimethyl-heptan-3-amine
   b. *N*-ethyl-3-methylcyclohexanamine

2.a. **B**
   b. **C**

3.a. **C**
   b. **B**

4.

a. Cl—⟨⟩—CN

b. ⟨⟩(CO₂⁻)(CO₂⁻) + Ph~NH₂

c. Cl-⟨⟩-I

d. ⟨⟩-CH₂-N(piperidine)

e. ⟨⟩ + ⟨⟩

5.

major

6.

Answers to Problems

25.1 The N atom of an ammonium salt is a stereogenic center when the N is surrounded by four different groups. All stereogenic centers are circled.

a.

b.

25.2

a. butan-2-amine
or
sec-butylamine

c. *N,N*-dimethylcyclohexanamine

e. *N*-ethylhexan-3-amine

b. dibutylamine

d. 2-methylnonan-5-amine

f. 2-methyl-*N*-propylcyclopentanamine

25.3 An **NH₂** group named as a substituent is called an **amino group.**

a. 2,4-dimethylhexan-3-amine

c. *N*-isopropyl-*p*-nitroaniline

e. *N,N*-dimethylethanamine

g. *N*-methylaniline

b. *N*-methylpentan-1-amine

d. *N*-methylpiperidine

f. 2-aminocyclohexanone

h. *m*-ethylaniline

25.4 Primary (1°) and 2° amines have higher bp's than similar compounds (like ethers) incapable of hydrogen bonding, but lower bp's than alcohols, which have stronger intermolecular hydrogen bonds. Tertiary amines (3°) have lower boiling points than 1° and 2° amines of comparable molecular weight because they have no N–H bonds with which to form hydrogen bonds.

alkane
**lowest
boiling point**

ether
**intermediate
boiling point**

amine
N–H can hydrogen bond.
highest boiling point

25.5 **The NH signal occurs between 0.5 and 5.0 ppm.** The protons on the carbon bonded to the amine nitrogen are deshielded and typically absorb at 2.3–3.0 ppm. The NH protons are not split.

molecular formula C₆H₁₅N
¹H NMR absorptions (ppm):
0.9 (singlet, 1 H) ⟶ NH
1.10 (triplet, 3 H) ⟶ CH₃ adjacent to CH₂
1.15 (singlet, 9 H) ⟶ (CH₃)₃C
2.6 (quartet, 2 H) ⟶ CH₂ adjacent to CH₃

25.6 The atoms of 2-phenylethanamine are in bold.

a.

LSD
lysergic acid diethyl amide

b.

codeine

25.7 S$_N$2 reaction of an alkyl halide with NH$_3$ or an amine forms an amine or an ammonium salt.

a.

$$\text{Cl} \xrightarrow[\text{excess}]{\text{NH}_3} \text{NH}_2$$

b.

$$\xrightarrow[\text{excess}]{\text{Br}}$$

25.8 The Gabriel synthesis converts an alkyl halide into a 1° amine by a two-step process: nucleophilic substitution followed by hydrolysis.

a.

b.

c. CH$_3$O ... NH$_2$ $\Longrightarrow$ CH$_3$O ... Br

25.9 The Gabriel synthesis prepares 1° amines from alkyl halides. Because the reaction proceeds by an S$_N$2 mechanism, the halide must be CH$_3$ or 1°, and X can't be bonded to an sp^2 hybridized C.

a.

aromatic
An S$_N$2 does not occur on
an aryl halide.
cannot be made by Gabriel
synthesis

b.

can be made by Gabriel synthesis

c.

2° amine
cannot be made by Gabriel
synthesis

d.

N on 3° C
An S$_N$2 does not occur on
a 3° RX.
cannot be made by
Gabriel synthesis

25.10 **Nitriles are reduced to 1° amines with LiAlH$_4$. Nitro groups are reduced to 1° amines** using a variety of reducing agents. **Primary (1°), 2°, and 3° amides are reduced to 1°, 2°, and 3° amines** respectively, using LiAlH$_4$.

a.

NH$_2$ $\Rightarrow$ NO$_2$ C≡N   NH$_2$

b.

NH$_2$ $\Rightarrow$ NO$_2$ C≡N   NH$_2$

c.

25.11 **Primary (1°), 2°, and 3° amides are reduced to 1°, 2°, and 3° amines** respectively, using LiAlH₄.

a.

b.

c.

25.12 Only amines with a CH₂ or CH₃ bonded to the N can be made by reduction of an amide.

a.

N bonded to benzene
cannot be made by reduction of
an amide

b.

N bonded to CH₂
can be made by reduction of
an amide

c.

N bonded to a 3° C
cannot be made by reduction of
an amide

d.

N with 2° C on both sides
cannot be made by reduction of
an amide

25.13 Reductive amination is a two-step method that converts aldehydes and ketones into 1°, 2°, and 3° amines. Reductive amination replaces a C=O by a C–H and C–N bond.

a.

b.

c.

d.

25.14 Reductive amination occurs using the ketone in **E** and the amine in **D**.

E

D

enalapril

25.15

a.

b.

25.16

a.

phentermine

Only amines that have a C bonded to a H and N atom can be made by reductive amination; that is, an amine must have the following structural feature:

In phentermine, the C bonded to N is not bonded to a H, so it cannot be made by reductive amination.

b. systematic name: 2-methyl-1-phenylpropan-2-amine

25.17
The pK_a of many protonated amines is 10–11, so the pK_a of the starting acid must be **less than 10** for equilibrium to favor the products. Amines are thus readily protonated by strong inorganic acids (e.g., HCl and H_2SO_4) and by carboxylic acids.

a.

c.

b.

25.18
An amine can be separated from other organic compounds by converting it to a water-soluble ammonium salt by an acid–base reaction. In each case, the extraction procedure would employ the following steps:
   • Dissolve the amine and either **X** or **Y** in CH_2Cl_2.
   • Add a solution of 10% HCl. The amine will be protonated and dissolve in the aqueous layer, while **X** or **Y** will remain in the organic layer as a neutral compound.
   • Separate the layers.

a.

• **soluble in H_2O**
• **insoluble in CH_2Cl_2**

• insoluble in H_2O
• soluble in CH_2Cl_2

b.

- soluble in H₂O
- insoluble in CH₂Cl₂

- insoluble in H₂O
- soluble in CH₂Cl₂

25.19 Primary (1°), 2°, and 3° alkylamines are more basic than NH₃ because of the electron-donating inductive effect of the R groups.

a.

2° alkylamine
CH₃ groups are electron donating.
stronger base

and NH₃

b.

1° alkylamine
stronger base

1° alkylamine
Cl is electron withdrawing.
weaker base

25.20 Arylamines are less basic than alkylamines because the electron pair on N is delocalized. Electron-donor groups add electron density to the benzene ring making the arylamine more basic than aniline. Electron-withdrawing groups remove electron density from the benzene ring, making the arylamine less basic than aniline.

a.

electron-
withdrawing group
least basic

arylamine
**intermediate
basicity**

electron-
donating group
most basic

b.

electron-
withdrawing group
least basic

arylamine
**intermediate
basicity**

alkylamine
most basic

25.21 Amides are much less basic than amines because the electron pair on N is highly delocalized.

amide
least basic

arylamine
intermediate basicity

alkylamine
most basic

25.22

a.   Electron pair on N occupies an *sp²* hybrid orbital.

sp² hybridized
more basic

This N is also *sp²* hybridized but the electron pair occupies a *p* orbital, so it can delocalize onto the aromatic ring. Delocalization makes this N less basic.

DMAP
4-(*N,N*-dimethylamino)pyridine

b.

sp³ hybridized N
stronger base

nicotine

sp² hybridized N
higher percent *s*-character
weaker base

25.23 HCl protonates the more basic N atom.

a.

delocalized electron pair that is part of an amide

stronger base
sp^3 hybridized N

b.

stronger base
sp^3 hybridized N
25% s-character

sp^2 hybridized N
33% s-character

25.24 Amines attack carbonyl groups to form products of nucleophilic addition or substitution.

a.

b.

c.

25.25 [1] Convert the amine (aniline) into an amide (acetanilide).
[2] **Carry out the Friedel–Crafts reaction.**
[3] **Hydrolyze the amide** to generate the free amino group.

a.

(+ ortho isomer)

b.

(+ para isomer)

25.26

a.

[1] CH$_3$I (excess)
[2] Ag$_2$O
[3] Δ

b.

[1] CH$_3$I (excess)
[2] Ag$_2$O
[3] Δ

c.

[1] CH$_3$I (excess)
[2] Ag$_2$O
[3] Δ

25.27 In a Hofmann elimination, the base removes a proton from the less substituted, more accessible β carbon atom, because of the bulky leaving group on the nearby α carbon.

a.

[1] CH$_3$I (excess)
[2] Ag$_2$O
[3] Δ

(+ Z isomer)   +   major product

b.

[1] CH$_3$I (excess)
[2] Ag$_2$O
[3] Δ

major product   +

c.

[1] CH$_3$I (excess)
[2] Ag$_2$O
[3] Δ

(3 β C's)
least substituted β carbon

(+ Z isomer)   +

major product, formed by removal of a H from the least substituted β C

25.28

a.

K$^+$ $^-$OC(CH$_3$)$_3$

c.

K$^+$ $^-$OC(CH$_3$)$_3$

b.

[1] CH$_3$I (excess)
[2] Ag$_2$O
[3] Δ

(E and Z)

d.

[1] CH$_3$I (excess)
[2] Ag$_2$O
[3] Δ

25.29

a. NH_2 (with CH₃ ortho) $\xrightarrow[\text{HCl}]{NaNO_2}$ $N_2^+\,Cl^-$ (with CH₃ ortho)

c. piperidine (NH) $\xrightarrow[\text{HCl}]{NaNO_2}$ N-nitrosopiperidine (N–NO)

b. CH₃CH₂–N(H)–CH₃ $\xrightarrow[\text{HCl}]{NaNO_2}$ CH₃CH₂–N(CH₃)–NO

d. sec-butylamine (NH₂) $\xrightarrow[\text{HCl}]{NaNO_2}$ $N_2^+\,Cl^-$ sec-butyl diazonium

25.30

a. NH_2 $\xrightarrow[\text{[2] CuBr}]{\text{[1] NaNO}_2,\ HCl}$ Br (benzene ring)

c. CH_3O—⟨benzene⟩—NH_2 $\xrightarrow[\text{[2] HBF}_4]{\text{[1] NaNO}_2,\ HCl}$ CH_3O—⟨benzene⟩—F

b. NH_2 (benzene, O_2N meta) $\xrightarrow[\text{[2] H}_2O]{\text{[1] NaNO}_2,\ HCl}$ OH (benzene, O_2N meta)

d. $N_2^+\,Cl^-$ (benzene, Cl ortho) $\xrightarrow[\text{[2] LiAlH}_4\ \ \text{[3] H}_2O]{\text{[1] CuCN}}$ CH_2NH_2 (benzene, Cl ortho)

25.31

a. benzene $\xrightarrow[\text{H}_2SO_4]{HNO_3}$ NO_2-benzene $\xrightarrow[\text{Pd-C}]{H_2}$ NH_2-benzene $\xrightarrow[\text{[2] HBF}_4]{\text{[1] NaNO}_2,\ HCl}$ F-benzene

b. NO_2-benzene (from a.) $\xrightarrow[\text{H}_2SO_4]{HNO_3}$ 1,3-dinitrobenzene (NO_2/NO_2) $\xrightarrow[\text{Pd-C}]{H_2}$ 1,3-diaminobenzene (NH_2/NH_2) $\xrightarrow[\text{[2] H}_2O]{\text{[1] NaNO}_2,\ HCl}$ 1,3-dihydroxybenzene (OH/OH)

c. NH_2-benzene (from a.) $\xrightarrow[\text{[2] NaI}]{\text{[1] NaNO}_2,\ HCl}$ I-benzene $\xrightarrow[\text{AlCl}_3]{CH_3Cl}$ o-iodotoluene (CH₃/I) (+ para isomer)

d. NH_2-benzene (from a.) $\xrightarrow[\text{FeCl}_3]{Cl_2\ \text{(excess)}}$ 2,4,6-trichloroaniline (NH_2, Cl/Cl/Cl) $\xrightarrow[\text{[2] H}_3PO_2]{\text{[1] NaNO}_2,\ HCl}$ 1,3,5-trichlorobenzene (Cl/Cl/Cl)

25.32

a. NH_2-benzene $\longrightarrow$ phenyl–N=N–⟨benzene⟩–NH_2 (4-(phenylazo)aniline)

b. HO–⟨benzene⟩ (phenol) $\longrightarrow$ phenyl–N=N–⟨benzene⟩–OH (4-(phenylazo)phenol)

c. HO—⟨benzene⟩—OH (hydroquinone) $\longrightarrow$ phenyl–N=N–⟨benzene⟩ with OH and HO substituents

25.33 To determine what starting materials are needed to synthesize a particular azo compound, always divide the molecule into two components: **one has a benzene ring with a diazonium ion, and one has a benzene ring with a very strong electron-donor group.**

a.

b.

25.34

a.

para red

b.

alizarine yellow R

25.35

a.

4,6-dimethylheptan-1-amine

b.

N,N-diethylcycloheptanamine

25.36

A
weaker base
(delocalized electron pair on N)

or

B
stronger base

25.37

a, b.

varenicline

most basic
only *sp*^{3} hybridized N

25.38

a. *N*-ethyl-2-methylbutan-1-amine

b. 4-ethyl-2-methyloctan-1-amine

c. *N*-ethyl-*N*-methylcyclohexanamine

d. *N*-*tert*-butyl-*N*-ethylaniline

e. 4-aminocyclohexanone

f. 2-ethylpyrrolidine

g. 2-methylhexan-3-amine

h. 3-ethyl-2-methylcyclohexanamine

25.39

a. *N*-isobutylcyclopentanamine

b. tri-*tert*-butylamine

N[C(CH$_3$)$_3$]$_3$

c. *N*,*N*-diisopropylaniline

d. *N*-methylpyrrole

e. *N*-methylcyclopentanamine

f. 3-methylhexan-2-amine

g. 2-*sec*-butylpiperidine

h. (*S*)-heptan-2-amine

25.40

2 stereogenic centers
4 stereoisomers

25.41

a.

delocalized
electron pair on N
least basic

sp^2 hybridized N
**intermediate
basicity**

sp^3 hybridized N
most basic

b.

electron-
withdrawing group
least basic

**intermediate
basicity**

electron-
donating group
most basic

25.42 The most basic N atom is protonated on treatment with acid.

most basic

a.

zolpidem

CH$_3$CO$_2$H

CH$_3$CO$_2^-$

a 3° alkylamine with an sp^3
hybridized N
most basic

b.

aripiprazole

CH$_3$CO$_2$H

CH$_3$CO$_2^-$

25.43 Pyrimidine has a second electronegative N atom that withdraws electron density from the other N, making the electron pair less available for electron donation.

This N pulls electron density
from the other N atom.

pyrimidine

25.44

a.

$N_b < N_a < N_c$

Order of basicity: $N_b < N_a < N_c$
N_b – The electron pair on this N atom is delocalized
  on the O atom; least basic.
N_a – The electron pair on this N atom is not
  delocalized, but is on an sp^2 hybridized atom.
N_c – The electron pair on this N atom is on an sp^3
  hybridized N; most basic.

b.

$N_b < N_a < N_c$

Order of basicity: $N_b < N_a < N_c$
N_b – The electron pair on this N atom is delocalized
  on the aromatic five-membered ring; least basic.
N_a – The electron pair on this N atom is not
  delocalized, but is on an sp^2 hybridized atom.
N_c – The electron pair on this N atom is on an sp^3
  hybridized N; most basic.

25.45 The para isomer is the weaker base because the electron pair on its NH$_2$ group can be delocalized onto the NO$_2$ group. In the meta isomer, no resonance structure places the electron pair on the NO$_2$ group, and fewer resonance structures can be drawn:

meta

para

25.46

A

pK_a of the conjugate acid = 5.2
stronger conjugate acid
weaker base
The electron pair of this arylamine
is delocalized on the benzene ring,
decreasing its basicity.

B

This two-carbon bridge makes it difficult
for the lone pair on N to delocalize on
the aromatic ring.

pK_a of the conjugate acid = 7.29
weaker conjugate acid
stronger base

Resonance structures that place a double bond between the N
atom and the benzene ring are destabilized. Since the electron
pair is more localized on N, compound **B** is more basic.

B

Geometry makes it difficult to have
a double bond here.

25.47

pyrrole
pK_a = 23
stronger acid

weaker conjugate base
The electron pair is delocalized, decreasing the
basicity. The N atom is sp^2 hybridized.

pyrrolidine
pK_a = 44
weaker acid

stronger conjugate base
The electron pair is not
delocalized on the ring. The N
atom is sp^3 hybridized.

25.48

a.

b.

or

c.

25.49 In reductive amination, one alkyl group on N comes from the carbonyl compound. The remainder of the molecule comes from NH_3 or an amine.

a.

b.

c.

25.50

a.

b.

c.

d.

25.51 Use the directions from Answer 25.18. Separation can be achieved because benzoic acid reacts with aqueous base and aniline reacts with aqueous acid according to the following equations:

benzoic acid
• soluble in CH_2Cl_2
• insoluble in H_2O

• soluble in H_2O
• insoluble in CH_2Cl_2

aniline
• soluble in CH_2Cl_2
• insoluble in H_2O

• soluble in H_2O
• insoluble in CH_2Cl_2

Toluene ($C_6H_5CH_3$), on the other hand, is not protonated or deprotonated in aqueous solution, so it is always soluble in CH_2Cl_2 and insoluble in H_2O. The following flow chart illustrates the process.

+ CH₂Cl₂ / 10% NaOH

H₂O / COO⁻Na⁺   CH₂Cl₂

10% HCl

H₂O / ⁺NH₃ Cl⁻   CH₂Cl₂

25.52

—NH₂   *p*-methylaniline

a. HCl → ⁺NH₃ Cl⁻

b. CH₃COCl →

c. (CH₃CO)₂O →

d. CH₃I (excess) → ⁺N(CH₃)₃I⁻

e. (CH₃)₂C=O →

f. CH₃COCl / AlCl₃ → ⁻AlCl₃ / ⁺NH₂

g. CH₃COOH → ⁺NH₃   CH₃COO⁻

h. NaNO₂ / HCl → N₂⁺Cl⁻

i. Step (b), then CH₃COCl / AlCl₃ →

j. CH₃CHO / NaBH₃CN → —NH

25.53

a. [1] CH₃I (excess) [2] Ag₂O [3] Δ → **major product** + (E + Z)

b. [1] CH₃I (excess) [2] Ag₂O [3] Δ → CH₂=CH₂ + + + **major product**

c.

[1] CH$_3$I (excess)
[2] Ag$_2$O
[3] Δ

major product

d.

β_1 β_2
β_3

[1] CH$_3$I (excess)
[2] Ag$_2$O
[3] Δ

β_1

major product

+

β_2

(E + Z)

+

β_3

(E + Z)

25.54

a.

one stereogenic center

benzphetamine

b. Amides that can be reduced to benzphetamine:

and

c. Amines + carbonyl compounds that form benzphetamine by reductive amination:

+

or

+

or

+

d.

β_1 α
β_2

[1] CH$_3$I
[2] Ag$_2$O
[3] Δ

+

+

major product
(elimination across α, β_2)

elimination across α, β_1

+

25.55

a. [cyclohexyl-CH₂CH₂—Cl] + NH₃ (excess) → [cyclohexyl-CH₂CH₂—NH₂]

g. [pyrrolidine] NH + NaNO₂ / HCl → N—N=O

b. [phthalimide] [1] KOH [2] (CH₃)₂CHCH₂Cl [3] ⁻OH, H₂O → (CH₃)₂CHCH₂NH₂ + [benzene-1,2-dicarboxylate, CO₂⁻ / CO₂⁻]

h. [pyrrolidine] NH + [benzaldehyde] + NaBH₃CN → N-benzyl pyrrolidine

c. Br—[benzene]—NO₂ + Sn / HCl → Br—[benzene]—NH₂

i. [piperidine, N–H] + [cyclohexanone, =O] → [N-cyclohexenyl piperidine]

d. [2-methylbutanenitrile, CN] [1] LiAlH₄ [2] H₂O → [amine, NH₂]

j. [amine, N with isopropyl and propyl, N–H] [1] CH₃I (excess) [2] Ag₂O [3] Δ → [propene] + [CH₃CH₂CH₂—N(CH₃)—] + [isopropyl-N(CH₃)₂]

e. [cyclohexanecarboxamide, C=O, HN–ethyl] [1] LiAlH₄ [2] H₂O → [cyclohexyl-CH₂–HN–ethyl]

f. [phenethylamine, NH₂] + [benzoic anhydride] → [PhCH₂CH₂–NH–C(=O)Ph amide] + [PhCH₂CH₂–NH₃⁺ ⁻O–C(=O)Ph]

25.56 NH₂ and H must be anti for the Hofmann elimination. Rotate around the C–C bond so the NH₂ and H are anti.

a. [Newman projection: front CH₃, C₆H₅, NH₂; back H, CH₂CH₃, C(CH₃)₃]
→ three steps → [alkene product with phenyl]

b. [Newman projection: front NH₂, H, CH₃; back C₆H₅, C(CH₃)₃, CH₂CH₃]
→ rotate 120° counterclockwise → [Newman projection: front NH₂, CH₃, CH₂CH₃; back C₆H₅, C(CH₃)₃, H]
→ three steps → [alkene product with phenyl]

c. [Newman projection: front C₆H₅, CH₃, H; back (CH₃)₃C, NH₂, CH₂CH₃]
→ rotate 120° counterclockwise → [Newman projection: front C₆H₅, H, CH₂CH₃; back (CH₃)₃C, NH₂, CH₃]
→ three steps → [alkene product with phenyl]

25.57

25.58

25.59

a.

b.

c.

25.60

25.61

a.

b.

25.62

25.63

Overall reaction:

The steps:

25.64 A nitrosonium ion ($^+$NO) is a weak electrophile, so electrophilic aromatic substitution occurs only with a strong electron-donor group that stabilizes the intermediate carbocation.

especially good resonance structure
All atoms have octets.

resonance-stabilized carbocation

25.65

a.

b.

(+ ortho isomer)

c. (from a.)

$$O_2N-C_6H_5 \xrightarrow[\text{Pd-C}]{H_2} H_2N-C_6H_5 \xrightarrow[\text{[2] CuCN}]{\text{[1] NaNO}_2,\text{ HCl}} NC-C_6H_5 \xrightarrow[\text{FeBr}_3]{Br_2} NC-C_6H_4-Br$$

d.

$$C_6H_6 \xrightarrow[\text{FeBr}_3]{Br_2} C_6H_5Br \xrightarrow[\text{H}_2\text{SO}_4]{HNO_3} (o\text{-NO}_2\text{-C}_6\text{H}_4\text{Br}) \text{ (+ para isomer)} \xrightarrow[\text{Pd-C}]{H_2} (o\text{-NH}_2\text{-C}_6\text{H}_4\text{Br}) \xrightarrow[\text{[2] H}_2\text{O}]{\text{[1] NaNO}_2,\text{ HCl}} (o\text{-OH-C}_6\text{H}_4\text{Br})$$

e. (from b.)

$$O_2N\text{-}C_6H_4\text{-CH}_3 \xrightarrow{\text{KMnO}_4} O_2N\text{-}C_6H_4\text{-COOH} \xrightarrow[\text{Pd-C}]{H_2} H_2N\text{-}C_6H_4\text{-COOH} \xrightarrow[\text{[2] H}_2\text{O}]{\text{[1] NaNO}_2,\text{ HCl}} HO\text{-}C_6H_4\text{-COOH}$$

f. (from c.)

$$C_6H_5\text{-NH}_2 \xrightarrow[\text{[2] H}_2\text{O}]{\text{[1] NaNO}_2,\text{ HCl}} C_6H_5\text{-OH} \xrightarrow[\text{(from c.)}]{C_6H_5N_2^+Cl^-} C_6H_5\text{-N=N-}C_6H_4\text{-OH}$$

25.66

a.

$$C_6H_5\text{-NH}_2 \xrightarrow[\text{[2] NaCN}]{\text{[1] NaNO}_2,\text{ HCl}} C_6H_5\text{-CN} \xrightarrow{H_3O^+} C_6H_5\text{-COOH} \xrightarrow[\text{[2] NH}_2\text{CH}_3]{\text{[1] SOCl}_2} C_6H_5\text{-CONH-CH}_3$$

b.

$$C_6H_5\text{-NH}_2 \xrightarrow{\text{CH}_3\text{COCl}} C_6H_5\text{-NHCOCH}_3 \xrightarrow[\text{FeBr}_3]{Br_2} (p\text{-Br-C}_6\text{H}_4\text{-NHCOCH}_3) \text{ (+ ortho isomer)} \xrightarrow[\text{AlCl}_3]{\text{CH}_3\text{Cl}} \xrightarrow[\text{H}_2\text{O}]{^-\text{OH}}$$

$$\xrightarrow[\text{[2] H}_3\text{PO}_2]{\text{[1] NaNO}_2,\text{ HCl}} (m\text{-Br-C}_6\text{H}_4\text{-CH}_3)$$

c.

$$H_2N\text{-}C_6H_5 \xrightarrow[\text{[2] H}_2\text{O}]{\text{[1] NaNO}_2,\text{ HCl}} HO\text{-}C_6H_5 \xrightarrow[\text{AlCl}_3]{\text{CH}_3\text{Cl}} HO\text{-}C_6H_4\text{-CH}_3 \xrightarrow[\text{(from a., Step [1])}]{C_6H_5N_2^+Cl^-} C_6H_5\text{-N=N-}C_6H_3(\text{OH})(\text{CH}_3) \text{ (+ ortho isomer)}$$

25.67

[1]

$$C_6H_6 \xrightarrow[\text{AlCl}_3]{\text{CH}_3\text{Cl}} C_6H_5\text{-CH}_3 \xrightarrow[h\nu]{Br_2} C_6H_5\text{-CH}_2\text{Br} \xrightarrow[\text{(excess)}]{NH_3} C_6H_5\text{-CH}_2\text{-NH}_2 \xrightarrow[\text{NaBH}_3\text{CN}]{\text{H}_2\text{C=O}} C_6H_5\text{-CH}_2\text{-NH-CH}_3$$

[2]

(from [1])

Benzyl bromide $\xrightarrow{^-OH}$ benzyl alcohol $\xrightarrow{PCC}$ benzaldehyde $\xrightarrow[NaBH_3CN]{CH_3NH_2}$ N-methylbenzylamine

[1] LiAlH$_4$
[2] H$_2$O

[3]

(from [1])

Toluene $\xrightarrow{KMnO_4}$ benzoic acid $\xrightarrow[\text{[2] } CH_3NH_2]{\text{[1] } SOCl_2}$ N-methylbenzamide $\xrightarrow[\text{[2] } H_2O]{\text{[1] } LiAlH_4}$ N-methylbenzylamine

[4]

Benzene $\xrightarrow{\substack{\text{[1] } HNO_3, H_2SO_4 \\ \text{[2] } H_2, Pd\text{-}C \\ \text{[3] } NaNO_2, HCl \\ \text{[4] } CuCN}}$ benzonitrile $\xrightarrow[\text{[2] } H_2O]{\text{[1] DIBAL-H}}$ benzaldehyde Then route [2].

[1] LiAlH$_4$
[2] H$_2$O

benzylamine Then route [1].

[5]

Benzene $\xrightarrow[FeBr_3]{Br_2}$ bromobenzene $\xrightarrow{Mg}$ phenylmagnesium bromide $\xrightarrow[\text{[2] } H_3O^+]{\text{[1] } CO_2}$ benzoic acid Then route [3].

[1] H$_2$C=O
[2] H$_2$O

benzyl alcohol Then route [2].

25.68

$\xrightarrow{mCPBA}$ epoxide $\xrightarrow[\text{[2] } H_2O]{\text{[1] } LiAlH_4}$ alcohol $\xrightarrow{PBr_3}$ bromide

PCC ↓

ketone $\xrightarrow[NaBH_3CN]{CH_3NH_2}$ MDMA [part (b)]

bromide $\xrightarrow{CH_3NH_2}$ MDMA [part (a)]

25.69

a.

Benzene $\xrightarrow[AlCl_3]{CH_3Cl}$ toluene $\xrightarrow[H_2SO_4]{HNO_3}$ p-nitrotoluene (+ ortho isomer) $\xrightarrow[Pd\text{-}C]{H_2}$ p-toluidine $\xrightarrow{\substack{\text{[1] } NaNO_2, HCl \\ \text{[2] } NaCN}}$ p-tolunitrile $\xrightarrow[\text{[2] } H_2O]{\text{[1] } LiAlH_4}$ amine

b.

c.

25.70

a.

b.

25.71

c. Probably a strong enough activator that the Friedel–Crafts reaction will still occur.

Make two parts:

25.72

Compound A: $C_8H_{11}N$
IR absorption at 3400 cm^{-1} → 2° amine
^1H NMR signals at (ppm):
- 1.3 (triplet, 3 H) CH$_3$ adjacent to 2 H's
- 3.2 (quartet, 2 H) CH$_2$ adjacent to 3 H's
- 3.6 (singlet, 1 H) amine H
- 6.8–7.2 (multiplet, 5 H) benzene ring

Compound B: $C_8H_{11}N$
IR absorption at 3310 cm^{-1} → 2° amine
^1H NMR signals at (ppm):
- 1.4 (singlet, 1 H) amine H
- 2.4 (singlet, 3 H) CH$_3$
- 3.8 (singlet, 2 H) CH$_2$
- 7.3 (multiplet, 5 H) benzene ring

Compound C: $C_8H_{11}N$
IR absorption at 3430 and
3350 cm^{-1} → 1° amine
^1H NMR signals at (ppm):
- 1.3 (triplet, 3 H) CH$_3$ near CH$_2$
- 2.5 (quartet, 2 H) CH$_2$ near CH$_3$
- 3.6 (singlet, 2 H) amine H's
- 6.7 (doublet, 2 H) ⎤ para disubstituted
- 7.0 (doublet, 2 H) ⎦ benzene ring

25.73

HO—CH₂CH₂—C≡N

Compound D:
Molecular ion at *m/z* = 71: C_3H_5NO (possible formula)
IR absorption at 3600–3200 cm⁻¹ → OH
                    2263 cm⁻¹ → CN
¹H NMR signals at (ppm):
  2.5 (triplet, 2 H) CH₂ adjacent to 2 H's
  3.1 (singlet, 1 H) OH
  3.8 (triplet, 2 H) CH₂ adjacent to 2 H's

HO—CH₂CH₂CH₂—NH₂

Compound E:
Molecular ion at *m/z* = 75: C_3H_9NO (possible formula)
IR absorption at 3600–3200 cm⁻¹ → OH
                    3636 cm⁻¹ → N–H of amine
¹H NMR signals at (ppm):
  1.6 (quintet, 2 H) CH₂ split by 2 CH₂'s
  2.5 (singlet, 3 H) NH₂ and OH
  2.8 (triplet, 2 H) CH₂ split by CH₂
  3.7 (triplet, 2 H) CH₂ split by CH₂

25.74 Guanidine is a strong base because its conjugate acid is stabilized by resonance. This resonance delocalization makes guanidine easily donate its electron pair; thus it's a strong base.

25.75 The compound with the most available electron pair or the compound with the highest electron density on an atom (N in this case) is the strongest base. Pyrrole is the weakest base because its lone pair is delocalized on the five-membered ring to make it aromatic. Both imidazole and thiazole contain sp^2 hybridized N atoms with electron pairs that are localized on N. Imidazole is a stronger base than thiazole, because its second N atom is more basic than thiazole's S atom, so it places more electron density on N by a resonance effect.

imidazole

This N atom is the strongest base.

thiazole

This form contributes less to the hybrid than the equivalent resonance structure in imidazole.

pyrrole
least basic

thiazole
intermediate basicity

imidazole
most basic

25.76

25.77

One possibility:

a.

(+ isomer)

albuterol

b.

A

(+ ortho)

25.78 CH_2=O reacts with the amine to form an intermediate imine, which undergoes an intramolecular Diels–Alder reaction.

X

Y
$C_{17}H_{23}NO$

one step

lupinine

epilupinine

Chapter 26 Carbon–Carbon Bond-Forming Reactions in Organic Synthesis

Chapter Review

Coupling reactions

[1] Coupling reactions of organocuprate reagents (26.1)

R'—X + R$_2$CuLi $\longrightarrow$ R'—R + RCu

X = Cl, Br, I   + LiX

- R'X can be CH$_3$X, RCH$_2$X, 2° cyclic halides, vinyl halides, and aryl halides.
- X may be Cl, Br, or I.
- With vinyl halides, coupling is stereospecific.

[2] Suzuki reaction (26.2)

$$R'—X \ + \ R-B\underset{Y}{\overset{Y}{\diagup}} \ \xrightarrow[\text{NaOH}]{\text{Pd(PPh}_3)_4} \ R'—R$$

X = Br, I

+ HO—BY$_2$
+ NaX

- R'X is most often a vinyl halide or aryl halide.
- With vinyl halides, coupling is stereospecific.

[3] Heck reaction (26.3)

$$R'—X \ + \ \diagup\hspace{-0.3em}\diagdown Z \ \xrightarrow[\text{Et}_3\text{N}]{\overset{\text{Pd(OAc)}_2}{\text{P(}o\text{-tolyl)}_3}} \ R'\diagdown\hspace{-0.3em}\diagup Z$$

X = Br or I

+ Et$_3$NH X$^-$

- R'X is a vinyl halide or aryl halide.
- Z = H, Ph, COOR, or CN
- With vinyl halides, coupling is stereospecific.
- The reaction forms trans alkenes.

Cyclopropane synthesis

[1] Addition of dihalocarbenes to alkenes (26.4)

- The reaction occurs with syn addition.
- The position of substituents in the alkene is retained in the cyclopropane.

[2] Simmons–Smith reaction (26.5)

- The reaction occurs with syn addition.
- The position of substituents in the alkene is retained in the cyclopropane.

Metathesis (26.6)

diene

- Metathesis works best when CH$_2$=CH$_2$, a gas that escapes from the reaction mixture, is formed as one product.

Practice Test on Chapter Review

1.a. Which functional groups react with lithium dialkyl cuprates?
   1. epoxides
   2. vinyl halides
   3. acid chlorides
   4. Compounds (1) and (2) both react with R_2CuLi.
   5. Compounds (1), (2), and (3) all react with R_2CuLi.

b. Which of the following statements is (are) true for the Suzuki reaction?
   1. Arylboranes can serve as one reactant.
   2. The reaction is stereospecific.
   3. The reaction occurs between a vinyl or aryl halide and an alkene in the presence of a palladium catalyst.
   4. Statements (1) and (2) are both true.
   5. Statements (1), (2), and (3) are all true.

c. Which of the following compounds can react with CH_2=CHCN in a Heck reaction?

   1.   4. Both (1) and (2) can react.

   2.   5. Compounds (1), (2), and (3) can all react.

   3. CH_2=CH_2

d. Which of the following compounds yield(s) a pair of enantiomers on reaction with the Simmons–Smith reagent?

   1.   3.

   2.   4. Compounds (1) and (2) both yield a pair of enantiomers.

                                     5. Compounds (1), (2), and (3) all yield a pair of enantiomers.

e. Which of the following compounds can be made by a ring-closing metathesis reaction?

   1.   3.

   2.   4. Compounds (1) and (2) can both be prepared.

                                     5. Compounds (1), (2), and (3) can all be prepared.

f. Which of the following compounds can be prepared from CH_3–C≡C–H by a Suzuki reaction? You may use other organic compounds or inorganic reagents.

1.

3.

2.

4. Compounds (1) and (2) can both be prepared.

5. Compounds (1), (2), and (3) can all be prepared.

2. Draw the product formed in each reaction. Indicate the stereochemistry around double bonds and stereogenic centers when necessary.

a.
[1] catecholborane
[2]
$_I$, Pd(PPh$_3$)$_4$, NaOH

e.
CHBr$_3$
KOC(CH$_3$)$_3$

b.
Br
[1] 2 Li
[2] 0.5 CuI
[3] Ph
Br

f.
Grubbs
cataylst

c.
Br
[1] 2 Li
[2] B(OCH$_3$)$_3$
[3]
Br, Pd(PPh$_3$)$_4$, NaOH

g.
Grubbs
cataylst

d.
OCH$_3$
CH$_3$O
I
+
CO$_2$CH$_3$
Pd(OAc)$_2$
P(o-tolyl)$_3$
Et$_3$N

3. What starting material is needed to synthesize each compound by ring-closure metathesis?

a.

b.
HO
OH   OH

c.
O   O

Answers to Practice Test

1.a. 5
  b. 4
  c. 4
  d. 2
  e. 3
  f. 4

2.

a.

b. Ph

c.

d. (OCH₃, CH₃O, CO₂CH₃)

e. (Br, Br)

f.

g.

3.

a.

b. (HO, OH, OH)

c.

Answers to Problems

26.1 A new C–C bond is formed in each coupling reaction.

a.

b.

c.

d.

26.2

C_{18} juvenile hormone

26.3

a.

or

b.

c.

26.4

a.

b.

c.

d.

26.5 The Suzuki reaction forms a new carbon–carbon bond between a vinyl halide and an arylborane.

26.6

a.

b.

c.

26.7

a. + →(Heck reaction)→

b. I + OCH₃ →(Heck reaction)→ OCH₃

c. CN + Br →(Heck reaction)→ CN

d. OCH₃ + Br →(Heck reaction)→ OCH₃

26.8 Locate the double bond with the aryl, COOR, or CN substituent, and break the molecule into two components at the end of the C=C not bonded to one of these substituents.

a. ⟹ +

b. ⟹ Br +

c. ⟹ +

26.9 Add the carbene carbon from either side of the alkene.

a. →(CHCl₃ / KOC(CH₃)₃)→ +
enantiomers

b. →(CHCl₃ / KOC(CH₃)₃)→ +
identical

c. →(CHCl₃ / KOC(CH₃)₃)→ +
enantiomers

26.10

a.

b.

c.
(from b.)

26.11

a.

b.

c.

26.12 The relative position of substituents in the reactant is retained in the product.

trans-hex-3-ene

two enantiomers of *trans*-1,2-diethylcyclopropane

26.13

a.
(*E* and *Z*)

c.

b.
(*E* and *Z*)

(CH₂=CH₂ is also formed in each reaction.)

26.14

cis-pent-2-ene

There are four products formed in this reaction including stereoisomers, and therefore, it is not a practical method to synthesize 1,2-disubstituted alkenes.

26.15

a.

b.

26.16

26.17 Cleave the C=C bond in the product, and then bond each carbon of the original alkene to a CH₂ group using a double bond.

a.

b.

c.

26.18 Inversion of configuration occurs with the substitution of the methyl group for the tosylate.

a.

b.

26.19

a.

b.

26.20

a.

b.

$B(OCH_3)_2$ + (pyridine)Br $\xrightarrow[\text{NaOH}]{\text{Pd(PPh}_3)_4}$

c.

$\xrightarrow[\substack{\text{P}(o\text{-tolyl})_3 \\ \text{Et}_3\text{N}}]{\text{Pd(OAc)}_2}$

d.

Cl $\xrightarrow{\substack{[1]\ \text{Li} \\ [2]\ \text{CuI} \\ [3]\ \text{Br}}}$

e.

Br + B $\xrightarrow[\text{NaOH}]{\text{Pd(PPh}_3)_4}$

f.

CH_3O—Br + CO_2CH_3 $\xrightarrow[\substack{\text{P}(o\text{-tolyl})_3 \\ \text{Et}_3\text{N}}]{\text{Pd(OAc)}_2}$ CH_3O—CO_2CH_3

g.

Br $\xrightarrow{\substack{[1]\ \text{Li} \\ [2]\ \text{B(OCH}_3)_3 \\ [3]\ \text{Br + Pd catalyst}}}$

h.

$\xrightarrow{\substack{[1]\ \text{H–B} \\ [2]\ \text{C}_6\text{H}_5\text{Br, Pd(PPh}_3)_4,\ \text{NaOH}}}$

26.21

a. $\Longrightarrow$ Br

b. $\Longrightarrow$ Br

c. CH_3O $\Longrightarrow$ CH_3O—Br

26.22

Each coupling reaction uses Pd(PPh$_3$)$_4$ and NaOH to form the conjugated diene.

ethynylcyclohexane

A

B

C

D

It is not possible to synthesize diene **D** using a Suzuki reaction with ethynylcyclohexane as starting material. Hydroboration of ethynylcyclohexane adds the elements of H and B in a syn fashion, affording a trans vinylborane. Since the Suzuki reaction is stereospecific, one of the double bonds in the product must be trans.

26.23 Locate the styrene part of the molecule, and break the molecule into two components. The second component in each reaction is styrene, C$_6$H$_5$CH=CH$_2$.

a.

styrene part

b.

styrene part

c.

styrene part

26.24

but-1-ene

[1] Li
[2] CuI

octane

26.25

A

+

B

Suzuki reaction

26.26

26.27 Add the carbene carbon from either side of the alkene.

26.28 The new three-membered ring has a stereogenic center on the C bonded to the phenyl group, so the phenyl group can be oriented in two different ways to afford two stereoisomers. These products are diastereomers of each other.

26.29 High-dilution conditions favor intramolecular metathesis.

26.30 Retrosynthetically break the double bond in the cyclic compound and add a new =CH$_2$ at each end to find the starting material.

a.

b.

c.

26.31 Alkene metathesis with two different alkenes is synthetically useful only when both alkenes are symmetrically substituted; that is, the two groups on each end of the double bond are identical to the two groups on the other end of the double bond.

a.

b.

This reaction is synthetically useful because it yields only one product.

c.

26.32

26.33 All double bonds can have either the E or Z configuration.

a.

b.

c.

26.34

a.

b.

c.

d.

e.

f.

g.

h.

26.35

26.36 This reaction follows the Simmons–Smith reaction mechanism illustrated in Mechanism 26.5.

26.37

26.38

26.39

a.

b. This suggests that the stereochemistry in Step [3] must occur with syn elimination of H and Pd to form **E**. Product **F** cannot form because the only H on the C bonded to the benzene ring is trans to the Pd species, so it cannot be removed if elimination occurs in a syn fashion.

26.40

(*Z*)-2-bromostyrene

26.41

26.42

26.43

a.

Synthesize these two components, and then use a Heck reaction to synthesize the product.

b.

26.44

a.

b.

(from a.)

26.45

a.

styrene

b.

26.46

a. $HC \equiv CH$

+ enantiomer

b.

26.47

a.

Either compound can be used to synthesize the organoborane, so two routes are possible.

Possibility [1]:

Possibility [2]:

b.

The acidic OH makes it impossible to prepare an organolithium reagent from this aryl halide, so this compound must be used as the aryl halide that couples with the organoborane from bromobenzene.

c.

This can't be converted to an organoborane reagent
via an organolithium reagent.

26.48

a.

Synthesis of starting material:

b.

Synthesis of starting material:

26.49

a.

Synthesis of starting material:

b.

Synthesis of starting material:

26.50

directed aldol
[1] $CH_3CHO + LDA \rightarrow {}^-CH_2CHO$
[2] H_3O^+

several steps

maytansine

A

26.51

a.

b.

c.

d.

26.52

a.

b.

26.53 There is more than one way to form **Z** by metathesis reactions. One possibility involves ring opening of the bicyclic alkene followed by successive ring closures to generate the five- and seven-membered rings.

26.54

26.55

26.56

a. Reaction of a terminal alkene with the catalyst forms a metal–carbene that undergoes an intramolecular reaction with the triple bond, generating a new metal–carbene. A second intramolecular reaction forms the bicyclic product.

b. Two products are possible because the cascade of reactions can begin at two different double bonds.

Chapter 27 Pericyclic Reactions

Chapter Review

Electrocyclic reactions (27.3)

Woodward–Hoffmann rules for electrocyclic reactions

Number of π bonds	Thermal reaction	Photochemical reaction
Even	Conrotatory	Disrotatory
Odd	Disrotatory	Conrotatory

Examples

The stereochemistry of a thermal electrocyclic reaction is opposite to that of a photochemical electrocyclic reaction.

(2E,4E)-hexa-2,4-diene
2 π bonds

+ enantiomer

- A thermal electrocyclic reaction with an even number of π bonds occurs in a conrotatory fashion.

- A photochemical electrocyclic reaction with an even number of π bonds occurs in a disrotatory fashion.

Cycloaddition reactions (27.4)

Woodward–Hoffmann rules for cycloaddition reactions

Number of π bonds	Thermal reaction	Photochemical reaction
Even	Antarafacial	Suprafacial
Odd	Suprafacial	Antarafacial

Examples

[1] A thermal [4 + 2] cycloaddition takes place in a suprafacial fashion with an odd number of π bonds. An antarafacial photochemical [4 + 2] cycloaddition to form a six-membered ring cannot occur, because of the geometrical constraints of forming a six-membered ring.

+
$CH_2 = CH_2$

$\xrightarrow{\Delta}$ suprafacial

cis product only

[2] A photochemical [2 + 2] cycloaddition takes place in a suprafacial fashion with an even number of π bonds. An antarafacial thermal [2 + 2] cycloaddition to form a four-membered ring cannot occur, because of the geometrical constraints of forming a four-membered ring.

Sigmatropic rearrangements (27.5)

Woodward–Hoffmann rules for sigmatropic rearrangements

Number of electron pairs	Thermal reaction	Photochemical reaction
Even	Antarafacial	Suprafacial
Odd	Suprafacial	Antarafacial

Examples

[1] A **Cope rearrangement** is a thermal [3,3] sigmatropic rearrangement that converts a 1,5-diene into an isomeric 1,5-diene.

1,5-diene   isomeric 1,5-diene

[2] An **oxy-Cope rearrangement** is a thermal [3,3] sigmatropic rearrangement that converts a 1,5-dien-3-ol into a δ,ε-unsaturated carbonyl compound, after tautomerization of an intermediate enol.

1,5-dien-3-ol   δ,ε-unsaturated carbonyl compound

[3] A **Claisen rearrangement** is a thermal [3,3] sigmatropic rearrangement that converts an unsaturated ether into a γ,δ-unsaturated carbonyl compound.

unsaturated ether   γ,δ-unsaturated carbonyl compound

Practice Test on Chapter Review

1. a. Which of the following pericyclic reactions is symmetry allowed and readily occurs?
    1. a photochemical conrotatory electrocyclic ring closure of a conjugated triene
    2. a disrotatory thermal electrocyclic ring opening of a substituted cyclohexadiene
    3. a thermal [2 + 2] cycloaddition
    4. Reactions (1) and (2) will both occur.
    5. Reactions (1), (2), and (3) will all occur.

   b. Which of the following reactions requires suprafacial stereochemistry to be symmetry allowed?
    1. a photochemical [1,5] sigmatropic rearrangement
    2. a thermal [8 + 2] cycloaddition
    3. a photochemical [4 + 2] cycloaddition
    4. Reactions (1) and (2) are both suprafacial.
    5. Reactions (1), (2), and (3) are all suprafacial.

   c. What product(s) are formed from the photochemical [2 + 2] cycloaddition of (E)-hex-3-ene?

        A   B   C

    1. A only
    2. B only
    3. C only
    4. A and B
    5. A, B, and C

   d. What product(s) are formed from the photochemical electrocyclic ring opening of cis-3,4-dimethylcyclobutene?

    1. (2E,4E)-hexa-2,4-diene
    2. (2E,4Z)-hexa-2,4-diene
    3. (Z)-hexa-1,3,5-triene
    4. Compounds (1) and (2) are both formed.
    5. Compounds (1), (2), and (3) are all formed.

e. What product(s) are formed by the [3,3] sigmatropic rearrangement of cyclodeca-1,5-dien-3-ol?

A   B   C

1. **A** only
2. **B** only
3. **C** only
4. **A** and **B**
5. **A, B,** and **C**

2. Consider the *p* orbitals of the terminal carbons of a conjugated polyene with like phases on the same side of the molecule (as in **A**) or opposite sides of the molecule (as in **B**), and answer each question.

A   B

a. Which drawing is consistent with the ground state HOMO of a conjugated triene?
b. Which drawing is consistent with the excited state LUMO of a conjugated diene?
c. Which drawing is consistent with the ground state LUMO for a conjugated tetraene?

3. What type of sigmatropic rearrangement is depicted in each reaction?

a.

b.

Answers to Practice Test

1. a. 4	2. a. **A**	3. a. [3,3]
b. 2	b. **B**	b. [1,3]
c. 4	c. **A**	
d. 1		
e. 3		

Answers to Problems

27.1   Use the following definitions:

* An **electrocyclic ring closure** is an intramolecular reaction that forms a cyclic product containing one more σ bond and one fewer π bond than the reactant. An **electrocyclic ring opening** is a reaction in which a σ bond of a cyclic reactant is cleaved to form a conjugated product with one more π bond.
* A **cycloaddition** is a reaction between two compounds with π bonds that forms a cyclic product with two new σ bonds.
* A **sigmatropic rearrangement** is a reaction in which a σ bond is broken in the reactant, the π bonds rearrange, and a σ bond is formed in the product.

a.

σ bond broken

2 π bonds

3 π bonds

electrocyclic reaction

b.

$H_2C=CH_2$

σ bond formed

σ bond formed

cycloaddition

c.

σ bond formed

electrocyclic reaction

4 π bonds

3 π bonds

d.

σ bond formed

sigmatropic rearrangement

1 π bond

σ bond broken

1 π bond

27.2

a. For a bonding molecular orbital, the number of bonding interactions is greater than the number of nodes.

b. For an antibonding molecular orbital, the number of bonding interactions is less than the number of nodes.

For buta-1,3-diene:

	Bonding	Nodes	Type of MO
ψ_1	3	0	bonding MO
ψ_2	2	1	bonding MO
ψ_3^*	1	2	antibonding MO
ψ_4^*	0	3	antibonding MO

27.3 The molecular orbitals of all conjugated dienes look similar.

a.

b. Ground state   c. Excited state

ψ_4^* ——   —— excited state LUMO

ψ_3^* ——   ⊥ excited state HOMO
ground state LUMO

ψ_2 ground state HOMO

ψ_1

27.4 a. There are 10 molecular orbitals from the 10 p orbitals of the five π bonds.
b. Five molecular orbitals are bonding and five molecular orbitals are antibonding.
c. The lowest energy molecular orbital (ψ_1) has zero nodes.

ψ_1

d. The highest energy molecular orbital (ψ_{10}^*) has nine nodes.

ψ_{10}^*

27.5 To draw the product of an electrocyclic reaction, use curved arrows and begin at a π bond. Move the π electrons to an adjacent carbon–carbon bond, and continue in a cyclic fashion.

a.

c.   re-draw   Δ

b.   Δ

27.6 Thermal electrocyclic reactions occur in a *disrotatory* fashion for a conjugated polyene with an *odd* number of π bonds, and in a *conrotatory* fashion for a conjugated polyene with an *even* number of π bonds.

27.7 For an *even* number of π bonds, thermal electrocyclic reactions occur in a *conrotatory* fashion.

27.8 Photochemical electrocyclic reactions occur in a *conrotatory* fashion for a conjugated polyene with an *odd* number of π bonds, and in a *disrotatory* fashion for a conjugated polyene with an *even* number of π bonds.

[The (*E,E*) diene is favored over the (*Z,Z*) diene.]

27.9 The photochemical electrocyclic reaction cleaves a six-membered ring to form a hexatriene.

7-dehydrocholesterol   provitamin D$_3$

27.10 Use the rules for electrocyclic reactions found in Answers 27.6 and 27.8.

b. re-draw

3 π bonds

Δ
disrotatory

+ enantiomer

hν
conrotatory

27.11 Use the rules for electrocyclic reactions found in Answer 27.6. A reaction with three π bonds and a disrotatory cyclization is thermal.

27.12 Count the number of π electrons in each reactant to classify the cycloaddition.

a.

2 π
electrons   2 π
electrons

(one possibility)

[2 + 2] cycloaddition

b.

4 π
electrons   2 π
electrons

[4 + 2] cycloaddition

c.

6 π
electrons   2 π
electrons

[6 + 2] cycloaddition

27.13 A thermal suprafacial addition is symmetry allowed in a [4 + 2] cycloaddition because like phases interact.

LUMO of the diene

Like phases interact.

HOMO of the alkene

27.14 A thermal [4 + 2] cycloaddition is suprafacial.

a.

(E,E) diene

b.

(E,Z) diene

27.15

a.

X

b.

diene HOMO

dienophile LUMO

X

The dienophile is under the diene, by the rule of endo addition (Section 16.13). The H's at the ring fusion are cis to each other, but trans to the CO_2CH_3 group.

27.16 A photochemical [2 + 2] cycloaddition is suprafacial.

a.

+ enantiomer

b.

+ enantiomer

27.17

a. The photochemical [6 + 4] cycloaddition involves five π bonds (the total number of π electrons divided by two) and is antarafacial.

b. A thermal [8 + 2] cycloaddition involves five π bonds and is suprafacial.

27.18 A photochemical [4 + 2] cycloaddition like the Diels–Alder reaction must proceed by an antarafacial pathway. This would require either the 1,3-diene or the alkene component to twist 180° in order for the like phases of the *p* orbitals to overlap. Such a rotation is not possible in the formation of a six-membered ring.

excited state HOMO of the diene

180° twist required

LUMO of the alkene

27.19 Locate the σ bonds broken and formed, and count the number of atoms that connect them.

a.

σ bond formed

σ bond broken

[1,3] sigmatropic rearrangement

b.

σ bond formed

σ bond broken

[3,3] sigmatropic rearrangement

27.20

a.

re-draw

[1,7]

b, c. The reaction involves four electron pairs (three π bonds and one σ bond), so it proceeds by an antarafacial pathway under thermal conditions, and by a suprafacial pathway under photochemical conditions.

27.21 Draw the products of each reaction.

a.

re-draw

[3,3]

b.

c.

27.22 Draw the product after protonation.

27.23 Re-draw geranial to put the ends of the 1,5-diene close together. Then draw three curved arrows, beginning at a π bond.

geranial

starting material for Cope rearrangement

27.24 Draw the product of Claisen rearrangement.

a.

c.

b.

27.25

a.

b.

metathesis

$+$ $CH_2{=}CH_2$

27.26 Predict the stereochemistry of each reaction using Table 27.4.
  a. A [6 + 4] thermal cycloaddition involves five electron pairs, making the reaction suprafacial.
  b. A photochemical electrocyclic ring closure of deca-1,3,5,7,9-pentaene involves five electron pairs, making the reaction conrotatory.
  c. A [4 + 4] photochemical cycloaddition involves four electron pairs, making the reaction suprafacial.
  d. A thermal [5,5] sigmatropic rearrangement involves five electron pairs, making the reaction suprafacial.

27.27 Use the rules found in Answers 27.6 and 27.8.

27.28 Draw the product of [3,3] sigmatropic rearrangement of each compound.

27.29 An electrocyclic reaction forms a product with one more or one fewer π bond than the starting material. A cycloaddition forms a ring with two new σ bonds. A sigmatropic rearrangement forms a product with the same number of π bonds, but the π bonds are rearranged. Use Table 27.4 to determine the stereochemistry.

a.

3 π bonds

Δ

thermal electrocyclic ring closure
• 3 π bonds
• disrotatory

2 π bonds

b.

2 π bonds

Δ

2 π bonds

H

$h\nu$

3 π bonds

thermal [1,5] sigmatropic rearrangement
• 3 electron pairs
  (2 π + 1 σ)
• suprafacial

photochemical electrocyclic ring opening
• 2 π bonds in diene formed
• disrotatory

c.

Δ

2 new σ bonds

thermal [6 + 4] cycloaddition
• 5 π bonds
• suprafacial

27.30 Use the rules for thermal electrocyclic reactions found in Answer 27.6.

a.

Δ
disrotatory

E
Z
Z

3 π bonds

b.

re-draw

Δ
conrotatory

2 π bonds

27.31 Use the rules for photochemical electrocyclic reactions found in Answer 27.8.

a.

$h\nu$
conrotatory

E
Z
E

3 π bonds

Although conrotatory ring opening could also form, at least in theory, an all-(*Z*) triene, steric hindrance during ring opening would cause the terminal CH₃'s to crash into one another, making this process unlikely.

b.

27.32 Use the rules found in Answer 27.8.

a.

4 π bonds

b.

4 π bonds

27.33 Use the rules found in Answers 27.6 and 27.8.

a, b.

2E
4Z
6Z
3 π bonds

c.

(one enantiomer)   3 π bonds

d.

(one enantiomer)   3 π bonds

27.34 The trans product is indicative of a disrotatory ring closure from the cyclic triene with the given stereochemistry at the double bonds. A disrotatory ring closure with a polyene having three π bonds must occur under thermal conditions.

trans

27.35 A disrotatory cyclization of a reactant with an even number of π bonds must occur under photochemical conditions.

27.36 Use the rules found in Answers 27.6 and 27.8.

27.37 Use the rules found in Answers 27.6 and 27.8.

(A 10-membered ring cannot contain two *E* double bonds.)

27.38 The reaction involves three π bonds in one reactant and two π bonds in the second reactant, so the reaction is a [6 + 4] cycloaddition. A suprafacial cycloaddition with five π bonds must proceed under thermal conditions.

27.39 The Diels–Alder reaction is a thermal, suprafacial [4 + 2] cycloaddition.

27.40 A photochemical [2 + 2] cycloaddition is suprafacial.

27.41 A thermal [4 + 2] cycloaddition is suprafacial.

27.42 Buta-1,3-diene can react with itself in a symmetry-allowed thermal [4 + 2] cycloaddition to form 4-vinylcyclohexene.

4-vinylcyclohexene

Cycloocta-1,5-diene would have to be formed from buta-1,3-diene by a [4 + 4] cycloaddition, which is not allowed under thermal conditions.

cycloocta-1,5-diene

27.43 A series of three [2 + 2] cycloadditions with *E* alkenes forms **X**.

27.44 Re-draw the reactant and product to more clearly show the relative location of the bonds broken and formed.

a.

b.

27.45 Draw the products of each reaction.

a.

[3,3]

b.

[3,3]

tautomerize

c.

[3,3]

27.46 a. Two [1,5] sigmatropic rearrangements occur.

[1,5]

[1,5]

5-methyl-
cyclopenta-1,3-diene

1-methyl-
cyclopenta-1,3-diene

2-methyl-
cyclopenta-1,3-diene

b.

5-methyl isomer → [1,3] (crossed out) → 2-methyl isomer

A [1,3] sigmatropic rearrangement requires photochemical conditions not thermal conditions, so 5-methylcyclopenta-1,3-diene cannot rearrange directly to its 2-methyl isomer by a [1,3] shift.

27.47

27.48 Re-draw **A** to put the ends of allyl vinyl ether close together, and use curved arrows to draw the Claisen product. Then re-draw the Claisen product to put the ends of the 1,5-diene close together to draw the product of the Cope rearrangement.

β-sinensal

27.49

The conversion of **B** to **C** is a [3,3] sigmatropic rearrangement of an intermediate enolate.

27.50 Use the definitions found in Answer 27.1.

an electrocyclic ring opening

2 π bonds → 3 π bonds

a [1,7] sigmatropic rearrangement

3 π bonds → 3 π bonds

an electrocyclic ring closure

3 π bonds → 2 π bonds

27.51

a.

b.

c.

d.

27.52

a.

b.

c.

d.

C_7H_8

27.53 The mechanism consists of sequential [3,3] sigmatropic rearrangements, followed by tautomerization.

27.54

27.55

conrotatory electrocyclic ring
opening forming
4 π bonds

disrotatory ring closure
involving only 3 of the 4
π bonds

27.56

[1,5] sigmatropic
rearrangement

re-draw

[4 + 2]
cycloaddition

E

27.57

electrocyclic ring opening
forming two π bonds

[4 + 2]
cycloaddition
suprafacial

27.58 The mechanism consists of a [4 + 2] cycloaddition, followed by intramolecular imine formation.

[4 + 2]

proton
transfer

+ H_3O^+

27.59

This bridged bicyclic system is cleaved.

[3,3]

This bond forms a new bridged ring system.

27.60

anionic oxy-Cope

protonation and tautomerization

B

C

D + ⁻OEt

27.61

+ HB⁺

+ ⁻OH

+ CO₂

+ H₂O

+ HO⁻

27.62 Conrotatory cyclization of **Y** using four π bonds forms **X**. Disrotatory ring closure of **X** can occur in two ways—on the top face or the bottom face of the eight-membered ring to form diastereomers.

Only this stereoisomer has the side chain close to the six-membered ring for Diels–Alder.

[4 + 2]

methyl ester of endiandric acid A

Endiandric acid A has a free COOH group instead of the CO_2CH_3 group.

endiandric acid E methyl ester

endiandric acid D methyl ester

27.63

a.

diene   dienophile

or

diene   dienophile

b.

diene   dienophile

or

diene

c.

$\xrightarrow[\text{cycloaddition}]{[4+2]}$

$\xrightarrow[\text{sigmatropic rearrangement}]{[3,3]}$

Chapter 28 Carbohydrates

Chapter Review

Important terms

- **Aldose** — A monosaccharide containing an aldehyde (28.2)
- **Ketose** — A monosaccharide containing a ketone (28.2)
- **D-Sugar** — A monosaccharide with the O bonded to the stereogenic center farthest from the carbonyl group drawn on the right in the Fischer projection (28.2C)
- **Epimers** — Two diastereomers that differ in configuration around one stereogenic center only (28.3)
- **Anomers** — Monosaccharides that differ in configuration at only the hemiacetal OH group (28.6)
- **Glycoside** — An acetal derived from a monosaccharide hemiacetal (28.7)

Acyclic, Haworth, and 3-D representations for D-glucose (28.6)

Reactions of monosaccharides involving the hemiacetal

[1] Glycoside formation (28.7A)

- Only the hemiacetal OH reacts.
- A mixture of α and β glycosides forms.

[2] Glycoside hydrolysis (28.7B)

- A mixture of α and β anomers forms.

Reactions of monosaccharides at the OH groups
[1] Ether formation (28.8)

- All OH groups react.
- The stereochemistry at all stereogenic centers is retained.

[2] Ester formation (28.8)

- All OH groups react.
- The stereochemistry at all stereogenic centers is retained.

Reactions of monosaccharides at the carbonyl group
[1] Oxidation of aldoses (28.9B)

- Aldonic acids are formed using:
  - Ag_2O, NH_4OH
  - Cu^{2+}
  - Br_2, H_2O
- Aldaric acids are formed with HNO_3, H_2O.

aldose → aldonic acid or aldaric acid

[2] Reduction of aldoses to alditols (28.9A)

aldose → (NaBH₄, CH₃OH) → alditol

[3] Wohl degradation (28.10A)

This C–C bond is cleaved.

CHO
H—²—OH

[1] NH₂OH
[2] Ac₂O, NaOAc
[3] NaOCH₃

CH₂OH → CHO ... CH₂OH

- The C1–C2 bond is cleaved to shorten an aldose chain by one carbon.
- The stereochemistry at all other stereogenic centers is retained.
- Two epimers at C2 form the same product.

[4] Kiliani–Fischer synthesis (28.10B)

CHO ... CH₂OH

[1] NaCN, HCl
[2] H₂, Pd-BaSO₄
[3] H₃O⁺

CHO
H—²—OH
CH₂OH

+

CHO
HO—²—H
CH₂OH

- One carbon is added to the aldehyde end of an aldose.
- Two epimers at C2 are formed.

Other reactions
[1] Hydrolysis of disaccharides (28.11)

This bond is cleaved.→

H₃O⁺

+ HO

A mixture of anomers is formed.

[2] Formation of N-glycosides (28.13B)

CH₂OH O H
H H
H OH
OH OH

RNH₂
mild H⁺

CH₂OH O H
H H
H NHR
OH OH

+

CH₂OH O NHR
H H
H H
OH OH

- Two anomers are formed.

Practice Test on Chapter Review

1.a. How are the following two representations related to each other?

A   B

1. **A** and **B** are anomers of each other.
2. **A** and **B** are epimers of each other.
3. **A** and **B** are diastereomers of each other.
4. Statements (1) and (2) are both true.
5. Statements (1), (2), and (3) are all true.

b. Which of the following statements is (are) true about monosaccharide **C**?

1. **C** is a D-sugar.
2. The β anomer is drawn.
3. **C** is an aldohexose.
4. Statements (1) and (2) are both true.
5. Statements (1), (2), and (3) are all true.

c. Which of the following are different representations for monosaccharide **D**?

CHO
H——OH
H——OH
HO——H
H——OH
CH₂OH

D

1.

2.

3.

4. Both (1) and (2) are representations for **D**.
5. Compounds (1), (2), and (3) all represent **D**.

d. Which aldoses give an optically active compound upon reaction with NaBH₄ in CH₃OH?

1.
CHO
H——OH
HO——H
HO——H
CH₂OH

2.
CHO
H——OH
HO——H
H——OH
CH₂OH

3.
CHO
H——OH
H——OH
H——OH
CH₂OH

4. Both (1) and (2) give an optically active product.
5. Compounds (1), (2), and (3) all give optically active products.

2. Answer each question about monosaccharide **D** as True (T) or False (F).

D

a. **D** is a D-sugar.
b. **D** is drawn as an α anomer.
c. **D** is an aldohexose.
d. Reduction of **D** with NaBH₄ in CH₃OH forms an optically inactive alditol.
e. Oxidation of **D** with Br₂, H₂O forms an optically active aldonic acid.

f. Oxidation of **D** with HNO₃ forms an optically active aldaric acid.
g. C2 has the *R* configuration.
h. Treatment of **D** with CH₃OH, HCl forms two products.
i. Treatment of **D** with Ag₂O, and CH₃I (excess) forms two products.
j. An epimer of **D** at C3 has an axial OH group.

3. Answer the following questions about the three monosaccharides (**A–C**) drawn below.

| | **A** | **B** | **C** |

a. Draw the α anomer of **A** in a Haworth projection.
b. Draw the β anomer of **B** in a three-dimensional representation using a chair conformation.
c. Convert **C** into the acyclic form of the monosaccharide using a Fischer projection.
d. Which two aldoses yield **A** in a Wohl degradation?

4. Draw the product of each reaction with the starting material D-xylose.

D-xylose

a. CH_3OH, HCl
b. $NaBH_4$, CH_3OH
c. Br_2, H_2O
d. [1] NaCN, HCl; [2] H_2, Pd-BaSO$_4$; [3] H_3O^+
e. Ac_2O, pyridine

Answers to Practice Test

1. a. 5
   b. 5
   c. 4
   d. 1

2. a. T
   b. F
   c. T
   d. F
   e. T
   f. T
   g. F
   h. T
   i. F
   j. T

3.

4.

Answers to Problems

28.1   A *ketose* is a monosaccharide containing a ketone. An *aldose* is a monosaccharide containing an aldehyde. A monosaccharide is called: a *triose* if it has three C's, a *tetrose* if it has four C's, a *pentose* if it has five C's, a *hexose* if it has six C's, and so forth.

a. a ketotetrose

$$CH_2OH$$
$$C=O$$
$$H-C-OH$$
$$CH_2OH$$

b. an aldopentose

$$CHO$$
$$H-C-OH$$
$$H-C-OH$$
$$H-C-OH$$
$$CH_2OH$$

c. an aldotetrose

$$CHO$$
$$H-C-OH$$
$$H-C-OH$$
$$CH_2OH$$

28.2   Rotate and re-draw each molecule to place the horizontal bonds in front of the plane and the vertical bonds behind the plane. Then use a cross to represent the stereogenic center in a Fischer projection formula.

a.

$$CH_3-C-OH$$ with HO–C=O top, CH₂CH₂OH bottom   =   CH₃ —|— OH (HO–C=O top, CH₂CH₂OH bottom)

b.   O=C–H top, HO–C–H, CH₃   --re-draw-->   HO–C–CH₃ (CHO top, H bottom) = HO —|— CH₃ (CHO top, H bottom)

c.   H, C(=O), HO, H   --re-draw-->   H–C–CH₂OH (CHO top, CH₂CH₃ bottom) = H —|— CH₂OH (CHO top, CH₂CH₃ bottom)

d.   H, C, CH₃, C=O   --re-draw-->   HO–C–CH₃ (CHO top, H bottom) = HO —|— CH₃ (CHO top, H bottom)

28.3   For each molecule:
[1] Convert the Fischer projection formula to a representation with wedges and dashes.
[2] Assign priorities (Section 5.6).
[3] Determine *R* or *S* in the usual manner. Reverse the answer if priority group [4] is oriented forward (on a wedge).

a.   Cl —|— CH₂Br (CH₂NH₂ top, H bottom) —[1]→ Cl–C–CH₂Br (CH₂NH₂ top, H bottom) —[2]→ 1 Cl–C–CH₂Br 2 (3 CH₂NH₂ top, H 4 bottom) —[3]→ 1 Cl–C–CH₂Br 2 (3 CH₂NH₂ top, H bottom)   **S configuration**

b.   Cl —|— H (CHO top, CH₂NH₂ bottom) —[1]→ Cl–C–H (CHO top, CH₂NH₂ bottom) —[2]→ 1 Cl–C–H 4 (2 CHO top, CH₂NH₂ 3 bottom) —[3]→ 1 Cl–C–H (2 CHO top, CH₂NH₂ 3 bottom) ← H forward   **S configuration**

c. The reactions [1], [2], [3] converting CHO/Cl—H/CH₂OH to **S configuration** — H forward

d. The reactions [1], [2], [3] converting COOH/Cl—CH₂Br/H to **S configuration**

28.4

The sequence: rotate → Convert staggered to eclipsed. → re-draw giving

CHO
HO—C—H
HO—C—H
HO—C—H
H—C—OH
CH₂OH

=

CHO
HO——H
HO——H
HO——H
H——OH
CH₂OH

28.5

CHO   *R*
H—C—OH   *S*
HO—C—H   *R*
H—C—OH   *R*
H—C—OH
CH₂OH

D-glucose

28.6 A D sugar has the OH group on the stereogenic center farthest from the carbonyl on the right. An L sugar has the OH group on the stereogenic center farthest from the carbonyl on the left.

a.

CHO
H——OH
H——OH
HO——H
CH₂OH

A

OH group on the left: **L sugar**

CHO
HO——H
H——OH
HO——H
CH₂OH

B

OH group on the left: **L sugar**

CHO
HO——H
HO——H
H——OH
CH₂OH

C

OH group on the right: **D sugar**

b. **A** and **B** are diastereomers.
A and **C** are enantiomers.
B and **C** are diastereomers.

28.7 There are 32 aldoheptoses; 16 are D sugars.

28.8 *Epimers* are two diastereomers that differ in the configuration around only one stereogenic center.

28.9
a. D-allose and L-allose: **enantiomers**
b. D-altrose and D-gulose: **diastereomers** but not epimers
c. D-galactose and D-talose: **epimers**
d. D-mannose and D-fructose: **constitutional isomers**
e. D-fructose and D-sorbose: **diastereomers** but not epimers
f. L-sorbose and L-tagatose: **epimers**

28.10

a. D-fructose, L-fructose, b. D-tagatose, c. L-sorbose

enantiomers

28.11

D-fructose, D-tagatose

28.12

28.13 Step [1]: Place the O atom in the upper right corner of a hexagon, and add the CH₂OH group on the first carbon counterclockwise from the O atom.

Step [2]: Place the anomeric carbon on the first carbon clockwise from the O atom.

Step [3]: Add the substituents at the three remaining stereogenic centers, clockwise around the ring.

28.14 To convert each Haworth projection into its acyclic form:
[1] Draw the C skeleton with the CHO on the top and the CH₂OH on the bottom.
[2] Draw in the OH group farthest from the C=O.
A CH₂OH group drawn up means a D sugar; a CH₂OH group drawn down means an L sugar.
[3] Add the three other stereogenic centers.
"Up" groups go on the left, and "down" groups go on the right.

28.15 To convert a Haworth projection into a 3-D representation with a chair cyclohexane:
[1] Draw the pyranose ring as a chair with the O as an "up" atom.
[2] Add the substituents around the ring.

28.16 Cyclization always forms a new stereogenic center at the anomeric carbon, so two different anomers are possible.

28.17

28.18

28.19

a. All circled O atoms are part of a glycoside.

b. Hydrolysis of rebaudioside A breaks each bond indicated with a dashed line and forms four molecules of glucose and the aglycon drawn.

rebaudioside A

Trade name Truvia

aglycon

Both anomers of glucose are formed, but only the β anomer is drawn.

28.20

28.21

D-tagatose → D-galactitol + D-talitol (NaBH₄, CH₃OH)

28.22 Carbohydrates containing a hemiacetal are in equilibrium with an acyclic aldehyde, making them reducing sugars. Glycosides are acetals, so they are not in equilibrium with any acyclic aldehyde, making them nonreducing sugars.

a. reducing sugar (hemiacetal) b. nonreducing sugar (acetal) c. lactose, reducing sugar (hemiacetal)

28.23

28.24 Molecules with a plane of symmetry are optically inactive.

a. D-erythrose → optically inactive b. D-lyxose → optically active c. D-galactose → optically inactive

28.25

D-idose   or   D-gulose   ⟶   D-xylose

28.26

a.   D-threose   ⟶   (two CHO structures) +

b.   D-ribose   ⟶   (two CHO structures) +

c.   D-galactose   ⟶   (two CHO structures) +

28.27

Possible optically inactive D-aldaric acids:

A'   ⟸   plane of symmetry   ⟶   ⟹   A''

This OH is on the **right** for a D sugar.

There are two possible structures for the D-aldopentose (**A'** and **A''**), and the Wohl degradation determines which structure corresponds to **A**.

This is **A**.

A'   —Wohl→   —[O]→   optically inactive

optically active   ←[O]—   B   ←Wohl—   A''

Since this compound has no plane of symmetry, its precursor is **B**, and thus **A'' = A**.

28.28

Optically inactive alditols formed from NaBH$_4$ reduction of a D-aldohexose.

Two D-aldohexoses (**A'** and **A''**) give optically inactive alditols on reduction. **A''** is formed from **B''** by Kiliani–Fischer synthesis. Because **B''** affords an optically active aldaric acid on oxidation, **B''** is **B** and **A''** is **A**. The alternate possibility (**A'**) is formed from an aldopentose **B'** that gives an optically inactive aldaric acid on oxidation.

28.29

α-D-glucose + β-D-glucose

The same products are formed on hydrolysis of the α and β anomers of maltose.

α anomer

28.30

β glycoside bond

cellobiose

Two possible anomers here. β OH is drawn.

28.31

a.

b.

dextran

28.32

chitin—a polysaccharide composed of NAG units

chitosan

28.33

a.

b.

28.34

28.35

a.

b.

28.36

a. Two purine bases (A and G) are both bicyclic bases. Therefore they are too big to hydrogen bond to each other on the inside of the DNA double helix.

b. Hydrogen bonding between guanine and cytosine has three hydrogen bonds, whereas between guanine and thymine there are only two. This makes hydrogen bonding between guanine and cytosine more favorable.

guanine
G

cytosine
C

guanine
G

thymine
T

28.37

a.

b.

28.38

A
β anomer   D-ribose

B
α anomer   D-allose

28.39 Label the compounds with *R* or *S* and then classify.

A
R

a.
R
identical

b.
R
identical

c.
S
enantiomer

28.40 Use the directions from Answer 28.2 to draw each Fischer projection.

a.

b.

c.

d.

28.41

a. D-arabinose | enantiomer | **b.** epimer (C3) | **c.** diastereomer (but not epimer) | **d.** constitutional isomer

28.42

A | B | C | D | E | F

a. **A** and **B**   epimers
b. **A** and **C**   diastereomers
c. **B** and **C**   enantiomers
d. **A** and **D**   constitutional isomers
e. **E** and **F**   diastereomers

28.43 Use the directions from Answer 28.13.

a. β-D-talopyranose
D-talose
farthest away C, OH on right = D sugar
[1] D sugar, CH₂OH is drawn up.
[2] β anomer OH is up for a D sugar.
[3]

b. α-D-galactopyranose
D-galactose
farthest away C, OH on right = D sugar
[1] D sugar, CH₂OH is drawn up.
[2] α anomer OH is down.
[3]

c. α-D-tagatofuranose

D-tagatose

CH₂OH
C=O
HO——H
HO——H
H——OH ← farthest away C, OH on right = D sugar
CH₂OH

[1] D sugar, CH₂OH is drawn up.

[2] α anomer OH is down.

[3]

28.44

CHO
H——OH ← C2
HO——H
H——OH
H——OH
CH₂OH

D-glucose

CHO
HO——H
HO——H
H——OH
H——OH
CH₂OH

epimer at C2

β anomer

28.45

a.

CHO
H——OH
H——OH
HO——H
H——OH ←
CH₂OH

farthest away C, OH on right = D sugar

α anomer (D sugar)

β anomer (D sugar)

b.

CHO
HO——H
H——OH
HO——H
H——OH ←
CH₂OH

farthest away C, OH on right = D sugar

α anomer (D sugar)

β anomer (D sugar)

28.46 Use the directions from Answer 28.14.

a.

CH₂OH is up = D sugar

"up" group on left

"down" group on right

"up" group on left

[1]

CHO
|
|
|
CH₂OH

[2]

CHO
|
|
H——OH ←
CH₂OH

OH on right = D sugar

[3]

CHO
H——OH
HO——H
HO——H
H——OH
CH₂OH

b.

"up" group
on left →

CH₂OH is down =
L sugar

CH₂OH is up =
D sugar

"down" group
on right

"up" group
on left

[1]   [2]   [3]

OH on left =
L sugar

OH on right =
D sugar

c.   =

d.   D sugar

28.47

D-arabinose

a.   β anomer   α anomer

b.   two anomers in the pyranose form

28.48

Two anomers of D-idose, as well as two conformations of each anomer:

equatorial CH₂OH group

α anomer

axial

4 axial substituents

4 equatorial OH groups

More stable conformation for the α anomer—the CH₂OH is
axial, but all other groups are equatorial.

β anomer

OH OH
OH
O
OH

←— equatorial CH₂OH group

⇌

OH — axial
OH
O
HO
HO
OH

— axial

OH

3 axial substituents

HO

3 equatorial OH groups

The more stable conformation for the β anomer—the CH₂OH is
axial, as is the anomeric OH, but three other OH groups are
equatorial.

28.49

a. CH₃I, Ag₂O

CH₃O
OCH₃
O
CH₃O
CH₃O OCH₃

d. The product in (a), then H₃O⁺

OCH₃ OCH₃
O
CH₃O
CH₃O OH

+ β anomer

b. CH₃OH, HCl

HO
OH
O
HO
OH OH

D-gulose

HO
OH
O
HO
HO OCH₃

+ β anomer

e. The product in (b), then Ac₂O, pyridine

OAc
OAc
O
OAc
OAc OCH₃

+ β anomer

c. Ac₂O, pyridine

OAc
OAc
O
OAc
OAc OAc

f. The product in (d), then C₆H₅CH₂Cl, Ag₂O

CH₃O
OCH₃
O
CH₃O
CH₃O OCH₂C₆H₅

+ β anomer

28.50

CHO
HO —— H
H —— OH
H —— OH
H —— OH
CH₂OH

D-altrose

a. (CH₃)₂CHOH, HCl

OH
OH
O
HO
OH OCH(CH₃)₂

+ β anomer

c. Br₂, H₂O

COOH
HO —— H
H —— OH
H —— OH
H —— OH
CH₂OH

b. NaBH₄, CH₃OH

CH₂OH
HO —— H
H —— OH
H —— OH
H —— OH
CH₂OH

d. HNO₃, H₂O

COOH
HO —— H
H —— OH
H —— OH
H —— OH
COOH

e. [1] NH$_2$OH
   [2] (CH$_3$CO)$_2$O, NaOCOCH$_3$
   [3] NaOCH$_3$

g. CH$_3$I, Ag$_2$O

+ β anomer

f. [1] NaCN, HCl
   [2] H$_2$, Pd-BaSO$_4$
   [3] H$_3$O$^+$

h. C$_6$H$_5$CH$_2$NH$_2$, mild H$^+$

+ β anomer

28.51

salicin

$\xrightarrow{\text{H}_3\text{O}^+}$

monosaccharide
(both anomers)

+

aglycon

solanine

$\xrightarrow{\text{H}_3\text{O}^+}$

monosaccharide
(both anomers)

+

aglycon

monosaccharide
(both anomers)

+

monosaccharide
(both anomers)

28.52

a.

$\Longrightarrow$

=

b.

$(\alpha + \beta)$

28.53

CHO
H——OH
HO——H
H——OH
H——OH
CH₂OH
D-glucose

CHO
HO——H
H——OH
H——OH
CH₂OH
D-arabinose

CHO
HO——H
HO——H
H——OH
H——OH
CH₂OH
D-mannose

28.54

a.

CHO
HO——H
HO——H
H——OH
CH₂OH

CHO
HO——H
HO——H
HO——H
H——OH
CH₂OH

+

CHO
H——OH
HO——H
HO——H
H——OH
CH₂OH

b.

CHO
HO——H
HO——H
HO——H
H——OH
CH₂OH

CHO
HO——H
HO——H
HO——H
HO——H
H——OH
CH₂OH

+

CHO
H——OH
HO——H
HO——H
HO——H
H——OH
CH₂OH

28.55

a.

$\xrightarrow{\text{CH}_3\text{OH}}_{\text{HCl}}$

$\xrightarrow{\text{I}}_{\text{Ag}_2\text{O}}$

+ α anomer   + α anomer   + α anomer

b.

CHO
H——OH
HO——H
H——OH
H——OH
CH₂OH

$\xrightarrow{\text{NaBH}_4}_{\text{CH}_3\text{OH}}$

CH₂OH
H——OH
HO——H
H——OH
H——OH
CH₂OH

$\xrightarrow{\text{CH}_3\text{I}}_{\text{Ag}_2\text{O}}$

CH₂OCH₃
H——OCH₃
CH₃O——H
H——OCH₃
H——OCH₃
CH₂OCH₃

c.

CHO
H——OH
HO——H
H——OH
H——OH
CH₂OH

$\xrightarrow{\text{Br}_2}_{\text{H}_2\text{O}}$

COOH
H——OH
HO——H
H——OH
H——OH
CH₂OH

$\xrightarrow{\text{Ac}_2\text{O}}_{\text{pyridine}}$

COOH
H——OAc
AcO——H
H——OAc
H——OAc
CH₂OAc

28.56 Molecules with a plane of symmetry are optically inactive.

D-ribose   D-xylose

28.57

a.

b.

c.

28.58

28.59

28.60

D-glucose

Protonation of this enolate can occur from two directions.

Protonation on O forms an enediol.

enediol
A

two protonation products

enediol
A

+ H₂O

Deprotonation of the OH at C2 of the enediol forms a new enolate that goes on to form the ketohexose.

28.61

Two D-aldopentoses (**A'** and **A"**) yield optically active aldaric acids when oxidized.

Optically active D-aldaric acids:

CHO	COOH		COOH	CHO

A' → [O] → optically active | optically active ← [O] ← A" (D-lyxose)

Only **A"** undergoes Wohl degradation to an aldotetrose that is oxidized to an optically active aldaric acid, so **A"** is the structure of the D-aldopentose in question.

28.62

Only two D-aldopentoses (**A'** and **A"**) yield optically inactive aldaric acids (**B'** and **B"**).

Only **A'** fits the criteria. Kiliani–Fischer synthesis of **A'** forms **C'** and **D'**, which are oxidized to one optically active and one optically inactive aldaric acid. A similar procedure with **A"** forms two optically active aldaric acids. Thus, the structures of **A–D** correspond to the structures of **A'–D'**.

28.63

Only two D-aldopentoses (**A'** and **A"**) are reduced to optically active alditols.

Product of Kiliani–Fischer synthesis:

Only **A"** fits the criteria. Kiliani–Fischer synthesis of **A"** forms **B"** and **C"**, which are oxidized to one optically inactive and one optically active diacid. A similar procedure with **A'** forms two optically active diacids. Thus, the structures of **A–C** correspond to **A"–C"**.

28.64 A disaccharide formed from two mannose units in a 1→4-α-glycosidic linkage:

α glycoside bond

28.65

a.

α glycoside bond

1→6-α–glycoside bond

reducing sugar
(hemiacetal)

b.

$$\xrightarrow[\text{Ag}_2\text{O}]{\text{CH}_3\text{I}}$$

C → **D**

$$\xrightarrow{\text{H}_3\text{O}^+}$$

E + **F** + CH₃OH

(Both anomers of **E** and **F** are
formed, but only one is drawn.)

28.66

a, b.

1→6-α-glycoside bond

1→6-α-glycoside bond

1→2-α-glycoside bond

stachyose

c.

$$\xrightarrow{\text{H}_3\text{O}^+}$$

identical

Two anomers of each monosaccharide are formed,
but only one anomer is drawn.

d. Stachyose is not a reducing sugar because it contains no hemiacetal.

e.

$$\xrightarrow[\text{Ag}_2\text{O}]{\text{CH}_3\text{I}}$$

f. product
in (e)

$$\xrightarrow{\text{H}_3\text{O}^+}$$

Two anomers of each monosaccharide are formed.

28.67

Isomaltose must be composed of two glucose units in an α-glycosidic linkage. Since it is a reducing sugar it contains a hemiacetal. The free OH groups in the hydrolysis products show where the two monosaccharides are joined.

isomaltose + α anomer

the hemiacetal

[1] CH₃I, Ag₂O
[2] H₃O⁺

(Both anomers are present.)

28.68

a.

b.

mannose   glucose

c.

d.

28.69

a.   OHC

rotate → re-draw

b. OH on left in Fischer
L-monosaccharide

Ring flip.

more stable chair

The α anomer has the CH₃ on C5 and the anomeric OH trans.

c. Fucose is unusual because it is an L-monosaccharide and it contains a CH₃ group rather than a CH₂OH group on its terminal carbon.

28.70

a.

CH$_2$OH
C=O
HO — H
H — OH
H — OH
CH$_2$OH

D-fructose

b.

CHO
H — OH
H — OH
H — OH
CH$_2$OH

D-ribose

c.

CHO
H — OH
HO — H
H — OH
H — OH
CH$_2$OH

D-glucose

d.

CHO
HO — H
H — OH
HO — H
HO — H
CH$_2$OH

L-glucose

28.71

Ignoring stereochemistry along the way:

Chapter 29 Amino Acids and Proteins

Chapter Review

Synthesis of amino acids (29.2)

[1] From α-halo carboxylic acids by S_N2 reaction

[2] By alkylation of diethyl acetamidomalonate

- Alkylation works best with unhindered alkyl halides—that is, with CH_3X and RCH_2X.

[3] Strecker synthesis

Preparation of optically active amino acids

[1] Resolution of enantiomers by forming diastereomers (29.3A)

- Convert a racemic mixture of amino acids into a racemic mixture of *N*-acetyl amino acids [(*S*)- and (*R*)-CH₃CONHCH(R)COOH].
- React the enantiomers with a chiral amine to form a mixture of diastereomers.
- Separate the diastereomers.
- Regenerate the amino acids by protonation of the carboxylate salt and hydrolysis of the *N*-acetyl group.

[2] Kinetic resolution using enzymes (29.3B)

[3] By enantioselective hydrogenation (29.4)

Rh* = chiral Rh hydrogenation catalyst

Summary of methods used for peptide sequencing (29.6)

- Complete hydrolysis of all amide bonds in a peptide gives the identity and amount of the individual amino acids.
- Edman degradation identifies the N-terminal amino acid. Repeated Edman degradations can be used to sequence a peptide from the N-terminal end.
- Cleavage with carboxypeptidase identifies the C-terminal amino acid.
- Partial hydrolysis of a peptide forms smaller fragments that can be sequenced. Amino acid sequences common to smaller fragments can be used to determine the sequence of the complete peptide.
- Selective cleavage of a peptide occurs with trypsin and chymotrypsin to identify the location of specific amino acids (Table 29.2).

Adding and removing protecting groups for amino acids (29.7)
[1] Protection of an amino group as a Boc derivative

[2] Deprotection of a Boc-protected amino acid

[3] Protection of an amino group as an Fmoc derivative

[4] Deprotection of an Fmoc-protected amino acid

[5] Protection of a carboxy group as an ester

methyl ester   benzyl ester

[6] Deprotection of an ester group

methyl ester   benzyl ester

Synthesis of dipeptides (29.7)
[1] Amide formation with DCC

[2] Four steps are needed to synthesize a dipeptide:
>   a. **Protect** the amino group of one amino acid using a Boc or Fmoc group.
>   b. **Protect** the carboxy group of the second amino acid using an ester.
>   c. Form the amide bond with **DCC.**
>   d. **Remove both protecting groups** in one or two reactions.

Summary of the Merrifield method of peptide synthesis (29.8)

[1] Attach an Fmoc-protected amino acid to a polymer derived from polystyrene.
[2] Remove the Fmoc protecting group.
[3] Form the amide bond with a second Fmoc-protected amino acid using DCC.
[4] Repeat steps [2] and [3].
[5] Remove the protecting group and detach the peptide from the polymer.

Practice Test on Chapter Review

1.a. Which statement is true about the peptide Ala–Gly–Tyr–Phe?
>   1. The N-terminal amino acid is Ala.
>   2. The N-terminal amino acid is Phe.
>   3. The peptide contains four peptide bonds.
>   4. Statements (1) and (3) are true.
>   5. Statements (2) and (3) are true.

  b. Which of the following peptides is hydrolyzed by trypsin?
>   1. Glu–Ser–Gly–Arg
>   2. Arg–Gln–Trp–Asp
>   3. Glu–Val–Leu–Lys
>   4. Peptides (1) and (2) are hydrolyzed.
>   5. Peptides (1), (2), and (3) are all hydrolyzed.

  c. In which types of protein structure is hydrogen bonding observed?
>   1. α-helix
>   2. β-pleated sheet
>   3. 3° structure
>   4. Hydrogen bonding is present in (1) and (2).
>   5. Hydrogen bonding is present in (1), (2), and (3).

2. Answer the following questions about peptides.

Ala   Val   Ser

  a. Draw the structure of the following tripeptide: Val–Ser–Ala.
  b. Give the three-letter abbreviation for the N-terminal amino acid.
  c. Give the three-letter abbreviation for the C-terminal amino acid.

3. Answer the following questions about the amino acid leucine (2-amino-4-methylpentanoic acid), which has pK_a's of 2.33 and 9.74 for its ionizable functional groups.
   a. Draw a Fischer projection for L-leucine and label the stereogenic center as R or S.
   b. What is the pI of leucine?
   c. Draw the structure of the predominant form of leucine at its isoelectric point.
   d. Draw the structure of the predominant form of leucine at pH 10.
   e. Is leucine an acidic, basic, or neutral amino acid?

4. What product is formed when the amino acid phenylalanine is treated with each reagent?
   a. PhCH$_2$OH, H$^+$
   b. Ac$_2$O, pyridine
   c. PhCOCl, pyridine
   d. (Boc)$_2$O
   e. C$_6$H$_5$N=C=S

5. Draw the amino acids and peptide fragments formed when the octapeptide Tyr–Gly–Ala–Lys–Val–Ser–Phe–Met is treated with each reagent or enzyme:
   a. chymotrypsin
   b. trypsin
   c. carboxypeptidase
   d. C$_6$H$_5$N=C=S

Answers to Practice Test

1.a. 1
 b. 2
 c. 5
2.a

 b. Val
 c. Ala
3.

 b. 6.04

4.

5.a. Tyr, Gly–Ala–Lys–Val–Ser–Phe, Met
 b. Tyr–Gly–Ala–Lys, Val–Ser–Phe–Met
 c. Tyr–Gly–Ala–Lys–Val–Ser–Phe, Met
 d. Tyr, Gly–Ala–Lys–Val–Ser–Phe–Met

Answers to Problems

29.1

L-isoleucine

29.2

a.   b.   c.   d.

29.3 In an amino acid, the electron-withdrawing carboxy group destabilizes the ammonium ion ($-NH_3^+$), making it more readily donate a proton; that is, it makes it a stronger acid. Also, the electron-withdrawing carboxy group removes electron density from the amino group ($-NH_2$) of the conjugate base, making it a weaker base than a 1° amine, which has no electron-withdrawing group.

29.4

zwitterionic form

29.5 The most direct way to synthesize an α-amino acid is by S_N2 **reaction of an α-halo carboxylic acid with a large excess of NH_3.**

a.   c.

b.

29.6

a.   c.

b.

29.7

$$[1]\ NaOEt$$
$$[2]\ CH_2=O$$
$$[3]\ H_3O^+,\ \Delta$$

serine

29.8

a.

valine

b.

leucine

c.

phenylalanine

29.9

a.

$$\xrightarrow[\text{large excess}]{NH_3}$$

c.

$$[1]\ NH_4Cl,\ NaCN$$
$$[2]\ H_3O^+$$

b.

$$[1]\ NaOEt$$
$$[2]\ (CH_3)_2CHCl$$
$$[3]\ H_3O^+,\ \Delta$$

d.

$$[1]\ NaOEt$$
$$[2]\ BrCH_2CO_2Et$$
$$[3]\ H_3O^+,\ \Delta$$

29.10 A chiral amine must be used to resolve a racemic mixture of amino acids.

a.

achiral

b.

achiral

c.

chiral
(can be used)

d.

chiral
(can be used)

29.11

To begin:

Convert the amino acids into *N*-acetyl amino acids (two enantiomers).

enantiomers

enantiomers

Step [1]:

React both enantiomers with the *R* isomer of the chiral amine.

proton transfer

(*R* isomer only)

diastereomers

These salts have the *same* configuration around one stereogenic center, but the *opposite* configuration about the other stereogenic center.

Step [2]:

Separate the diastereomers.

separate

Step [3]:

Regenerate the amino acid by hydrolysis of the amide.

H₂O, ⁻OH

H₂O, ⁻OH

(*R*)-leucine

(*S*)-leucine

The chiral amine is also regenerated.

The amino acids are now separated.

29.12

[1] (CH₃CO)₂O

[2] acylase

(mixture of enantiomers)

(*S*)-leucine

N-acetyl-(*R*)-leucine

29.13

a.

c.

b.

29.14 Draw the peptide by joining adjacent COOH and NH₂ groups in amide bonds.

a.

Val   Glu   Val–Glu

b.

Gly   His   Leu   Gly–His–Leu

c.

M   A   T   T

M–A–T–T

29.15

a.

Arg–Asn–Val
R–N–V

b.

Lys–His–Gln
K–H–Q

29.16 There are six different tripeptides that can be formed from three amino acids (A, B, C): A–B–C, A–C–B, B–A–C, B–C–A, C–A–B, and C–B–A.

29.17

leu-enkephalin

29.18

a.

glutathione

b.

The peptide bond beween glutamic acid and its adjacent amino acid (cysteine) is formed from the COOH in the R group of glutamic acid, not the α COOH.

α COO⁻   This comes from the amino acid glutamic acid.

This carboxy group is used to form the amide bond in the peptide, not the α COOH, as is usual. That's what makes glutathione's structure unusual.

glutamic acid

29.19

a. from Ala   b. from Val

29.20 Determine the sequence of the octapeptide as in Sample Problem 29.2. Look for overlapping sequences in the fragments.

common amino acids

Ala–Leu–Tyr

Tyr–Leu–Val–Cys

Val–Cys–Gly–Glu

Answer:

Ala–Leu–Tyr–Leu–Val–Cys–Gly–Glu

29.21 Trypsin cleaves peptides at amide bonds with a carbonyl group from Arg and Lys. Chymotrypsin cleaves at amide bonds with a carbonyl group from Phe, Tyr, and Trp.

a. [1] Gly–Ala–Phe–Leu–Lys + Ala
   [2] Phe–Tyr–Gly–Cys–Arg + Ser
   [3] Thr–Pro–Lys + Glu–His–Gly–Phe–Cys–Trp–Val–Val–Phe
b. [1] Gly–Ala–Phe + Leu–Lys–Ala
   [2] Phe + Tyr + Gly–Cys–Arg–Ser
   [3] Thr–Pro–Lys–Glu–His–Gly–Phe + Cys–Trp + Val–Val–Phe

29.22

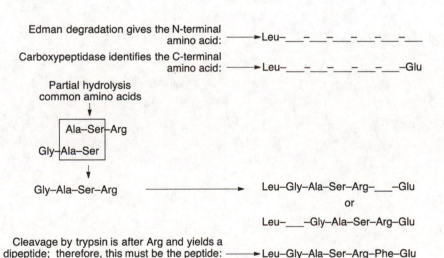

Edman degradation gives the N-terminal amino acid: →Leu–___–___–___–___–___–___–___

Carboxypeptidase identifies the C-terminal amino acid: →Leu–___–___–___–___–___–___–Glu

Partial hydrolysis common amino acids

Ala–Ser–Arg

Gly–Ala–Ser

Gly–Ala–Ser–Arg   →   Leu–Gly–Ala–Ser–Arg–___–Glu
                          or
                         Leu–___–Gly–Ala–Ser–Arg–Glu

Cleavage by trypsin is after Arg and yields a dipeptide; therefore, this must be the peptide: →Leu–Gly–Ala–Ser–Arg–Phe–Glu

29.23

a.

b.

29.24 The dipeptide depicted in the 3-D model has alanine as the N-terminal amino acid and cysteine as the C-terminal amino acid.

29.25

| All Fmoc-protected amino acids are made by the following general reaction: |

The steps:

29.26 In a *parallel* β-pleated sheet, the strands run in the *same* direction from the N- to C-terminal amino acid. In an *antiparallel* β-pleated sheet, the strands run in the *opposite* direction.

parallel

antiparallel

29.27

a. Ser and Tyr

side chains with
OH groups
hydrogen bonding

c. 2 Phe residues

van der Waals forces

b. Val and Leu

side chains with only
C–C and C–H bonds
van der Waals forces

29.28 a. The R group for glycine is a hydrogen. The R groups must be small to allow the β-pleated sheets to stack on top of each other. With large R groups, steric hindrance prevents stacking.

b. Silk fibers are water insoluble because most of the polar functional groups are in the interior of the stacked sheets. The β-pleated sheets are stacked one on top of another, so few polar functional groups are available for hydrogen bonding to water.

29.29

29.30

a. N-terminal amino acid: alanine
C-terminal amino acid: serine

b. A–Q–C–S

c. Amide bonds are **bold** (not wedges).

Ala–Gln–Cys–Ser
A–Q–C–S

29.31 The dipeptide is composed of phenylalanine and leucine.

29.32

a. (R)-penicillamine   (S)-penicillamine   b.

29.33 The electron pair on the N atom not part of a double bond is delocalized on the five-membered ring, making it less basic.

When this N is protonated...   ...the ring is no longer aromatic.
                                      sp³ hybridized N atom

When this N is protonated...   ...the ring is still aromatic.
                                      6 π electrons

preferred path

29.34

no p orbital on N

This electron pair is delocalized on the bicyclic ring system (giving it 10 π electrons), making it less available for donation, and thus less basic.

The ring structure on tryptophan is aromatic because each atom contains a p orbital. Protonation of the N atom would disrupt the aromaticity, making this a less favorable reaction.

29.35 At its isoelectric point, each amino acid is neutral.

a.

alanine

b.

methionine

c.

aspartic acid

d.

lysine

29.36

a. **threonine**
$pI = 5.06$
(+1) charge at pH = 1

b. **methionine**
$pI = 5.74$
(+1) charge at pH = 1

c. **aspartic acid**
$pI = 2.98$
(+1) charge at pH = 1

d. **arginine**
$pI = 5.41$
(+2) charge at pH = 1

29.37

a. **valine**
$pI = 6.00$
(−1) charge
at pH = 11

b. **proline**
$pI = 6.30$
(−1) charge
at pH = 11

c. **glutamic acid**
$pI = 3.08$
(−2) charge
at pH = 11

d. **lysine**
$pI = 9.74$
(−1) charge
at pH = 11

29.38 The terminal NH_2 and COOH groups are ionizable functional groups, so they can gain or lose protons in aqueous solution.

a.

Ala

A–A–A

(drawn with all uncharged atoms)

b. At pH = 1

c. The pK_a of the COOH of the tripeptide is higher than the pK_a of the COOH group of alanine, making it less acidic. This occurs because the COOH group in the tripeptide is farther away from the $-NH_3^+$ group. The positively charged $-NH_3^+$ group stabilizes the negatively charged carboxylate anion of alanine more than the carboxylate anion of the tripeptide because it is so much closer in alanine. The opposite effect is observed with the ionization of the $-NH_3^+$ group. In alanine, the $-NH_3^+$ is closer to the COO^- group, so it is more difficult to lose a proton, resulting in a higher pK_a. In the tripeptide, the $-NH_3^+$ is farther away from the COO^-, so it is less affected by its presence.

29.39

a.

b.

c.

d.

29.40

a. Asn

b. His

c. Trp

29.41

29.42

a.

b.

29.43

Br₂ / CH₃COOH → A

[1] NaOEt
[2] Cl–CH₂CH₂–S–CH₃ → C

[1] NaOH, H₂O
[2] H₃O⁺, Δ → D

29.44

[1] NaOEt → + EtOH

[2] → H–OH₂⁺

[3] → + H₂O

H₃O⁺ / Δ → glutamic acid

29.45

CN → [1] DIBAL-H [2] H₂O → A (CHO)

Ph₃P=CHCO₂Et → B (CO₂Et) (+ cis isomer)

H₂ / Pd–C → C (CO₂Et)

KOH / H₂O → D (CO₂H)

DCC → E

29.46

enantiomers

proton
transfer

diastereomers

separate

−OH

−OH

R isomer

S isomer

The chiral sulfonic
acid is
regenerated.

29.47

To begin:
Convert the amino acids into amino acid esters (two enantiomers).

Step [1]:
React both enantiomers with the *R* isomer of mandelic acid.

These salts have the *same* configuration around one stereogenic center, but the *opposite* configuration about the other stereogenic center.

Step [2]:
Separate the diastereomers.

Step [3]:
Regenerate the amino acids by hydrolysis of the esters.

29.48

To begin:
Convert the amino acids into *N*-acetyl amino acids (two enantiomers).

Step [1]:
React both enantiomers with the *R* isomer of the chiral amine.

proton transfer

brucine

diastereomers

Step [2]:
Separate the diastereomers.

separate

Step [3]:
Regenerate the amino acid by hydrolysis of the amide.

H_2O, ^-OH

H_2O, ^-OH

(*S*)-phenylalanine

(*R*)-phenylalanine

The amino acids are now separated.

The chiral amine is also regenerated.

29.49

a.

racemic mixture

b.

29.50

L-dopa

29.51

a.

Phe–Ala

c.

Lys–Gly

b.

Gly–Gln

d.

R–H

29.52 Amide bonds are bold lines (not wedges). The structure is drawn with all uncharged atoms.

C-terminal

N-terminal

Asp–Arg–Val–Tyr
D–R–V–Y

29.53 Name a peptide from the N-terminal to the C-terminal end.

a.

Gly–Asp–Glu
G–D–E

b.

Ala–Gly–Arg
A–G–R

29.54 The unusual features of gramicidin S are the presence of the uncommon enantiomer of phenylalanine (D-Phe) in the molecule, as well as the presence of an uncommon amino acid, labeled **X.**

gramicidin S

Amino acids:

Val   Leu

Pro   D-Phe

X

29.55

a. A–P–F + L–K–W + S–G–R–G
b. A–P–F–L–K + W–S–G–R + G
c. A–P–F–L–K–W–S–G–R + G
d. A + P–F–L–K–W–S–G–R–G

29.56

angiotensin I

Three-letter abbreviation: Asp–Arg–Val–Tyr–Ile–His–Pro–Phe–His–Leu

a. Trypsin cleavage products: Asp–Arg + Val–Tyr–Ile–His–Pro–Phe–His–Leu

b. Chymotrypsin cleavage products: Asp–Arg–Val–Tyr + Ile–His–Pro–Phe + His–Leu

c. ACE cleavage products: Asp–Arg–Val–Tyr–Ile–His–Pro–Phe (angiotensin II) + His–Leu

29.57

common amino acids

a.
Gly–Ala
Ala–His
His–Tyr

Answer:

Gly–Ala–His–Tyr

common amino acids

b.
Lys–His
His–Gly–Glu
Gly–Glu–Phe

Answer:

Lys–His–Gly–Glu–Phe

29.58 Gly is the N-terminal amino acid (from Edman degradation), and Leu is the C-terminal amino acid (from treatment with carboxypeptidase). Partial hydrolysis gives the rest of the sequence.

common amino acids

Gly–Ala–Phe–His
Phe–His–Ile
Ile–His–Leu

Answer:

Gly–Gly–Ala–Phe–His–Ile–His–Leu

29.59 Edman degradation data give the N-terminal amino acid for the octapeptide and all smaller peptides.

A: Glu–Arg–Val–Tyr
B: Ile–Leu–His–Phe
C: Glu–Arg
D: Val–Tyr

Glu–Arg–Val–Tyr–Ile–Leu–His + Phe

29.60

a.

Ph CH₂OH, H⁺

b. (Boc)₂O, Et₃N

c. product in (a) + product in (b) → DCC

d. H₂, Pd-C

e. product in (d) → CF₃COOH

f. + Fmoc–Cl → Na₂CO₃, H₂O

29.61

a.

Gly

Boc₂O / Et₃N → A

Ala + HO–CH₂Ph / H⁺ → B

A + B → DCC → HBr / CH₃COOH → Gly–Ala

b.

Ile

Boc₂O / Et₃N → A

Ala + HO–CH₂Ph / H⁺ → B

A + B → DCC → H₂ / Pd-C → C

Phe + HO–CH₂Ph / H⁺ →

C / DCC → HBr / CH₃COOH → Ile–Ala–Phe

29.62 Make all the Fmoc derivatives as described in Problem 29.25.

a.

(Fmoc-Phe)

[1] base
[2] Cl—POLYMER

[1] piperidine
[2] DCC

(Fmoc-Phe)

[1] piperidine
[2] DCC +

(Fmoc-Leu)

[1] piperidine
[2] DCC

(Fmoc-Ala)

[1] piperidine
[2] HF

Ala–Leu–Phe–Phe

+

F—POLYMER

b.

29.63

29.64

a. A *p*-nitrophenyl ester activates the carboxy group of the first amino acid to amide formation by converting the OH group into a good leaving group, the *p*-nitrophenoxide group, which is highly resonance stabilized. In this case the electron-withdrawing NO_2 group further stabilizes the leaving group.

p-nitrophenoxide

The negative charge is delocalized
on the O atom of the NO_2 group.

b. The *p*-methoxyphenyl ester contains an electron-donating OCH_3 group, making $CH_3OC_6H_4O^-$ a poorer leaving group than $NO_2C_6H_4O^-$, so this ester does *not* activate the amino acid to amide formation as much.

29.65

Fmoc-protected amino acid

proton transfer

29.66 Amino acids commonly found in the interior of a globular protein have nonpolar or weakly polar side chains: isoleucine and phenylalanine. Amino acids commonly found on the surface have COOH, NH_2, and other groups that can hydrogen bond to water: aspartic acid, lysine, arginine, and glutamic acid.

29.67 The proline residues on collagen are hydroxylated to increase hydrogen bonding interactions.

The new OH group allows more hydrogen bonding interactions between the chains of the triple helix, thus stabilizing it.

29.68

valine

(Racemic valine and leucine are formed as products, but the synthesis of the tripeptide is drawn with one enantiomer only.)

leucine

29.69 Perhaps using a chiral amine R*NH₂ (or related chiral nitrogen-containing compound) to make a chiral imine, will now favor formation of one of the amino nitriles in the Strecker synthesis. Hydrolysis of the CN group and removal of R* would then form the amino acid.

29.70 This reaction is similar to the reaction of penicillin with the glycopeptide transpeptidase enzyme discussed in Section 22.14. Serine has a nucleophilic OH, which can open the strained β-lactone to form a covalently bound, inactive enzyme.

orlistat

Three operations occur: nucleophilic addition; loss of the alkoxide leaving group; proton transfer.

29.71

thiazolinone

proton transfer

proton transfer

proton transfer

N-phenylthiohydantoin

Chapter 30 Synthetic Polymers

Chapter Review

Chain-growth polymers—Addition polymers

[1] Chain-growth polymers with alkene starting materials (30.2)

- General reaction:

- Mechanism—three possibilities, depending on the identity of Z:

Type	Identity of Z	Initiator	Comments
[1] radical polymerization	Z stabilizes a radical. Z = R, Ph, Cl, etc.	A source of radicals (ROOR)	Termination occurs by radical coupling or disproportionation. Chain branching occurs.
[2] cationic polymerization	Z stabilizes a carbocation. Z = R, Ph, OR, etc.	H–A or a Lewis acid (BF$_3$ + H$_2$O)	Termination occurs by loss of a proton.
[3] anionic polymerization	Z stabilizes a carbanion. Z = Ph, COOR, COR, CN, etc.	An organolithium reagent (R–Li)	Termination occurs only when an acid or other electrophile is added.

[2] Chain-growth polymers with epoxide starting materials (30.3)

- The mechanism is S$_N$2.
- Ring opening occurs at the less substituted carbon of the epoxide.

Examples of step-growth polymers—Condensation polymers (30.6)

Polyamides	Polyesters

nylon 6

Kevlar

polyethylene terephthalate

copolymer of
glycolic and lactic acids

Polyurethanes	Polycarbonates

a polyurethane

Lexan

Structure and properties

- Polymers prepared from monomers having the general structure $CH_2=CHZ$ can be **isotactic, syndiotactic,** or **atactic,** depending on the identity of Z and the method of preparation (30.4).
- **Ziegler–Natta catalysts** form polymers without significant branching. Polymers can be isotactic, syndiotactic, or atactic, depending on the catalyst. Polymers prepared from 1,3-dienes have the E or Z configuration, depending on the monomer (30.4, 30.5).
- Most polymers contain ordered crystalline regions and less ordered amorphous regions (30.7). The greater the crystallinity, the harder the polymer.
- **Elastomers** are polymers that stretch and can return to their original shape (30.5).
- **Thermoplastics** are polymers that can be molded, shaped, and cooled such that the new form is preserved (30.7).
- **Thermosetting polymers** are composed of complex networks of covalent bonds, so they cannot be melted to form a liquid phase (30.7).

Practice Test on Chapter Review

1.a. Which of the following statements is (are) true about chain-growth polymers?
   1. The reaction mechanism involves initiation, propagation, and termination.
   2. The reaction may occur with anionic, cationic, or radical intermediates.
   3. Epoxides can serve as monomers.
   4. Statements (1) and (2) are both true.
   5. Statements (1), (2), and (3) are all true.

b. Which of the following alkenes is likely to undergo anionic polymerization?
1. $CH_2=CHCO_2CH_3$
2. $CH_2=CHOCH_3$
3. $CH_2=CHCH_2CO_2CH_3$
4. Both (1) and (2) will react.
5. Compounds (1), (2), and (3) will all react.

c. Which of the following compounds can serve as an initiator in cationic polymerization?
1. butyllithium
2. $(CH_3)_3COOC(CH_3)_3$
3. BF_3
4. Both (1) and (2) can serve as initiators.
5. Compounds (1), (2), and (3) can all serve as initiators.

d. Which of the following statements is (are) true about step-growth polymers?
1. A small molecule such as H_2O or HCl is extruded during synthesis.
2. Polycarbonates are an example of a step-growth polymer.
3. Step-growth polymers are also called addition polymers.
4. Statements (1) and (2) are both true.
5. Statements (1), (2), and (3) are all true.

2. Label each statement as True (T) or False (F).
a. A polyester is the most easily recycled polymer.
b. Natural rubber is a polymer of repeating isoprene units in which all double bonds have the E configuration.
c. A syndiotactic polymer has all Z groups bonded to the polymer chain on the same side.
d. A polyether can be formed by anionic polymerization of an epoxide.
e. An epoxy resin is a chain-growth polymer.
f. A branched polymer is more amorphous, giving it a higher T_m.
g. Polystyrene is a thermoplastic that can be melted and molded in shapes that are retained when the polymer is cooled.
h. A polyurethane is a condensation polymer.
i. Ziegler–Natta catalysts are used to form highly branched chain-growth polymers.
j. Using a feedstock from a renewable source is one method of green polymer synthesis.

3. What monomer(s) are needed to synthesize each polymer?

Answers to Practice Test

1.a. 5   2.a. T   f. F   3.
b. 1   b. F   g. T
c. 3   c. F   h. T
d. 4   d. T   i. F
   e. F   j. T

a.

b.

c.

d.

e.

Answers to Problems

30.1 Place brackets around the repeating unit that creates the polymer.

poly(vinyl chloride)

nylon 6,6

30.2 Draw each polymer formed by chain-growth polymerization.

a.

b.

c.

d.

30.3 Draw each polymer formed by radical polymerization.

a.

b.

30.4 Use Mechanism 30.1 as a model of radical polymerization.

Initiation:

Propagation:

Repeat Step [3] over and over.

Termination:

(Ac = CH₃CO–)

30.5 Radical polymerization forms a long chain of polystyrene with phenyl groups bonded to every other carbon. To form branches on this polystyrene chain, a radical on a second polymer chain abstracts a H atom. Abstraction of H_a forms a resonance-stabilized radical **A'**. The 2° radical **B'** (without added resonance stabilization) is formed by abstraction of H_b. Abstraction of H_a is favored, therefore, and this radical goes on to form products with 4° C's (**A**).

30.6 Cationic polymerization proceeds via a carbocation intermediate. Substrates that form more stable 3° carbocations react more readily in these polymerization reactions than substrates that form less stable 1° carbocations. $CH_2=C(CH_3)_2$ will form a more substituted carbocation than $CH_2=CH_2$.

30.7 Cationic polymerization occurs with alkene monomers having substituents that can stabilize carbocations, such as alkyl groups and other electron-donor groups. Anionic polymerization occurs with alkene monomers having substituents that can stabilize a negative charge, such as COR, COOR, or CN.

a. electron-withdrawing group **anionic polymerization**
b. alkyl group **cationic polymerization**
c. an electron-donating resonance effect **cationic polymerization**
d. electron-withdrawing group **anionic polymerization**

30.8 Use Mechanism 30.4 as a model of anionic polymerization.

30.9 Styrene ($CH_2=CHPh$) can by polymerized by all three methods of chain-growth polymerization because a benzene ring can stabilize a radical, a carbocation, or a carbanion by resonance delocalization.

* = •, +, or −

30.10 Draw the copolymers formed in each reaction.

a.

b.

30.11

ABS

30.12

a.

b.

30.13

neoprene

All double bonds have the *Z* configuration.

E configuration of each double bond

Two higher priority groups (1's)
are on the same side of the
double bond - - - > *Z* configuration.

30.14

A

The resonance-stabilized radical can
react at two carbons.

B

30.15

a.

b.

c.

30.16

furandicarboxylic acid · · · · · ethylene glycol · · · · · PEF

30.17

Repeat for all ester bonds.

30.18

Repeat this process for all other CO bonds.

$+ \quad 2\ C_6H_5OH$

30.19

1,4-dihydroxybenzene

epichlorohydrin
(excess)

A

B

30.20

30.21

cardinol

30.22 Chemical recycling of HDPE and LDPE is not easily done because these polymers are both long chains of CH_2 groups joined together in a linear fashion. Because there are only C–C bonds and no functional groups in the polymer chain, there are no easy methods to convert the polymers to their monomers. This process is readily accomplished only when the polymer backbone contains hydrolyzable functional groups.

30.23

a.

b.

30.24

a.

b.

30.25

a.

b.

30.26

a.   b.

30.27 Draw the polymer formed by chain-growth polymerization as in Answer 30.2.

a.   c.

b.   d.

30.28 Draw the copolymers.

a.   c.

b.   d.

30.29

a.

b.

c.

d.   and

30.30

a.

b.

c.

d.

30.31 An **isotactic polymer** has all Z groups on the same side of the carbon backbone. A **syndiotactic polymer** has the Z groups alternating from one side of the carbon chain to the other. An **atactic polymer** has the Z groups oriented randomly along the polymer chain.

a.

b.

c.

30.32

from ethylene oxide

30.33

a.

and

b.

and

c.

and

d.

30.34

a.

+

Quiana

b.

+

Nomex

30.35

30.36

Kevlar

30.37

polyester **A**	PET	nylon 6,6
$T_g = <0\ °C$	$T_g = 70\ °C$	$T_g = 53\ °C$
$T_m = 50\ °C$	$T_m = 265\ °C$	$T_m = 265\ °C$

a. Polyester **A** has a lower T_g and T_m than PET because its polymer chain is more flexible. There are no rigid benzene rings, so the polymer is less ordered.

b. Polyester **A** has a lower T_g and T_m than nylon 6,6 because the N–H bonds of nylon 6,6 allow chains to hydrogen bond to each other, which makes the polymer more ordered.

c. The T_m for Kevlar would be higher than that of nylon 6,6, because in addition to extensive hydrogen bonding between chains, each chain contains rigid benzene rings. This results in a more ordered polymer.

30.38

A   dibutyl phthalate

Diester **A** is often used as a plasticizer in place of dibutyl phthalate because it has a higher molecular weight, giving it a higher boiling point. **A** should therefore be less volatile than dibutyl phthalate, so it should evaporate from a polymer less readily.

30.39

Initiation:

Propagation:

Repeat Step [3] over and over to form gutta-percha.

Termination:

30.40

a highly resonance-stabilized carbocation

Repeat Steps [3] and [4]. ⟶ **A**

new C–C bond

A is the major product formed due to the 1,2-H shift (Step [3]) that occurs to form a resonance-stabilized carbocation.
B is the product that would form without this shift.

A
major product

B

30.41

$F_3\bar{B}-O^+\overset{H}{\underset{H}{}}$ ⟶   This carbocation is unstable because it is located next to an electron-withdrawing CN group that bears a δ+ on its C atom. This carbocation is difficult to form, so $CH_2=CHCN$ is only slowly polymerized under cationic conditions.

This 2° carbocation is more stable because it is not directly bonded to the electron-withdrawing CN group. As a result, it is more readily formed. Thus, cationic polymerization can occur more readily.

30.42

Initiation: Bu—Li + [Ph alkene] $\xrightarrow{[1]}$ Bu—CH₂—CH⁻Ph Li⁺

Propagation: Bu... Ph + [Ph alkene] $\xrightarrow{[2]}$ Bu... Ph Ph $\xrightarrow[\text{over and over.}]{\text{Repeat Step [2]}}$

Termination: ... Ph Ph + $\ddot{O}=C=\ddot{O}$ $\xrightarrow{[3]}$ ... Ph Ph carboxylate

30.43 The substituent on styrene determines whether cationic or anionic polymerization is preferred. When the substituent stabilizes a carbocation, cationic polymerization will occur. When the substituent stabilizes a carbanion, anionic polymerization will occur.

a. [styrene—OCH₃]
cationic
polymerization

b. [styrene—NO₂]
anionic
polymerization

c. [styrene—CF₃]
anionic
polymerization

d. [styrene—ethyl]
cationic
polymerization

30.44 The rate of anionic polymerization depends on the ability of the substituents on the alkene to stabilize an intermediate carbanion: the better a substituent stabilizes a carbanion, the faster anionic polymerization occurs.

least   most

increasing ability to undergo anionic polymerization

30.45 The reason for this selectivity is explained in Figure 9.9. In the ring opening of an unsymmetrical epoxide under acidic conditions, nucleophilic attack occurs at the carbon atom that is more able to accept a δ+ in the transition state; that is, nucleophilic attack occurs at the more substituted carbon. The transition state having a δ+ on a C with an electron-donating CH₃ group is more stabilized (lower in energy), permitting a faster reaction.

Repeat Steps [4] and [5] over and over.

30.46

30.47

30.48

a urethane

30.49

a.

b.

c.

d.

e.

f.

g.

h.

30.50 Polyethylene bottles are resistant to NaOH because they are hydrocarbons with no reactive sites. Polyester shirts and nylon stockings both contain functional groups. Nylon contains amides and polyester contains esters, two functional groups that are susceptible to hydrolysis with aqueous NaOH. Thus, the polymers are converted to their monomer starting materials, creating a hole in the garment.

30.51

30.52

a.

vinyl alcohol → poly(vinyl alcohol)

Poly(vinyl alcohol) cannot be prepared from vinyl alcohol because vinyl alcohol is not a stable monomer. It is the enol of acetaldehyde (CH_3CHO), and thus it can't be converted to poly(vinyl alcohol).

b.

vinyl acetate $\xrightarrow[\text{radical polymerization}]{ROOR}$ poly(vinyl acetate) $\xrightarrow[\text{hydrolysis}]{^-OH \quad H_2O}$ poly(vinyl alcohol) + $CH_3CO_2^-$

c.

poly(vinyl alcohol) $\xrightarrow{H^+}$ poly(vinyl butyral)

an acetal

30.53

benzene $\xrightarrow[\text{AlCl}_3]{CH_3Cl}$ toluene $\xrightarrow[\text{AlCl}_3]{CH_3Cl}$ xylene $\xrightarrow{KMnO_4}$ terephthalic acid

$CH_2{=}CH_2$ $\xrightarrow[\text{[2] NaHSO}_3\text{, H}_2\text{O}]{\text{[1] OsO}_4}$ ethylene glycol

or

$\xrightarrow{mCPBA}$ (epoxide) $\xrightarrow{H_2O, \ ^-OH}$ ethylene glycol

30.54

phenol + $CH_2{=}O$ ⟶

Bakelite

Since phenol has no substituents at any ortho or para position, an extensive network of covalent bonds can join the benzene rings together at all ortho and para positions to the OH groups.

p-cresol

Since *p*-cresol has a CH₃ group at the para position to the OH group, new bonds can be formed only at two ortho positions, so that a less extensive three-dimensional network can form.

30.55

a.

ε-caprolactone → polycaprolactone

b.

p-dioxanone → polydioxanone

30.56

poly(ester amide) **A**

leucine

the benzyl ester of lysine

30.57

benzyl salicylate
(2 equiv)

+

sebacoyl chloride

PolyAspirin

salicylic acid
(2 equiv)

+

sebacic acid

30.58

30.59

a.

b. Abstraction of the H is more facile than abstraction of the other H's because the H atom that is removed is six atoms from the radical. The transition state for this intramolecular reaction is cyclic, and resembles a six-membered ring, the most stable ring size. Other H's are too far away or the transition state would resemble a smaller, less stable ring.

30.60